Water Studies

WIT*PRESS*

WIT Press publishes leading books in Science and Technology.
Visit our website for the current list of titles.
www.witpress.com

WIT*eLibrary*

Home of the Transactions of the Wessex Institute.
The WIT electronic-library provides the international scientific community with immediate and permanent access to individual papers presented at WIT conferences.
http://library.witpress.com

Water Studies

Editor

C.A. Brebbia
Wessex Institute, UK

WIT*PRESS* Southampton, Boston

Editor:

C.A. Brebbia
Wessex Institute, UK

Published by

WIT Press
Ashurst Lodge, Ashurst, Southampton, SO40 7AA, UK
Tel: 44 (0) 238 029 3223; Fax: 44 (0) 238 029 2853
E-Mail: witpress@witpress.com
http://www.witpress.com

For USA, Canada and Mexico

Computational Mechanics International Inc
25 Bridge Street, Billerica, MA 01821, USA
Tel: 978 667 5841; Fax: 978 667 7582
E-Mail: infousa@witpress.com
http://www.witpress.com

British Library Cataloguing-in-Publication Data

A Catalogue record for this book is available
from the British Library

Library of Congress Catalog Card Number: 2018932135

ISBN: 978-1-78466-323-0
eISBN: 978-1-78466-324-7

The texts of the papers in this volume were set individually by the authors or under their supervision. Only minor corrections to the text may have been carried out by the publisher.

Printed in Great Britain by Lightning Source, UK.

The material contained herein is reprinted from special editions of the International Journal of Sustainable Development and Planning, Volume 13, Number 3 and the International Journal of Environmental Impact, Volume 1, Number 3 published by WIT Press.

Preface

Water is essential for sustaining life on our planet and its uneven distribution is a source of permanent conflict. The growth of the human population combined with the irregularity in precipitation and water availability may restrict even further the access to water in certain regions of the world. This problem is made more severe by anthropogenic activities that affect its quality. Agriculture and industry, major users of water, are at the same time those that contribute to its contamination. Water distribution networks in urban areas, as well as soiled water collection systems, present serious problems in response to a growing population as well as the need to maintain ageing infrastructures.

Changes in the landscape, use of the land and climate conditions lead to a continuous revaluation of water resources management objectives. This requires the development of better measuring tools as well as the use of increasingly accurate computer software. Catastrophic events such as floods and associated landslides, erosion and sedimentation can have serious effects not only on life and property but also on the regions ecology. Frequently these problems are aggravated by the unforeseen consequences of man made changes.

The papers in this volume are written by professionals and academics involved in water studies to exchange knowledge and gain an insight into the state of the art in the current technology, techniques and solutions in sustainable water management as they have been developed and applied in different countries. Contributions include a wide variety of stakeholders from research and academia, to industrial sectors as well as government organisations.

The Editors
2018

Contents

EMPOWERMENT VERSUS CHARITY IN BRINGING MICROFLUSH TOILETS TO THE POOR: SUSTAINABLE LOCAL ECONOMIES FOR THE COMMON GOOD

S. MECCA[†] & A. AYALA
S-Lab, Department of Engineering-Physics-Systems, Providence College, USA.
†Member of the Global Sustainable Aid Project (GSAP), globalsustainableaidproject.org.

ABSTRACT

The GSAP (Grand Challenges Explorations Award) Microflush toilet system, a locally sourced-locally fabricated toilet that features a macro-organism enhanced aerobic filter-digester and an innovative valve that flushes on just 150 cc of water has proven to be an effective sustainable sanitation solution for developing world tropical communities. This sustainable, locally sourced, locally fabricated technology has been deployed in 18 countries around the world through trained MAKERs creating for them a local business opportunity. The household toilet averages $300 of which $100 is profit for the MAKER. For the lowest income quintile households, a microloan is often required to effect toilet ownership and, when such credit is unavailable or inaccessible, GSAP's LENDER model, operated by the MAKER or by another community player becomes a viable solution. Recent data from this intervention suggests a more powerful investment model to empower multiple local MAKERs with a sustainable sanitation credit fund (SANCRED). GSAP's SANCRED approach as it evolved from its LENDER model is described. Early experience with essential elements of the model deployed with teams of Maasai women MAKERs is presented. The potential of the models, which emphasize economic development for the common good, solving the menacing condition of sanitation in the world and meeting Goal #6 and several others of the UN Sustainable Development Goals (SDGs) are noted.

Keywords: closed systems, community development, economy for the common good, GSAP Microflush toilets, off-grid, revolving sanitation loan fund, rural wash, sanitation credit fund, sanitation, sustainable development goals, sustainable systemstoilets.

1 BACKGROUND

Under a Grand Challenges Explorations Award from the Bill & Melinda Gates Foundation, GSAP (in Ghana known as the Ghana Sustainable Aid Project) successfully prototyped a marriage of two technologies: a filter-digester and the Microflush valve [1]. The valve, which was designed in the S-Lab in the Department of Engineering-Physics-Systems at Providence College flushes on as little as 150 cc of reused water from the previous user's hand washing and results in an isolation of waste from human contact, eliminating odors and disease-bearing flies. GSAP has incorporated the valve into multiple designs including a generic macro-organism enhanced aerobic digester. The resulting GSAP Microflush (GMF) system has been designed in a model that is fabricated using local skills and local materials. So far, GSAP has trained and empowered local artisans as toilet MAKERs in 18 countries around the world. The entrepreneurial model is effecting reliable off-grid odor-free fly-free sanitation for thousands of households while creating sustainable enterprises for the MAKERs. Some of the trained MAKERs were masons and owned sufficient tools to begin their toilet making operations. Some were men and women, who did not even have the basic tools to begin their enterprise. Some of the latter were part of local cooperatives, who provided startup support and others were assisted by donor 'scholarships/grants/loans' from individuals or service organizations, e.g. Rotary. The typical grant is $500, $100 of which provides the required tools and $400 purchases materials for the first two toilets. Upon demonstrated sale and completion of three toilets, the grant/loan is forgiven, and the new toilet MAKER has

DOI: 10.2495/SDP-V13-N3-373-381

begun a sustainable enterprise of toilet sales with a balance of cash sales and sales with a loan. In communities with a preponderance of households in the bottom quintile of the household income spectrum, growth of the MAKER's enterprise (and satisfaction of community demand for the toilets) is limited by the proportion of MAKER proceeds that can be given over to small household toilet loans. In these conditions, the GSAP's LENDER or SANCRED programs offer great promise.

2 REVIEW OF THE LENDER PROGRAM

Soon after the publication of GSAP's LENDER model [2] readers asked what the minimum requirement is for funding a sustainable LENDER operation in a small community. The questions were provocative and the analyses that followed were revealing. The program can be placed in the hands of the MAKER or a separate entity; in either case the pro-formas are similar but depend on the expected mix of outright sales versus loan sales for the community. Paralleling the MAKER scholarship program, a $2000 grant, of which $300 takes care of some needed tools–a tablet or notebook computer and some paper forms, e.g. simple loan templates. The balance of the grant becomes the LENDER revolving loan fund from which toilet loans are made. The pro-forma depends on a number of factors especially the terms of the sale and household loan. For the model results shown in Fig. 1, the following assumptions were made:

Household microloan interest rate per annum	0.2
Toilet Price to the Household	$325
Gross Origination Fees per toilet to the LENDER	$50
Gross Profit per toilet to the MAKER	$75
Toilet Down Payment	$125
Loan amount	$200
Toilet Loans per month	1.40

We use parameters consistent with programs in Ghana. The 20% annual interest rate is about half of prevailing bank rates. However, in this model, there is a $25 loan origination fee charged bringing the price of the toilet to $325 (versus $300 for a cash sale) and the MAKER foregoes $25 of his/her profit as an incentive to the LENDER. A $125 down payment is received and 24-month loan is written with monthly payments of $10.25 of which the LENDER receives $1.25 in compensation. This small initial return requires patience on the part of the LENDER until assets and loans reach significant levels to assure a larger self-payment.

The assets shown in Figure 1 maintain a 1.4 loans per month rate for the duration of the months shown. By month 50, some 70 loans have been made bringing toilets to over 1000 of the neediest people in the community. The more likely situation is that the new loan creation rates would gradually increase beginning around month 20 to take advantage of the growth of assets in the loan fund. A 3% growth in toilet loans per month beginning in month 21 will result in a 5 year total of about 140 toilets serving more than 2000 people. LENDER profits which began at $83 per month in month 1 will have reached $925 by month 60, well above median and average monthly salaries in the country. Bear in mind that the MAKER will have made outright cash sales in addition to sales involving the aforementioned loans. The 5-year results under this loan growth scenario are shown in Figure 2.

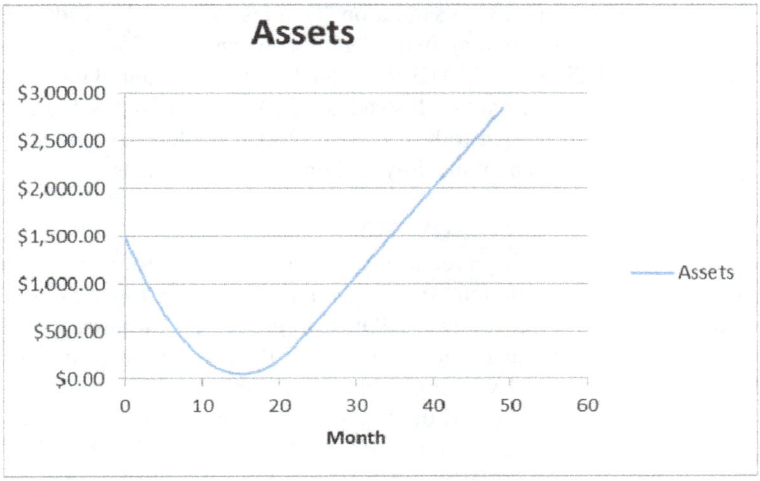

Figure 1: Fifty-month LENDER asset change for the conditions stated in the model (see text).

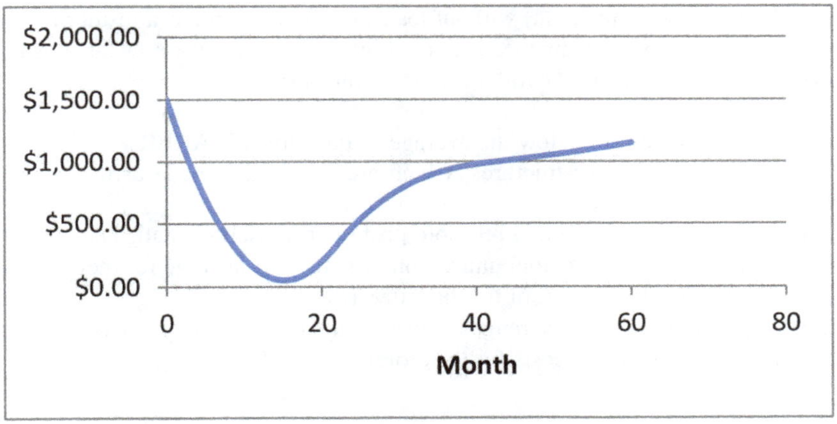

Figure 2: Five year asset changes for a 3% increase in loan creation rate beginning in month 21.

With the help of donors, GSAP has created similar MAKER and LENDER models in rural communities in Ghana, Kenya, Uganda and Bolivia. Early reports suggest that the models are working. One 3 year program in Ghana has resulted in a large MAKER-LENDER enterprise that won a Changemaker award in the country. Two other very recent ones have resulted in multiple toilets per month. In Kenya, a 3 year old program has resulted in over 50 toilets with startup MAKER_LENDER total funding of just $1500. In Kenya this program has spun off to create a MAKER_LENDER program with the Maasai in the southern part of the country. In Bolivia a similar program has begun with the Quechuan women in which GSAP has partnered with Procasha, an organization fostering women's cooperatives among this indigenous people. In the sections that follow, we will look more closely at the Maasai as a striking example of empowerment over charity and explore a final model for building local economies for the common good [3] in sanitation. That lesson should already be obvious in the foregoing examples. Startup funding in the foregoing example is $500 for a MAKER and $2000 for a LENDER or $2500 for a MAKER_LENDER. That amount of money would have built eight toilets for needy households. Instead the MAKER-LENDER builds 140 toilets over 5 years all the while providing toilets for over 2000 people who otherwise would be practicing open defecation or using unhealthy undignified open latrines.

3 LESSONS FROM THE MAASAI WOMEN

In the first quarter of 2016 following a request from a spokesperson for women of the Maasai lands in southern Kenya made through GSAP's country director in Kenya, GSAP secured a multiphase grant to fund a MAKER and LENDER program for 14 Maasai women (7 teams of 3). The three phase project began immediately with the class training and acquisition of tools sets for each team, some of which are shown in Figure 3. This was followed by the identification of a field training site and the 5 day training of each team was carried out by GSAP's country director, who made multiple trips to the Maasai lands. Phase I-II of the program, which is described here, involved a $3750 grant, the activities and cash flow from which are summarized in Tables 1 and 2, respectively.

Teams 1 through 5 were active in making loans with the funds they had and with some of their toilet sale proceeds; teams 6 & 7 completed 10 school block stalls; among teams 1–5, 4 toilets were built and sold for (cash) without loans. Going forward, as a result of their cooperation on the school block, teams 6 & 7 merged into one (which we now label as group 6). We observe the following from the toilet activities thus far:

- Average material costs are below the average of our global MAKERs. This is a result of lower costs for some superstructures, which are sometimes made using local customs from mud, dung and grass.
- The Maasai women are making a bit more profit per household stall. They are charging similar amounts for the household units, a bit less than the average for their school block stalls and realizing a bit less profit for the latter as well.
- Overall, they are well within the ranges of prices, and profits being realized elsewhere and a bit below on their material costs for the aforementioned reason.

Figure 3: A few of the Maasai women toilet MAKERs receiving their tool sets.

Table 1: The activities of the teams of Maasai women toilet MAKERs during the first quarter of their grant program.

Group/ Team	Toilets Sold (no loan)	Toilets Sold (w/loan)	Sale Amount	Material Cost	Loan amount	Gross Profit	Payments to date
1		1	$300	$137	$200	$163	$115
2		1	$300	$137	$200	$163	$92
3		3	$900	$411	$600	$489	$136
4		2	$300	$274	$400	$26	$180
5		1	$300	$137	$200	$163	$46
1–7 as coop	4	0	$1,400	$548	$0	$852	n/a
School stalls	10		$3,800	$1,750	$0	$2,050	n/a
Totals	14	8	$7,300	$3,394	$1,600	$3,906	$569
Average			$332	$154	$200	$178	$71

- It is interesting to note the collaboration of teams 6 & 7 resulting in a single entity.
- We did not expect the Maasai teams to be in a position to make any loans. The Kenya director encouraged them to set aside 75% of their profits for the loan fund. While this may not be the best long-term strategy (and in fact they may have had to pocket more than 25% of their profits), it is clear that they understand the revolving loan concept and the need for some restraint early on as they establish their enterprises. This is evident in the 8 toilet loans they made even at this early stage.
- We note from photos taken, the emerging high quality of the Maasai fabrications. As noted above, they make their own blocks from mud, dung and grass and their craftsmanship is remarkable looking at the final product.

It should be noted that the Maasai women are for the most part illiterate; few have more than a year or two of primary school education. Yet they have recognized the opportunity in their enterprise and are seizing on this accordingly. This is evident in both the aforementioned activities and in the cash flow shown in Table 2.

Again, a couple of things stand out:

- The expenditure for tool sets exceeds the average ($100 per MAKER) for our programs in other countries. The carrying distance for heavy materials (cement, sand) is such that wheelbarrows were needed for each team; also, a drill and extra hex saw blades were purchased for each team.
- The profits and balance of funds in the Maasai account total ~$5520. The balance of funds is $775 so the 7 teams total return during this period was $4745 or $431 per team.

There are many lessons from this experience. First, if this followed the pattern of the long-standing model of humanitarian service, a charitable contribution for community sanitation of $3750 (the phase I-II sum) could have given a school a 10 stall toilet block OR it could have purchased toilets for a dozen households. Instead it gave a business to 21 women,

Table 2: The cash flow against the Phase I–II funding is shown in the table below.

	Summary Transactions	Balance
Phase l grant	$3,750	$3,750
Geoffrey travel	–$250	$3,500
Tools	–$855	$2,645
Materials	–$3,394	–$749
Toilet sales	$7,600	$6,851
Loans	–$1,600	$5,521
Payments	$569	$5,820
Maasai profits	–$4,206	$1,614

provided toilets for 12 households AND 10 school block stalls. There is still $775 in the Maasai account (~110 $/team) and they have realized $431 per team in profits. In fact, the worth of the enterprise already exceeds the phase I-II grant. Another lesson is that all of this is being accomplished by a collective of 21 illiterate indigenous women. GSAP has trained many women in several of the 18 countries in which it has worked and, while women MAKERs often organize themselves into teams (this helps in some of the more strenuous efforts–lifting slabs and such), their craftsmanship is excellent and their business skills (in spite of the lack of education) are responsible and amazing. The power of the coop also comes through in this experience and others like it. The message is clear as the social value expectations are being realized.

Phase III of the program, currently in place, is the establishment of a revolving loan fund in the sum of $14,300. The fund will be administered through GSAP's Center in Kitale, Kenya but will eventually pass to the Maasai teams as they develop experience and training in lending and managing a loan fund. Investors in the Maasai and other GSAP MAKER-LENDER programs, given the success of the programs, have now wondered about developing a sustainable investment model. We describe this in the next section.

4 SUSTAINING A SANITATION CREDIT FUND (SANCREF) MODEL

It is fair to ask whether or not Phase III monies are really needed for a project like the Maasai described in the previous section. Could the balance in their account and a portion of their proceeds sustain their individual enterprises? It is clear that this could be done but the rate of growth would be extremely slow and the demand for toilets would go unmet among poor households unless there is access to loan funds. One can envision a period where such loans are made from a Phase III grant loan and interest proceeds expand the fund, which continues to make household loans until such time as the MAKERs have sufficient resources and experience to make their own loans and realize the interest from same themselves. Enter the GSAP SANCREF model currently under development.

The SANCREF dynamics are shown in Fig. 4.

The fund would be expected to support the $500 MAKER training grants previously discussed. MAKERs would sell toilets for cash and sell toilets with loans from the SANCREF fund underwritten to the households and co-signed by the MAKERs. One loan payment would be withheld from the MAKER to be returned as the last payment of the

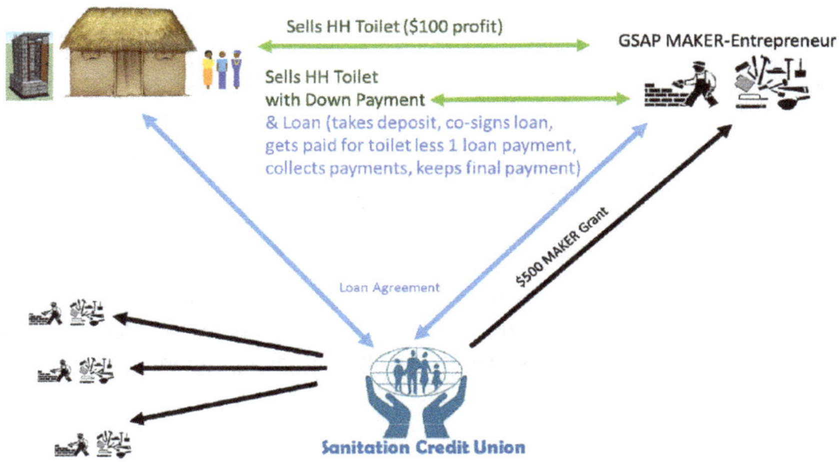

Figure 4: GSAP'ss SANCREF Model. See text.

loan. As the MAKERs' own assets grow, they would be expected to make their own loans and gradually leave the SANCREF. As the SANCREF's assets allow, it would fuel yet other MAKERs and the cycle would begin anew building a sustainable core from which nearby communities would realize GSAP's sustainable sanitation solution through new MAKERs with access to the SANCREF. While potential investors have suggested a return of the SANCREF's assets to the investor, GSAP believes the successive use of these funds within the region/country would create more social value in the successive programs than the value resulting from the extraction of those funds for re-investment in another region of the world. Further, an investment of just $2500 in a SANCREF suggests that the additional management and tracking required for the eventual extraction of the fund creates effort and expense not unwarranted for such a small investment. GSAP expects to test this model against the more standard Maasai MAKER-LENDER program by rolling it out with women in the Samburu region in northern Kenya this year. This will provide a side by side long term comparison of the two systems each with teams of women from indigenous tribes in similar contexts.

5 A PRACTICAL ISSUE

There are many challenges to be overcome in implementing the SANCRED model, one of which is local tracking of toilet sales, loans and asset cash flow. The S-Lab at Providence College has worked on creating a web centric tracking system using an on-line, mobile-enabled, secure relational database having flexible field definition features to allow different types of data including images/photos, location, and bar code identification (if necessary). The system is depicted in Fig. 5 along with the type of data being tracked and reports supported.

While the system [5] is web-centric and mobile enabled, connection to the internet during the collection of data is not required. Collected data can be uploaded to the SANCRED cloud storage when the client (mobile device-smart phone) eventually does see internet service. This feature is important as many household systems are sold in regions where internet service is lacking.

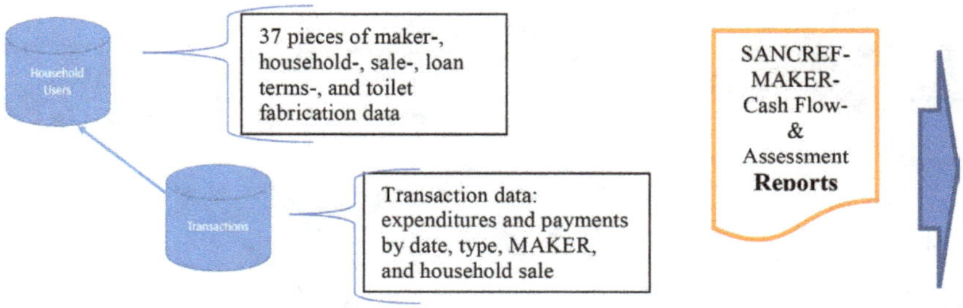

Figure 5: Two-file relational database tracking. 1. Household toilet sale and loan information and 2. Financial transactions. System reports include Detailed and Summary SANFREF- MAKER- Cash Flow- performance and program assessments.

6 CONCLUDING REMARKS AND AN INTERESTING SPECULATION

There is already evidence that GSAP's MAKER-LENDER models represent a sustainable way to bring its sustainable off-grid microflush toilet to developing communities around the world. The proposed SANCREF program has the potential to create a nuclear-breeder type solution to whole regions of communities with minimal financial outlay. It is interesting to speculate on the investment required in SANCREF programs to solve the menacing condition of sanitation in the world, meeting the UN's SDG Goal #6 by 2030. An effective MAKER program can build 10–12 toilets per month serving 150–180 people with an effective household toilet. That's about 2000 people per year per MAKER. Assuming a need to serve half of a world population assumed to be ~ 8B by 2025 means we need toilets for 4B people; spread over 10 years, we need about 2 M toilet MAKERs; at $500 per MAKER and another $2000 in SANCREF funding per first MAKER and realizing the sustainable (2–5 year) roll-over nature of SANCREF funds, $1B is needed for MAKERs and conservatively another $2B (assuming the 5 year roll-over) or a total of $3B to meet SDG #6 with a global deployment of this model. Alternatively, giving a toilet to needy households of the world, the old charity model, will require $300/toilet × 4B people/15 people/toilet or $120 billion! Instead with the SANCREF program, at the end of 10 years the residual SANCREF funds will exceed the $3B total investment that was made and it will have empowered 2 million MAKER enterprises- all local economies for the common good maintaining household and school toilets in the communities they serve. Of course, there are many factors left out of such ballpark estimates, but the fact that the sanitation problem appears solvable with an investment of less than $1 per person to be served gives us hope in the empowerment model and encouragement to endure in our efforts to solve the menacing sanitation problem facing billions of people with an economy for the common good.

ACKNOWLEDGEMENTS

The authors acknowledge the efforts of GSAP Country Director in Kenya, Geoffrey Nyakundi, and express our gratitude to the nonprofit organizations: Global Sustainable Aid Project for its overall support and Women in Social Entrepreneurship for its financial support of the Maasai project.

REFERENCES

[1] Mecca, S., Davis, H. & Davis, A., The Microflush/Biofil System: Results to Date of Prototype Installations in Ghana. *Fecal Sludge Management FSM2 Conference*, Durban, 2012.

[2] Mecca, S., Davis, H. & Davis, A., Application of GSAP Microflush toilets: a sustainable development approach to rural and peri-urban sanitation. *WIT Transactions Ecology and the Environment*, 175, 2013.

[3] Felber, C., *Change Everything: Creating an Economy for the Common Good*, (English translation from the German original), Zed Books Ltd, Unit 28 The Foundry, London, 2015.

[4] United Nations Sustainable Development Goals. Available at: http://www.un.org/sustainabledevelopment/sustainable-development-goals, September 15, 2015.

[5] EpiCollect.net is a Beta database product developed by Imperial College London.

ASSESSING THE EFFECTIVENESS OF WATER AND SANITATION SECTOR GOVERNANCE NETWORKS IN DEVELOPING COUNTRIES: A POLICY ANALYSIS FRAMEWORK

RIGOBERTO E. ALVARADO & LISA M. BORNSTEIN
School of Urban Planning, McGill University, Canada.

ABSTRACT

In developing countries, water and sanitation services for rural and peri-urban areas often are provided by networks comprised of governmental and non-governmental actors. The resulting governance systems are rarely evaluated, in part because the methods to do so are complex and unclear. This paper builds on network governance theory to (a) propose a new framework for the assessment of the effectiveness of Water and Sanitation governance networks in developing countries and (b) apply it through field research in Honduras. Network theory suggests that, since the sum of the network is greater than its individual parts, the effectiveness of a network should be evaluated based on the performance of the overall network rather than that of its individual network actors. The proposed assessment framework starts with this premise and evaluates overall network effectiveness in the four stages of the policy process: policy development; policy decisions; implementation; and monitoring & evaluation. For the case of Honduras, performance indicators were specified for each policy stage, and an assessment conducted of the overall network's performance. Key findings from the assessment relate to the importance of meta-governance coordination functions, dramatic expansion of services, and key gaps in network integration. The research, and the assessment framework, will be of interest to those concerned with the effective delivery of basic services, particularly to secondary cities of the developing world where, as in Honduras, governance network commonly provide services and data for assessment are not yet compiled.

Keywords: evaluation, network governance, performance indicators, policy cycle, water and sanitation, water boards, water quality, water supply.

1 INTRODUCTION

Governments and institutions worldwide provide water and sanitation services to the population through the traditional, formal government commitments and, increasingly, through networks and partnerships. These networks and partnerships may involve diverse actors – for example, international agencies, national regulatory bodies, non-governmental organizations (NGOs), municipal government and water boards (local resident associations) – that work together to define policy, pool resources and deliver services (or other goods). Countries using a network and partnership approach have often seen the rapid expansion of water and sanitation facilities, especially in rural and sub-urban areas [1, 2]. The challenge for evaluation, and the overall effectiveness of the water and sanitation sector, is that while assessment approaches are known and established for traditional service provision, they are lacking for network and partnership approaches [3].

The value of monitoring and evaluation is known. In the case of water and sanitation, evaluation can lead to better use of resources and better quality service, whether in terms of those reached, the service provided or equity outcomes [4]. Periodic evaluations by governments or civil society could likely accelerate and improve the effectiveness of service provision, including for the poorest members of society. Undertaking regular evaluations is essential for continuous performance improvement and for the most effective use of limited resources [5]. In the case of governance networks, assessment is complicated by the highly dynamic nature

DOI: 10.2495/SDP-V13-N3-382-393

of the network itself and the involvement of multiple stakeholders, who have differing levels of power and influence, operate at multiple geographical scales and have different roles and engagement levels in the network.

Our research takes on this challenge, developing a new approach to the assessment of network governance. The approach is both theoretical–conceptual, drawing on existing literature on network governance to design an evaluation framework around the policy cycle, and empirical in that the specific performance indicators are tailored to a local context and the evaluation approach is tested within that context. Our case study setting is Tela, a medium-size municipality on the northern coast of Honduras, where network governance in the water and sanitation sector (WSS) has risen over the past decade.

The development of an evaluation approach for WSS governance networks entailed five components: a review of literature on governance networks and their evaluation to develop the policy analysis framework (PAF); identification of the governance network for a case study municipality; distribution of resulting research tools to stakeholders in that network; analysis of those results to reach conclusions about the effectiveness of the network in the Municipality of Tela's WSS; and a meta-analysis of the usefulness of the PAF evaluation framework. The water and sanitation sector of the Municipality of Tela was selected as a case study because it was one of the first municipalities in Honduras to operate under the new water and sanitation framework law of 2003. A qualitative line of inquiry was used given the exploratory aspect of the research [6]. This article reports specifically on the development and application of this new policy analysis framework (PAF) to the assessment of the effectiveness of Water and Sanitation Sector, an approach targeted to the governance networks increasingly present in the sector in medium-sized municipalities of developing countries. The material is presented in five parts: this introduction, the theoretical and empirical foundations of an evaluation approach for governance networks, the application of the approach using context-specific performance indicators, application of the policy analysis framework, and concluding comments.

2 THEORETICAL AND EMPIRICAL FOUNDATIONS

Water and sanitation programs in developing countries are increasingly handled through a network mode of governance. As such, it becomes urgent to assess these networks. However, the evaluation of WSS governance networks is a complex task, in part because of the number of different stakeholders and also due to the multi-level nature of the task, the highly dynamic nature of the network, the potential for external influence, and the multiple perspectives from which one could design and launch legitimate assessments [7]. For example, evaluations can be initiated by international development organizations, central government institutions, municipalities, participating communities or external evaluators, each of whom may have their own criteria for effectiveness. Implicit in this statement is that assessments can be done from multiple perspectives; obviously, evaluating from all perspectives is not realistic or practical. The question thus arises of what can pragmatically be undertaken to assess effectiveness and decision making within the network context.

Appropriate theoretical-based methods for the assessment of governance networks were sought in the literature; the field is incomplete and highly fragmented. Moreover, it is not evident how to assess effectiveness or what assessment methods should be applied [8]. This situation emerges in part because the evaluation community, over the past decades, has focused on evaluating projects and programs, with little guidance provided for the assessment of institutions and networks [9]. Moreover, organizations and States, democratic or not,

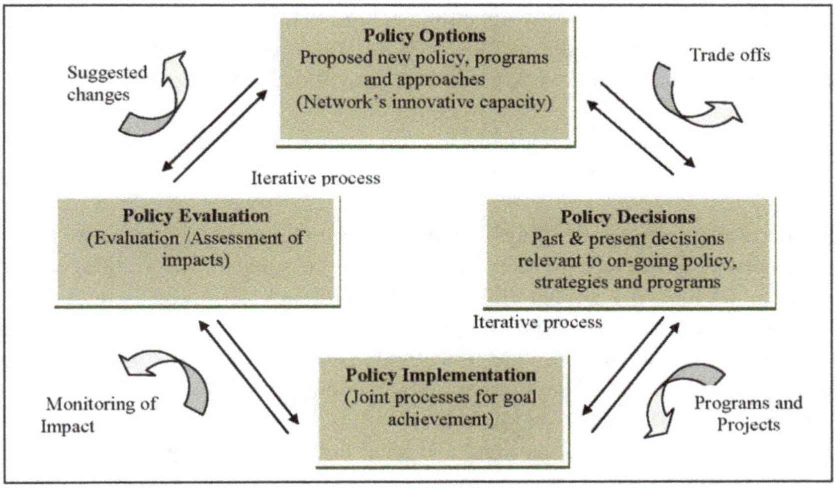

Figure 1: The policy analytical framework (PAF).

Note: For simplicity the framework is called PAF, but pragmatically it is a tool for the analysis of policy and related programs and projects. *Source:* Authors

dislike receiving unwelcome truths, thus administrative structures frequently undermined well-intentioned assessment exercises [10]. Thus, it should come as no surprise that the assessment of institutions as 'governance networks' is at a relatively early stage of development [11, 12]. As a consequence, there is considerable ongoing debate among academics and public administrators concerning appropriate evaluation frameworks [13–15].

Despite the debates, we identified several useful directions. Literature on network theory suggests that instead of devising different sets of criteria for different stakeholders, it is better, as indicated by Sorensen and Torfing [16], to relate the measurement of performance of networks to the different phases of the policy process. This observation was the starting point for our approach; we elaborated performance indicators for each of the four stages of the policy cycle and applied them both to individual stakeholders and, via integration of the results, to the network as a whole. Additional considerations in evaluating the effectiveness of the WSS governance network included: Meyer [17] perspective that new ways of framing and addressing an issue, new learning processes and new visions are vital outputs of a network; and Turrini *et al.* [14] emphasis on the network's ability to reach stated goals, capacity for innovation and change, and its longer-term sustainability.

The overall approach to network governance defined, we still had the challenges of specifying what constituted the governance network for a specific territory and how to empirically assess its effectiveness. These tasks were tackled using a two phase design. The first phase identified the WSS governance network by detecting the critical stakeholders; the second phase made an approximation of governance network effectiveness through the assessment of network processes and outputs following the proposed four stage analytical framework (Fig. 1).

2.1 Identification of the governance network

To identify the critical stakeholders in the WSS network, we used a method developed by the World Bank's Urban Management Program: following specific issues and programs to establish the involvement of various stakeholders [18, 19]. Individual entities' strategies and actions were documented and the overall network was mapped. Those entities, which are identified as 'working in the network form of organization,' were then studied, as described below.

2.2 The assessment of network effectiveness

In the second research phase, the Policy Analytical Framework (PAF) was used to assess the governance network's outputs. At each of the four developmental-procedural stages of the PAF, an evaluation tool was applied (Table 1). This approach permitted an integrated assessment, one that considered the concrete outputs of the network as a whole, such as the number of water systems constructed, as well as less tangible elements such as visions, strategies and on-going processes [17, 20]. The approach also generated valuable information about the network's structure, composition and linkages.

The PAF tool was given to the national and local institutions in the WSS governance network in the case study area. Representatives of each institution were asked to complete the form by responding to the following open-ended guide questions: (1) What relevant decisions have been taken? (2) What has been implemented to accomplish which goals? (3) What assessments have been made? (4) What new policy options have been suggested or adopted to strengthen goal achievement?

2.3 Evaluation Criteria for the WSS governance network

A great variety of performance indicators specific to the water and sanitation sector have been reported in the literature. Many of these standard indicators relate to specific programs and projects [2, 21–23]. Little work has been done in the overall assessment of WSS governance networks, perhaps because deep understanding of networks requires collection of data on multiple institutions, which can be time consuming and costly [24]. To evaluate the effectiveness of Tela's WSS network, customized performance indicators were needed.

Criteria for the selection of performance indicators included correspondence to one of the four steps of the PAF and relevance to the network's attainment of its stated long-term

Table 1: Policy analysis framework tool (to be completed via field assessment with stakeholders)

Policy decisions		
Past & present decisions relevant to on-going policy, strategies & programs	Stakeholder(s)	Comments
Implementation		
Projects & actions performed	Stakeholder(s)	Comments
Evaluation		
Evaluations/assessments performed	Stakeholder(s)	Comments
Policy options		
Proposed new policy, programs, approaches & practices	Stakeholder(s)	Comments

objectives. Performance indicators to assess the WSS governance network were specified for each stage of the policy process (Table 2). Twenty-three indicators were selected, covering important issues related directly to WSS delivery (e.g. water availability and quality) as well as to governance, such as coordination capacity, conflict resolution, and organizational culture. Four indicators refer to the policy decision step, six to implementation, six to assessment, and seven to new policy options. Data collected on the indicators provide a basis for assessing network performance vis-à-vis stated goals for both network outputs and processes.

Table 2: Performance indicators.

Policy decision indicators		Implementation indicators	
1	A long-term engagement with the local water sanitation sector	5	Strengthening local sector institutions
2	An institution/organization working towards the development of sustainable and cost-effective WSS systems	6	Strengthening water boards and service providers
3	Coordination with other stakeholders of the network/number of links with other network partners	7	Constructing WSS systems
4	New WSS projects are including supervision and control methodology (e.g. ERSAPS/USCL, see Figure 2)	8	Working toward micro-watershed protection
		9	Establishment and management of WSS information system
		10	Participating in national or local forum for information exchange (e.g. RASHON, see Figure 2)
Assessment Indicators		**New Policy Options Indicators**	
11	Supervision of the construction of new WSS systems	17	New approaches to WSS management considering climate change
12	Assessments of program or project implementation by funding organization	18	Development of new financial approaches for the WSS
13	External third-party assessments during program development and at project completion	19	Advocacy towards strengthening of a local WSS leader (Policy & Strategy) (e.g. COMAS, see Figure 2)
14	Ex-post assessment one year after project completion	20	Advocacy for strengthening local WSS technical assistance partners (e.g. SANAA, see Figure 2)
15	Assessment and/or supervision of existing WSS system (on-going process)	21	Advocacy for the installation of household micro-water meters
16	Assessments and control of quality of water (on-going process)	22	Addressing gender inequality

In addition, the research framework allowed investigation of the network's capacity for longer-term sustainability, coordination among members, resolution of conflict, development of linkages, spaces for discourse and meta-governance functions. Cultural factors facilitating or impeding network outputs and processes were also explored. Network theory literature identifies all these factors as crucial to governance effectiveness.

3 EVALUATION WITH SPECIFIC PERFORMANCE INDICATORS

Eight entities – indicated in grey on Figure 2 – were found to constitute the WSS network for Tela; they were asked to complete the questions listed in Table 1. Data were obtained for each stage of the policy framework via document review, semi-structured interviews (with key informants of government, bilateral and multilateral agencies, partner NGOs, community-based organizations (CBOs), and water boards), and a three-week participatory observation in the case study area [25]. Illustrative findings per policy stage follow.

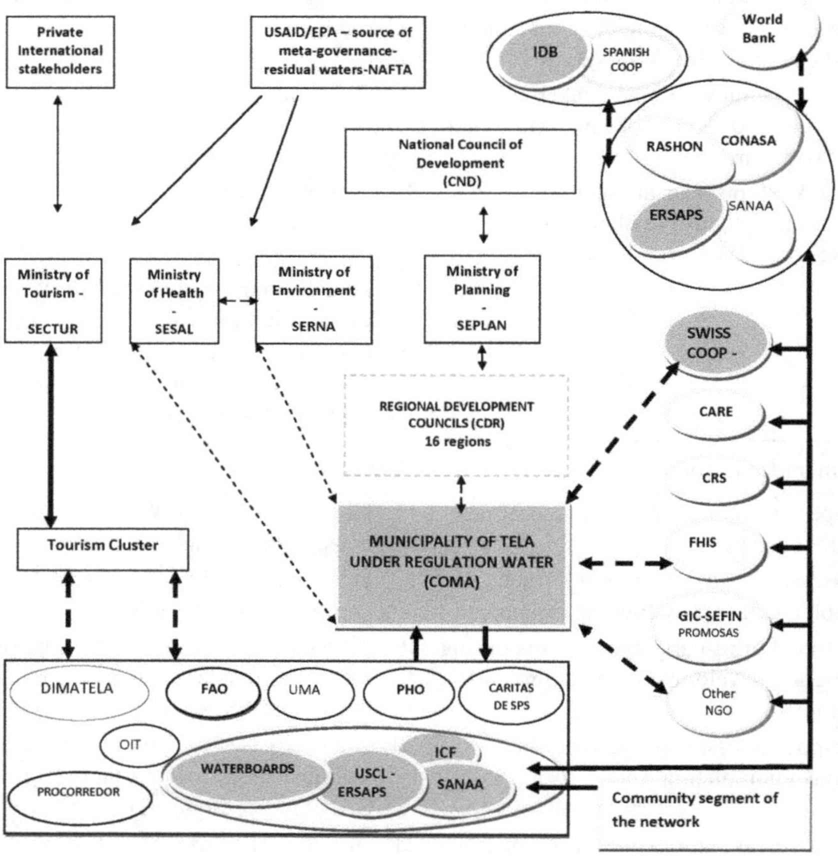

Figure 2: Conformation of the WSS governance network.

Note: The thickness of the arrow indicates strength of a linkage – An intermittent arrow indicates an occasional linkage. *Source:* authors

3.1 Policy decisions issues (Performance indicators 1–4)

When combined, the data from the individual stakeholders suggest the network, as constituted, functions reasonably well with respect to policy decision issues. Policy decision indicators show that the network has a long-term engagement with the WSS sector and that it is oriented toward the development of sustainable cost-effective systems. Additionally, indicators reveal a good level of coordination, collaboration and trust among its local members, with a trio of stakeholders – the *Unidad de Supervisión y Control Local, Servicio Autónomo Nacional de Acueductos y Alcantarillados,* and *Instituto Nacional de Conservación Forestal y Desarrollo Forestal, Areas Protegidas y Vida Silvestre* (listed as USCL-SANAA-ICF in Figure 2) – operating effectively through informal agreements.

3.2 Implementation issues (Performance indicators 5–10)

With respect to implementation, evidence shows uneven performance, with the network performing well in some areas and not in others. Indicators show that three stakeholders are working to strengthen local institutions and the water boards. These actions have resulted in a network fully engaged in regulating 48 water boards with plans to expand coverage to other communities in the municipality. Some water boards are requesting to join the regulated system, reflecting recognition that the network strengthens those that participate. Moreover, the research found that the overall network is very active in micro-watershed protection, though some participants have failed to incorporate such protection into the design and implementation of their WSS programs.

3.3 Monitoring and assessment issues (Performance indicators 11–16)

As with the other performance areas, the network shows mixed results with respect to monitoring and evaluation. The international segment of the network consistently monitors progress and prepares project completion reports. However, the network does not seem to favor external third party assessment at project completion or appear concerned with ex-post assessment one year after project completion as only one stakeholder of the network is consistently applying these methods.

A unique characteristic of the community segment of the network is its interest in the assessment and control of the quality of water, a concern not expressed by ERSAPS, the national regulation authority. Local actions to improve water quality focus on the monitoring of micro-watersheds, chlorination of water reservoirs and water sampling for coliform count (although the effectiveness of the chlorination protocols appear questionable as some test results indicated high coliform counts even after chlorination).

3.4 New policy issues (Performance indicators 17–23)

The network shows mixed results with respect to policy options and new approaches. Important issues of local interest include: the signing of a formal agreement between water and health authorities to jointly supervise of water quality. Issues of climate change and new financial approaches generated little interest among local stakeholders.

4 EVALUATION WITH THE PAF CYCLE

Water and sanitation for the municipality of Tela is managed through two approaches. The first focuses on improving conditions in the City of Tela, and the second focuses on improving conditions at the rural level.

The first approach places the responsibility for urban service provision exclusively in the hands of the Municipal Division of Waters of Tela (DIMATELA), which reports to the mayor. DIMATELA was not active in the WSS network. Observers said, '*DIMATELA resources are channeled to purchase products and services unrelated to water and sanitation, while many water problems remain unresolved due to lack of funding*.'

The second approach of the Municipality of Tela focuses on managing water and sanitation in the rural areas. Empirical evidence from the research indicates that rural provision is managed through a WSS governance network working actively in the region. To assess effectiveness, analysis was done of its on-going processes and outputs at each of the four discrete stages of the PAF framework. Discussion of key aspects per stage follows.

The first stage of analysis, which refers to policy decisions, shows that the community segment of the network (Fig. 1) as a rule does not create, propose or implement new policies or suggest new approaches. It follows policies that are generated elsewhere. But exceptions exist: the community segment decided that the water supply of the water boards should be monitoring for coliform count and this constitutes an example of the community segment participating in decision making.

The second stage of analysis, which refers to implementation capabilities, indicates that the network is relatively active in overall program and project development. However, its construction of new water systems is at a very low level of activity (less than 10), although many communities need new and/or improved systems. The lack of a network member with water system construction capabilities partially accounts for low performance in this area, as does lack of dedicated funding. The research found that the network was addressing both constraints: inclusion of an experienced construction entity in the network and fresh funds made available by the Inter-American Development Bank (IDB) for new rural systems, of which at least six are being planned for the Tela region.

The third stage of the analysis relates to monitoring, evaluation and other forms of assessment. The network had 48 regulated water boards in 2012. Each board is supervised quarterly for adequacy of tariffs, frequency of service, volume of water supplied, system disinfection, water quality, and status of sanitation. Great care is taken in the monitoring and control of the finances of the water boards to the point that the culture of paying a monthly fee for water provision is now deeply rooted among the regulated water boards. Reports indicate that the average tariffs collected in the municipality of Tela is 28% over its operating costs, with a low level of arrears at the water board level [26]. This approach has improved the quality of life of the participating communities, and is a model of excellence within a country context that does not have a culture of supervision or regulation.

The fourth stage of the cycle, which refers to the network's contribution to the generation of new policies or processes, shows that the higher-level governance portion of the network played a major role in the enactment of the WSS framework law of 2003. This law resulted in significant advancement of the rural WSS agenda of the Municipality of Tela. Meanwhile, the community segment of the network has also been active, advocating for improved coordination with Public Health authorities – to help supervise water quality – and expanded protection of micro-watersheds up-stream of water reservoirs.

4.1 Linkages, meta-governance and culture issues of the WSS governance network

It was determined that the Municipality of Tela has an active local WSS governance network composed of four stakeholders strictly working at the community level, with another four stakeholders operating at higher levels of governance. The three local entities (ICF, USCL and SANAA) and the water boards constitute the community segment of the network, and the Municipal Corporation (COMAS), the Swiss Cooperation, and the two meta-governors, ERSAPS and IDB, comprise the higher-level segment of the network (Fig. 2). The four members of the community segment have excellent horizontal linkages and operate very smoothly – intra-network conflicts are almost non-existent, possibly because of transparency and convivial relations among the members.

4.2 Highlights of the WSS governance network

Overall the network made significantly contributions to advancing the rural WSS agenda of the municipality of Tela, a development made possible because higher-level governance portions of the network is permanently engaged in policy and institutional development, and played a major role in the enactment of the new WSS framework law of 2003. It was then that the country embarked on the development of the water sector from a municipal perspective. This municipal focus greatly benefited the Municipality of Tela as it was one of the first to receive technical assistance and funds for capacity building of its water sector.

The high-level portion of the network has supported and continues to support institutions, organizations and networks in the Water and Sanitation sector in the country. In fact, the World Health Organization (WHO) identifies Honduras as one of the three Latin American countries that has most advanced in water and sanitation provision since the 1990s [27]. How was this advancement accomplished? The research suggests that the leadership of international development organizations engaged with water and sanitation was a pivotal factor. An unfortunate lack of Honduran state leadership for WSS [28] meant that, as our research shows, the WSS governance network of the Municipality of Tela turned to international stakeholders and ERSAP (Fig. 2) for meta-governance of the network.

Furthermore, the research points toward policy and organizational dynamics important to future water and sanitation decision-making. For instance, the research indicates that, overall, the Water and Sanitation country sector is steered by two meta-governors, namely the Inter-American Development Bank and the World Bank, with strong support provided by ERSAP and aid from other strategic institutions such as the Swiss Cooperation for Development (SCD). To improve the effectiveness of all the institutions and networks working in the water sector in the country, more support could be directed to local entities. Some are already active in the network and promoting WSS nationally: ERSAPS (the national regulatory institution for water supply and sanitation) is providing some WSS leadership nationally; and RASHON (Honduras Water and Sanitation Network), a public/private legally chartered institution, acts as a forum for information-sharing on water and sanitation issues in Honduras. Two national-level institutions, the National Committee of Water and Sanitation (CONASA) and National Autonomous Service of Water and Sewage (SANAA), were not considered members of the WSS governance network but could, if transformed, potentially be important stakeholders. Fieldwork completed in 2010–2012, revealed, for instance, that CONASA had conducted minimal work during a 10-year period, mainly because of the lack of leadership and ministerial support. Similarly, lack of resources accounted, at least partially, for SANAA's overall disengagement from its technical assistance responsibilities [28, 29].

5 CONCLUSIONS AND SUGGESTIONS FOR FUTURE RESEARCH

This research proposes and applies a new framework for the assessment of the effectiveness of Water and Sanitation Sector (WSS) governance networks in developing countries. The framework is grounded on the theoretical perspective that, rather than applying different criteria for each of the stakeholders, it is better to relate the overall performance of a network to the different phases of the policy process, [16]. The resulting approach, based on network theory, could improve water and sanitation provision while advancing knowledge in the emerging field of governance network assessment.

The proposed two-phase analytical method first identifies the WSS governance network, and then makes an approximation of effectiveness through the policy analysis framework (PAF), which evaluates effectiveness based on network outputs and on-going processes (Fig. 1). The framework, following leading network theories, relates the overall performance of the network to the different phases of the policy process [16]. Each of the four stages of the policy process is supported by the application of performance indicators custom-designed to assess the WSS governance network (Table 2). The chosen indicators not only cover a wide range of important water issues, such as quality and availability, but also important network aspects, such as coordination capacity, conflict resolution capacity, and culture.

The proposed PAF assessment methodology, with its related family of custom indicators, opens up many areas for future research [30]. Other WSS governance networks can be evaluated by applying or slightly modifying the 23 indicators designed for this research or, alternatively, the WSS network could be evaluated from other perspectives such as efficiency, relevance, financial viability, and gender participation, restructuring the indicators to suit. In addition, future research is not restricted to WSS governance networks. Based on the general applicability of the methodology, it can be replicated to assess networks operating in other sectors such as education, environment, green growth and food security, again incorporating a set of relevant performance indicators.

This new analytical framework for the assessment of the effectiveness of water and sanitation sector (WSS) governance networks is a simple and useful tool for informed decision making for those working to improve water and sanitation provision through network forms of organization, particularly for the benefit of communities or society at large. The investigation also advances governance network evaluation methodology.

ACKNOWLEDGEMENTS

This research was made possible thanks to the financial support of the FQRSC (Fonds de Recherche sur la Sociélé et la Culture du Quebec), IDRC (International Development Research Center), the Association of AUCC/CIDA (Sustainable tourism project – Honduras) and the School of Urban Planning, McGill University, Montreal, Canada.

REFERENCES

[1] Gray, T. & Stewart, A., *International Library of African Studies: Governance of Water and Sanitation in Africa: Achieving Sustainable Development Through Partnerships*, I.B. Tauris: London, GBRp. 110, 2009.

[2] White, H. & Black, R., *Targeting Development: Critical Perspectives on the Millennium Development Goals,* Taylor & Francis, 2003.

[3] Parker, R., Berg, G., Sieburger, M. & Heuser, S. (eds), Water and development: taking lessons from evaluation - Summary of discussions at the Berlin Conference. *The Berlin Conference on Water and Development;* 2010; Berlin: The Independent Evaluation Group of the World Bank (IEG) and the Evaluation Department of KfW Entwicklungsbank (development bank).

[4] Tesselaar, R. Engelbertink, J., Kemp, A.D. & Jorritsma, H. (eds), *From Infrastructure to Sustainable Impact: Policy Review of the Dutch Contribution to Drinking Water and Sanitation (1990–2011)*, The Hague, The Netherlands: Netherlands, Ministry of Foreign Affairs, Policy and Operations Review Department (IOB), 2012.

[5] Cotton, A., Adams, J. & Shaw, D., Improving water supply and sanitation programme effectiveness: Lessons from WaterAid's outcome evaluation studies. *Water and Environment Journal,* **27**, pp. 1–9, 2013.
https://doi.org/10.1111/j.1747-6593.2012.00317.x

[6] Yin, RK., *Case Study Research: Design and Methods -Third Edition - Applied Social Research Methods Volume 5*, Leonard Bickman DJR, editor: Sage Publications, 2003.

[7] Albrecht, M., Elbe, J., Elbe, S. & Meyer, W., Analyzing and evaluating regional governance networks: three challenges for applications. *Evaluation*, **20**, pp. 58–74, 2014.
https://doi.org/10.1177/1356389013518457

[8] Torfing, J. & Sørensen, E., The European debate on governance networks: towards a new and viable paradigm? *Policy and Society*, **33**, pp. 329–344, 2014.
https://doi.org/10.1016/j.polsoc.2014.10.003

[9] Rojas, K. & Lusthaus C., The evolution of institutional and organizational assessment. In G. Anderson (ed), *Shaping International Evaluation*, Montreal, 2010.

[10] Moran, M., Rein, M. & Goodin RE., The public and its policies. *The Oxford Handbook of Public Policy*, Oxford University Press, 2006.

[11] Hertting, N. & Vedung, E., Purposes and criteria in network governance evaluation: How far does standard evaluation vocabulary takes us? *Evaluation*, **18**, pp. 27–46, 2012.
https://doi.org/10.1177/1356389011431021

[12] Lusthaus, C., Garven, K. & Grandi, S., Chapter four: evaluating partnerships in the not for profit sector. In Universalia, editor. *Shaping International Evaluation*, Universalia Management Group, 2010.

[13] Span, K., Luijkx, K., Schols, J. & Schalk, R., The relationship between governance role and performance in local public interorganizational networks: A conceptual analysis. *The American Review of Public Administration,* **42**(2), pp. 186–201, 2012.
https://doi.org/10.1177/0275074011402193

[14] Turrini, A., Cristofoli, D., Frosini, F. & Nasi, G., Network literatute about determinants of network effectiveness. *Public Administration*, **88**(2), pp. 528–550, 2010.
https://doi.org/10.1111/j.1467-9299.2009.01791.x

[15] Nance, E.B., Multistakeholder evaluation of condominial sewer services. *American Journal of Evaluation*, **26**(4), pp. 480–500, 2005.
https://doi.org/10.1177/1098214005281324

[16] Sorensen, E. & Torfing, J., Making governance networks effective and democratic through metagovernance. *Public Administration*, **87**, pp. 243–258, 2009.
https://doi.org/10.1111/j.1467-9299.2009.01753.x

[17] Meyer, M.W., *Rethinking Performance Measurement*, Cambridge University Press, 2002.

[18] Brown, D. & Mooleedhar, T., The response of public authorities to a local emergency: the Demera Road incident. *Human Systems Management*, **17**, pp. 49–62, 1998.

[19] Bartone, C., Bernstein, J. & Leitman, J., *Toward Environmental Strategies for Cities, Programme UM*, editor, Washington: The World Bank, 1995.

[20] Wiek, A. & Larson, K., Water, People, and sustainability—A systems framework for analyzing and assessing water governance regimes. *Water Resources Management*, **26**(11), pp. 3153–3171, 2012.
https://doi.org/10.1007/s11269-012-0065-6

[21] Water and Sanitation Program - Central American Region. *Poverty and Sanitation - An analysis of the linkage between poverty and access to basic sanitation in Honduras*, available at www.wsp.org, 2008.

[22] Gurria, A., IV. Governance: at the root of the problem, at the hearth of the solution. *Water: Towards a Culture of Responsibility*, University of New Hampshire Press, 2011.

[23] ERSAPS. *Indicadores 2010 del Sector Agua y Saneamiento en Honduras*, ERSAPS, 2010.

[24] Provan, K.G. & Kenis, P., Modes of network governance: Structure, management, and effectiveness. *Journal of Public Administration Research and Theory*, **18**, pp. 229–252, 2007.
https://doi.org/10.1093/jopart/mum015

[25] Jorgensen, D.L., *Participant observation: A methodology for human studies - Applied Social Research Methods Series,* Volume 15, Sage Publications, 1989.

[26] ERSAPS. *Indicadores 2011 del sector Agua Potable y Saneamiento en Honduras.* ERSAPS, 2011.

[27] Organización Mundial de la Salud. En 24 años, 17% de latinoamericanos tienen acceso a agua potable y saneamiento, available at www.americaeconomia.com/politica-sociedad/mundo/en-24-anos-17-mas-de-latinoamericanos-tienen-acceso-agua-potable-y-saneamien (accessed 8 January 2017)

[28] Inter-American Development Bank. Informe de Terminacion de Projecto: Programa en Inversión en Agua Potable y Saneamiento HO0072, available at www.idbdocs.iadb.org/wsdocs/getdocument.aspx?docnum=37318279 (accessed 8 January 2017)

[29] Gentes, I., Agua potable, saneamiento y calidad en Honduras: Desde una vision consensuada a la accion concertada, available at www.ersaps.hn/documentos/interes/PAPSAC_Sintesis.pdf: Programa de Apoyo Presupuestario Sectorial en Agua y Calidad (PAPSAC), convenio de financiación no Reembolsable No. DCI-ALA/2011/22-863 de la Union Europea, 2014 (accessed 9 January 2017)

[30] Alvarado, R.E., Assessing the effectiveness of water and sanitation sector governance networks in developing countries: a policy analysis framework and its application to the governance network for the Municipality of Tela, Honduras, Ph D thesis, Mcgill University, 2015, available at http://digitool.library.mcgill.ca/R/TTJTQV5LC4T9YTJDMDT459C1KUBLLIDUH269AGKFGFPA9D4QQE-00476?func=collections-result&collection_id=1575

SUSTAINABILITY OF BASIN LEVEL DEVELOPMENT UNDER A CHANGING CLIMATE

IBRAHIM. ALAMEDDINE, ABBAS FAYYAD, MAJDI ABOU NAJM & MUTASEM EL-FADEL
Department of Civil & Environmental Engineering, Faculty of Engineering & Architecture, American University of Beirut, Beirut, Lebanon.

ABSTRACT

The potential impacts of projected future climate change scenarios on the hydrologic response of a water-stressed Mediterranean river basin (Upper Litani River Basin in Lebanon) are quantified and assessed using the Water Evaluation and Planning (WEAP) model. Projected basin-level changes in water availability are then compared to multi-sector demands estimated under six basin-level development scenarios. The sustainability under these scenarios and the resilience of the system in the face of the projected climatic changes are then assessed in terms of a water resources index, demand reliability, demand satisfaction index, demand reliability index and the average duration of failure. The results reveal that the basin is expected to experience significant alteration in its hydrologic cycle and that current plans envisioning an increase in irrigated areas within the basin, is non-sustainable and will lead to a highly water stressed system. A conservative basin-level plan that integrates both supply- and demand-side measures is proposed in an effort to achieve a more sustainable system.

Keywords: Climate Change, Water Stress Indices, Watershed Management, WEAP

1 INTRODUCTION

Climate change has been identified as one of the main drivers of water availability in the 21st century [1, 2]. Its impacts are known to have wide ranging effects on the socio-economic, political, water [2, 3] and agricultural [3, 4] sectors jeopardizing the sustainability of environmental systems [3, 5]. The relationship between climate and water resources, in particular, is recognized with potential perturbations in surface runoff, groundwater recharge, lake levels, and water quality [3]. It is projected that climatic changes will increase the percentage of the world's population currently living in water stressed regions, from one-third to two-thirds over the next few decades [4, 6]. In the face of potential shortages, additional water resources will need to be developed to meet the future water demand. As a result, most river basins are expected to face increased pressures [6, 7]. Mediterranean basins are identified as one of the most vulnerable freshwater systems [2, 3]. In this context, social, economic and technological changes will need to be implemented to alleviate or minimize water scarcity challenges [8–11].

Unfortunately, many current national and/or regional water plans fail to appropriately account for projected climate change impacts and often disregard the importance of the regional and local context [8, 12]. This oversight stresses river systems, reducing their reliability and resilience and increasing their vulnerability and average duration of failure. To avert these shortcomings, basin level adaptation measures have to be properly formulated and their impacts assessed under future climatic conditions. In this context, future alternatives for water resources management under a changing climate were assessed at a river basin scale (the Upper Litani River Basin in Lebanon) taking into account demographic, economic, technological, and management constraints and opportunities. The efficacy of proposed alternatives was tested against a set of future climatic scenarios that represent Intergovernmental Panel on Climate Change (IPCC) defined story lines executed under several global circulation models (GCMs). Water and sustainability performance metrics [13–15] were then used for performance assessment and quantification of the combined impacts associated with

DOI: 10.2495/SDP-V13-N3-394-405

climate change and management alternatives on the water resources system in the basin. The paper concludes with arguing the need to integrate both supply- and demand-side measures for setting a more sustainable watershed management plan.

2 METHODS

2.1 Study Area

Extending over an area of ~2,186 km^2, the Litani river basin is the most important renewable freshwater system in Lebanon, a country that is expected to receive less precipitation and experiences higher temperatures under most future climatic scenarios coupled with a projected increase in water demand. Geomorphologically, the basin is divided into two sub-basins by the Qaraoun dam. The upper part (Fig. 1), the pilot area considered in this study, stretches from the Northern Bekaa plain to the dam. It covers an area of 1,545 km^2 with a population of ~660,000 in 2010 that is expected to reach 1.45 million by 2100. The upper Litani river basin (ULRB) is also the most agricultural productive area in Lebanon, with over 800 km^2 of productive agricultural land [16], of which 290 km^2 are intensively cultivated and irrigated. Future plans aim at increasing irrigated areas to ~540 km^2 by the late 2030 early 2040.

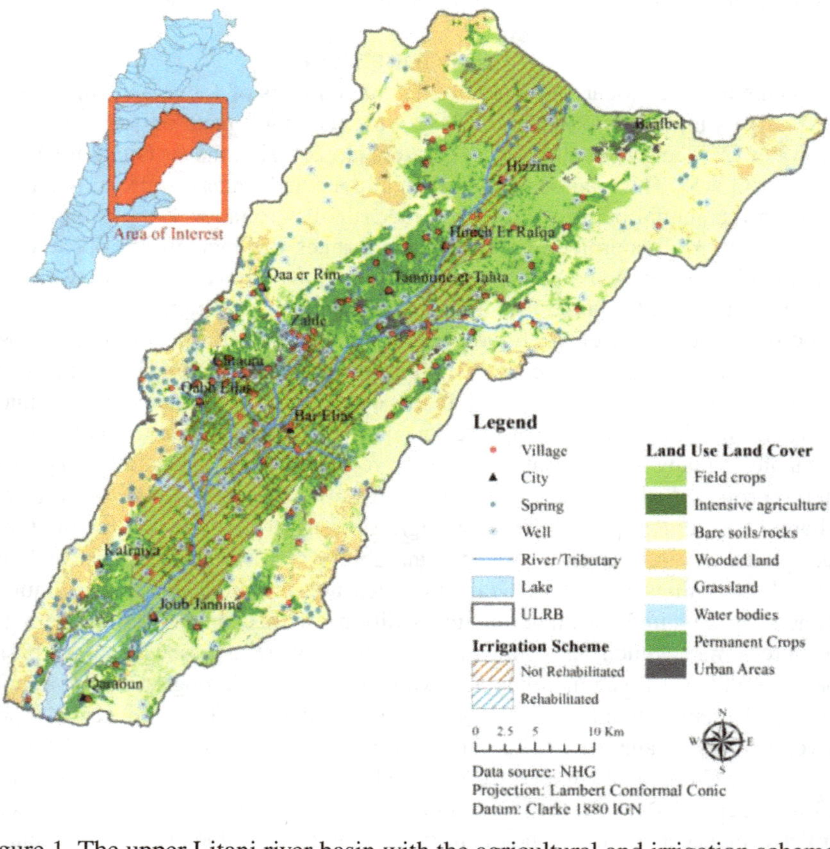

Figure 1. The upper Litani river basin with the agricultural and irrigation schemes.

Surface water resources in the ULRB are estimated at 290 MCM/year, of which 200 million cubic meter (MCM)/year are currently used. In addition, groundwater abstraction in the basin is significant (~190 MCM) and largely non-regulated. The water stored in the Qaraoun Lake, a reservoir with an estimated area of ~12 km^2 and a capacity of 220 MCM, is also used for hydropower and irrigation at a rate of 160 MCM/year [17, 18]. While at present only 15 MCM/year are used for irrigation, future plans envision using between 70 and 80 MCM/year of the reservoir's water for irrigation. The basin's water infrastructure is relatively old and inefficient, with an average supply service less than 10 hours a day and more than 50% of the transmission and distribution networks installed is more than 40 years ago. Future climatic prediction for the region foresee decreased precipitation and increased temperatures [19] resulting in lower surface runoff and groundwater recharge rates coupled with increased evapotranspiration [20]. Similarly, socio-economic pressures (population growth, shifts in agriculture practices and changes in water management) are expected to reduce the availability of water resources in the basin.

2.1.1 Future Climate Scenarios

Twenty four future climatic scenarios were used to assess the impact of climate change on the hydrology and water resources in the ULRB for the time period 2011–2100 [19]. 24 scenarios were constructed by coupling the four IPCC Special Report on Emissions Scenarios (SRES) emission scenarios (i.e. A1FI, A2, B2 and B1) with the five commonly used GCMs (i.e. Coupled Global Climate Model (CGCM2), Commonwealth Scientific and Research Organisation (CSIRO2), European Centre for Medium Range Weather Forecasts and Hamburg (ECHAM4), Hadley Centre Coupled Model (HadCM3) and Parallel Climate Model (PCM)). In addition, four ensemble scenarios were defined that averaged emission scenarios over the different GCM models [21–24]. The spatial resolution of the gridded climatic data was 0.5 degree (~50 km) [19]. Assessing the future climatic data over the study area indicated that the most pessimistic scenario is the fossil intensive ECHAM4-A1FI, while the CSIRO2-B2 (ecologically friendly) was found to be the most optimistic scenario. In this study, we present our results in terms of the four ensemble scenarios (A1FI, A2, B2 and B1) along with the most pessimistic and optimistic scenarios (ECHAM4-A1FI and CSIRO2-B2, respectively).

2.1.2 Water Resources Management Alternatives

To investigate the impacts of future climate change on the hydrology of the ULRB, six future management alternatives were defined namely, the Reference ('Do Nothing') (RA), the Non-Conservative (NCA), the Conservative Agricultural (CAA), the Management-Based Adaptation (MAA), the Structural Focused (SFA) and the Full Development (FDA). All six alternatives account for population growth in addition to different sets of proposed demand and supply-side mitigation measures. The RA represents a future with the same current water resources practices, while accounting for the increase in population and for a slight change in current agricultural practices. The NCA accounts for population growth and an expansion in agricultural practices coupled with no adaptation measures. This alternative represents mostly closely the proposed governmental plan for the basin. The CAA and the MAA account for limited expansion in agricultural practices, while considering different demand side adaptation measures. While the CAA focuses primarily on reducing agricultural water consumption by changing cropping patterns, the MAA projects a future where major losses from the supply system are reduced through enhanced connectivity and network rehabilitation. On the other hand, the SFA accounts for increases in population and agricultural practices, with emphasis on supply-side adaptation measures. Finally, the FDA assumes the implementation of a full suite of potential demand and supply-side adaptation measures at the basin level.

Table 1 summarizes the adaptation measures for each alternative with corresponding estimated costs. Table 2 outlines the key constraints and model variables that were defined for each alternative. Note that policy and nonphysical measures proposed in the alternatives were implemented in the Water Evaluation And Planning (WEAP) model [25, 26] through a reduction of areas (e.g. agricultural areas) and/or changes in the supply and return flow constraints. On the other hand, legislative and legal framework measures were accounted for by introducing constraints on permissible water abstraction rates (e.g. groundwater abstraction).

Table 1: Adaptation measures under future water resources management alternatives.

		Responses		
Alternatives	Adaptation Measure	*Increase supply*	*Demand reductions*	Cost
RA	None			
NCA	None			
CAA	Reduce agricultural water consumption (Improved Irrigation)		++++	+
	Change in type of crops		++	++
	Enhanced connectivity	+	++	+
	Improvement in agricultural technology & research		+	+
MAA	Enhanced connectivity		+++	+/++
	Network rehabilitation		+	+
	Reduced agricultural water consumption (Improved irrigation)		++	+
	Change in crop types		+	+
	Water conservation/ Sustainable water use (domestic water consumption)		++	+
SSA	Enhanced connectivity		+++	+/++
	Network rehabilitation & expansion (infrastructure/ distribution)	+++	+	++++
	Wastewater treatment	+	++	++++
	Increased water-use efficiency (Water reuse, water recycling)	+/++	+	+++
FDA	Enhanced connectivity		+++	+/++
	Network rehabilitation & expansion (infrastructure/ distribution)	+++	+	++++
	Wastewater treatment	+	++	++++
	Reduce agricultural water consumption (Improved Irrigation)		++++	+

(Continued)

Table 1: (*Continued*)

Alterna-tives	Adaptation Measure	Responses: *Increase supply*	Responses: *Demand reductions*	Cost
	Change in type of crops		++	++
	Improvement in agricultural technology & research		+	+
	Environmental sustainability & protection	++		++
	Increased water-use efficiency (Water reuse, water recycling)	+/++	+	+++
	Water conservation/ Sustainable water use (domestic water consumption)	+	+	+

Predicted water savings: < 10 MCM = +; 10–25 MCM = ++; 25–50 MCM = +++; > 50 MCM = ++++; Estimated implimentation costs: < 1 $/m^3$ = +; 1 – 2 $/m^3$ = ++; 2 – 3 $/m^3$ = +++; > 3 $/m^3$ = +++

Table 2: Summary of alternatives, drivers and key adaptation measures.

	Key drivers and adaptation measures
RA	• Population in basin moderately increasing ~1.45 million by 2100 • Economic development stable: water demand ~10% of urban demand by 2100 • Water consumption per capita 160 l/capita/day • Agricultural area stable (290 km^2 irrigated, 800 km^2 total agricultural lands (TAL)) • System losses 52.5%; irrigation efficiency 63% • Crop type distribution (55% mixed vegetables, 25% fruits, and 20% wheat) • Receives up to 15 MCM of water from Qaraoun Lake
NCA	• Population in basin moderately increasing ~1.45 million by 2100 • Moderate economic development: water demand 30% of urban demand by 2100 • Water consumption per capita 180 l/capita/day by 2100 • Increasing agricultural area (520 km^2 irrigated by 2040 and stabilizing; 900 km^2 TAL) • System losses 52.5%; irrigation efficiency 63% • Crop type distribution: (56% mixed vegetables, 19% fruits, and 25% wheat) • Receives up to 15 MCM of water from Qaraoun Lake
CAA	• Population growth moderately increasing ~1.45 million in 2100 • Economic development stable: water demand ~10% of total urban demand by 2100 • Water consumption per 160 l/capita/day • Increasing agricultural area (390 km^2 irrigated by 2040 and stabilizing, 800 km^2 TAL) • System losses 42.5%; irrigation efficiency 73.25% • Crop type distribution (40% vegetables, 15% fruits, and 45% wheat) • Receives up to 85 MCM of water from Qaraoun Lake

(*Continued*)

Table 2: (*Continued*)

	Key drivers and adaptation measures
MAA	• Population moderately increasing reaches ~1.45 million in 2100 • Moderate economic development: water demand 20% of total urban demand by 2100 • Water consumption per capita 140 l/capita/day • Increasing agricultural area (390 km^2 irrigated by 2040 and stabilizing, 800 km^2 TAL) • System losses 37.5%; irrigation efficiency 70% • Crop type distribution (45% vegetables, 20% fruits, and 35% wheat) • Receives up to 45 MCM of water from Qaraoun Lake
SSA	• Population moderately increasing reaches 1.45 million in 2100 • Moderate economic development: water demand 30% of total urban demand in 2100 • Water consumption per capita 180 l/capita/day by 2100 • Increasing agricultural area (390 km^2 irrigated by 2040 and stabilizing, 800 km^2 TAL) • System losses 20%; irrigation efficiency 69% • Crop type distribution (45% vegetables, 20% fruits, and 35% wheat) • Wastewater treatment and increased water-use efficiency • Receives up to 45 MCM of water from Qaraoun Lake
FDA	• Population in the basin moderately increasing reaches 1.45 million in 2100 • Moderate economic development: water demand 30% of urban demand in 2100 • Water consumption per capita 58.4 m^3/capita/year (160 l/capita/day) • Increasing agricultural area (390 km^2 irrigated by 2045 and stabilizing, 800 km^2 TAL) • System losses 20%; irrigation efficiency 73.25% • Crop type distribution: (40% vegetables, 15% fruits, and 45% wheat) • Wastewater treatment and increased water-use efficiency • Receives up to 85 MCM of water from Qaraoun Lake

2.2 Hydrological modelling, water stress indices, and performance metrics

A calibrated WEAP integrated hydrologic and water management model for the URLB [20] was used to investigate future climate impacts on water resources availability and to examine the effects of future management plans. The model generated monthly hydrologic basin responses (river flow, groundwater recharge, and evapotranspiration) that were used to define the amount of available water. These volumes were compared with demands under the various water resources management alternatives in order to determine the magnitudes of future water scarcities in the basin. The estimation of available water sources was conducted on a monthly time basis over the time period from 1961 to 2100. Seven water stress indices were used to assess the impact of management alternatives on the ULRB's water resources system:

C_R: Reliability that measures the time frequency when the system is able to supply enough water to meet the demand requirement; it is expressed in percent time;

C_{RS}: Resilience that measures the speed at which the supply system is able to recover after a failure (in %);
V_T: Time vulnerability that measures the maximum duration of system's failure in meeting its demands (in months);
V_V: Volumetric vulnerability that measures the maximum volume of the system's failure in meeting its demands (in MCM);
D_{SI}: Demand satisfaction representing reliability of volumetric supply of water (in % deficit in volume)
D_{RI}: Demand reliability that represents the total volume of supply that met the demand under a condition of no failure divided by the total water demand volume (in % deficit in volume);
F_{DI}: Failure deficiency that is the average duration of failure (in months)

More detailed discussions on the adopted metrics can be found in Hashimoto *et al.* [27], Zongxue *et al.* [28], Fowler *et al.* [29], Martin-Carrasco and Garrote [30] and Pulido-Velazquez *et al.* [31] to name a few. All indices were used to examine the reliability, resilience, demand satisfaction and reliability, and deficit vulnerability of the ULRB water sources system under the baseline (1961–2010) and future projections of climatic conditions and basin management alternatives (2011–2100).

3 RESULTS AND DISCUSSIONS

Simulation results showed that between 1961 and the mid 1990's the ULRB was able to meet almost all its demands. Starting in the mid-90's, the basin underwent significant agricultural expansion that coincided with several dry years. As such, the system experienced decreased water availability and increased water shortages. Current (averaged over 2001–2010) water demands in the basin, estimated at ~390 MCM/year, exceed the long-time average water supply of ~250 MCM/year (averaged over 1961–2010). Projected water demands under future water management alternatives were estimated independently of climatic conditions. Under the do-nothing RA management alternative, the average yearly demand reached 500 MCM/year by 2100. The increase in demand is mostly attributed to population increase. Under the NCA, the water demand was estimated at ~740 MCM/year by 2100, exceeding by far the annual renewable water resources (i.e. surface flow and recharge). Under the CAA, water demand reached 440 MCM/year by 2040, before decreasing to ~400 MCM/year by 2100. The increase in demand is instigated by the planned implementation of new irrigation schemes as well as changes in crop types between 2011 and 2040. The drop in demand is attributed to the change in crop types and a greater efficiency in agricultural water use. The MAA and SCA had average water demands of 456 and 500 MCM/year (average 2011–2100), respectively. Under these alternatives, upgrading the agricultural, distribution network, and wastewater systems were able to offset the agricultural expansion planned within the basin. Finally, under FDA, the annual water demand was estimated to range between ~420 MCM/year and 360 MCM/year between 2011 and 2100, respectively. The reduction in annual water requirement is attributed to the implementation of water saving measures as well as to the greater use of the stored water from the Qaraoun reservoir.

The available water supply in the basin under the different climate change scenarios was determined. Under ECHAM4-A1FI, the worst-case future climate change scenario, the projected average water supply, from both surface and groundwater sources, was estimated at 320 MCM/year (average 2011–2100). The mean supply reached 370 MCM/year under the most optimistic scenario, CSIRO2-B2. Meanwhile, under the ensemble scenarios the yearly water supply ranged between 337 and 345 MCM/year, with a standard deviation of ~60 MCM/year. As expected the variability between scenarios increases as time progresses.

Comparing projected water supply under future climate scenarios with the predicted demands under the defined water management alternatives provides the expected future water deficits in the ULRB. Under the RA, the impact of climate change for the first 30 years of the simulation is a deficit of ~120 MCM/year. The impact of climate change and increased demand becomes more apparent after the 2050's, whereby deficits consistently exceed 200 MCM/year. Under the NCA management alternative, the water deficit in the basin is dramatic. By 2030, the magnitude of the deficit is expected to surpass the basin's annual renewable resources. Under the CAA alternative, the proposed demand-side agricultural mitigation measures were able to reduce the water deficits to levels slightly higher than the long-term average for the basin. The water deficits under the MAA and SSA supply-side alternatives were in general very close to the deficits recorded for the RA, highlighting the limited ability of supply-side measures to reduce deficits. Note that the MAA, which incorporates some elements of demand-side mitigation measures, resulted in slightly lower deficits when compared with SSA. Under the FDA alternative, which accounts for all adaptation measures, the ULRB is projected to have minimal water shortages, ~20 MCM under the best CSIRO2-B2 climate scenario, and ~50 MCM/year under the worst ECHAM4-A1FI climate scenario, and between 28 and 35 MCM/year under the four ensemble models.

The water indicators for the ULRB were calculated for the baseline period between 1961 and 2010 as well as under future climatic scenarios (2011–2100). Between 1961 and 2010, the performance metrics were acceptable. The overall system's reliability (C_R) reached a high of ~97%, the average duration of failure (F_{DI}) of a deficient system was around 2.85 months, and it's maximum duration (V_T) was 4 months, while the system's resilience (C_{RS}) reaching 35% recovery/per month, the system demand satisfaction (D_{SI}) was around 95%, and the demand reliability (D_{RI}) reached 93%. Water stress indices for all management alternatives were calculated under the various climate change scenarios for 2011–2100 (Table 3). The results further corroborate the fact that the do nothing scenario will reduce the reliability (C_R) of the system to as low as 66% and increase the duration of system failure to around 4.2 months under the worst ECHAM4-A1FI. With the NCA the system's reliability (C_R) is not expected to go above 63% even under the most optimistic climatic scenario. In fact, under the NCA the ULRB would be able to supply the total water demand only 50% of the times. Under the CAA, the system's reliability will drop to between 73% and 88% and the duration of failure is expected to increase up to 4 months. Water reliability under the MAA and SSA were largely similar with the R_A varying between 65 and 74%. The demand satisfaction index was slightly higher for MAA, while the SSA had slightly shorter F_{DI} values and a slightly faster recovery. Under FDA, the overall system reliability ranged between 78 and 92%. The ULRB would thus be able to supply between 86 to 94% of its water demands. If the FDA mitigation measures are implemented in the URLB, it is projected that the annual water deficit would not exceed the 150 MCM/year threshold even under the driest year modeled under the worst-case climate change scenario (ECHAM4-A1F1).

Table 3: Performance metrics under all climatic scenarios and water management alternatives (2011–2100).

		ECHAM4-A1FI	Ensemble-A1FI	Ensemble-A2	Ensemble-B1	Ensemble-B2	CSIRO2-B2
RA	C_R (%)	66	69	70	71	71	74
	D_{SI}; D_{RI}	64; 52	68; 56	69; 58	70; 60	70; 60	75; 64
	C_{RS} (%)	24	26	27	27	27	30
	V_V	350	322	314	315	312	283
	F_{DI} ; V_T	4.2; 7	3.8; 7	3.7; 7	3.7; 6	3.7; 6	3.3; 5
NCA	C_R (%)	57	60	60	61	61	63
	D_{SI}; D_{RI}	44; 33	47; 36	48; 38	48; 38	48; 38	52; 41
	C_{RS} (%)	19	21	21	21	22	23
	V_V	646	627	620	620	619	587
	F_{DI} ; V_T	5.3; 8	4.8; 7	4.8; 7	4.8; 7	4.5; 7	4.3; 6
CAA	C_R (%)	73	78	80	81	81	88
	D_{SI}; D_{RI}	80; 69	84; 76	86; 78	86; 79	86; 79	91; 87
	C_{RS} (%)	25	29	29	29	30	32
	V_V	186	169	154	154	152	135
	F_{DI} ; V_T	4; 7	3.4; 6	3.4; 6	3.4; 6	3.3; 6	3.1; 5
MAA	C_R (%)	65	69	69	70	70	74
	D_{SI}, D_{RI}	69; 57	73; 62	74; 63	75; 63	75; 63	80; 70
	C_{RS} (%)	23	25	25	26	26	29
	V_V	272	257	249	249	247	212
	F_{DI} ; V_T	4.3; 7	4; 6	4; 6	3.8; 6	3.8; 6	3.4; 6
SSA	C_R (%)	66	68	68	69	69	73
	D_{SI}, D_{RI}	63; 48	66; 52	68; 53	68; 53	68; 53	72; 60
	C_{RS} (%)	24	26	26	26	26	30
	V_V	329	307	297	297	295	261
	F_{DI} ; V_T	4.2; 7	3.8; 6	3.8; 6	3.8; 6	3.8; 6	3.3; 6
FDA	C_R (%)	78	84	85	86	87	92
	D_{SI}, D_{RI}	86; 76	90; 82	91; 84	91; 85	91; 85	94; 90
	C_{RS} (%)	29	33	33	34	35	37
	V_V	148	142	141	143	142	129
	F_{DI} ; V_T	3.4; 7	3; 6	3; 5	2.9; 5	2.9; 5	2.7; 4

4 CONCLUSION

The potential impacts of future climate change on the ULRB were assessed with regards to six watershed management alternatives that incorporated different supply- and demand-side measures. The model simulations showed that the basin is highly vulnerable to projected future climate changes, even if no future development projects are implemented within the basin. The results also highlighted the risk of current plans aiming to increase agricultural areas in the basin up to 900 km^2, of which 520 km^2 would be intensively irrigated. If implemented, it will decrease the ULRB reliability to less than 63% even under the most optimistic future climatic scenario. With such a low reliability and increased vulnerabilities to the system, socio-economic hardships within the basin are expected along with an increase in water conflicts between various demand sectors. Moreover, with the total system demand exceeding supply the environmental situation in the basin –in terms of environmental flows and water quality- is expected to degrade significantly. This bleak situation is shared with other basins in the Middle East, many of which are trans-boundary river systems (Nile, Euphrates-Tigris, Orontes, Jordan and El-Kebir basins). There system failure will only strain existing precarious trans-boundary water conflicts. The future water scarcity in all basins in the Middle East are tightly linked to remediating poor water management in the agricultural sector. The model results highlighted the limitation of implementing supply-side mitigation measures alone. It is only when demand-side agricultural adaptation measures are implemented that the basin will achieve higher sustainability levels. Other adaptation measures targeting other economic sectors can help; but are ineffective alone to alleviate shortages within the basin. As such, priority across the Middle East should be given towards improving irrigation methods, exploring crops with lower water consumption, and increasing the efficiency of the distribution network.

ACKNOWLEDGEMENTS

The authors would like to acknowledge the support of the US Agency for International Development under the PEER initiative (Grant Number USAID-OAA-A-I1-00012).

REFERENCES

[1] Frederick, K.D., *Water resources and climate change* Edward Elgar Publishing, 2002.
[2] Bates, B., et al., *Climate change and water.* Intergovernmental Panel on Climate Change (IPCC), 2008.
[3] Parry, M. L., *Climate Change 2007: impacts, adaptation and vulnerability: contribution of Working Group II to the fourth assessment report of the Intergovernmental Panel on Climate Change.* Vol. 4. Cambridge University Press, 2007.
[4] Rijsberman, F.R., Water scarcity: fact or fiction? *Agricultural Water Management*, **80**(1), pp. 5–22, 2006.
https://doi.org/10.1016/j.agwat.2005.07.001
[5] Smakhtin, V., Revenga, C. & Döll, P., A pilot global assessment of environmental water requirements and scarcity. *Water International*, **29**(3), pp. 307–317, 2004.
https://doi.org/10.1080/02508060408691785
[6] Engelman, R., Cincotta, R.P., Dye, B., Gardner-Outlaw, T. & Wisnewski, J., *People in the Balance: population and natural resources at the turn of the millennium.* Population Action International, Washington DC, 2000.
[7] Vedwan, N., Ahmad, S., Miralles-Wilhelm, F., Broad, K., Letson, D. & Podesta, G., Institutional evolution in Lake Okeechobee management in Florida: characteristics, impacts, and limitations. *Water Resources Management*, **22**(6), pp. 699–718, 2008.
https://doi.org/10.1007/s11269-007-9187-7
[8] Vörösmarty, C.J., Global water resources: vulnerability from climate change and population growth. *Science*, **289**(5477), p. 284, 2000.
https://doi.org/10.1126/science.289.5477.284

[9] Moss, R.H., et al., The next generation of scenarios for climate change research and assessment. *Nature*, **463**(7282), pp. 747–756.
https://doi.org/10.1038/nature08823

[10] Cooley, H., et al., *Understanding and reducing the risks of climate change for transboundary waters*. Pacific Institute Oakland, 2009.

[11] Bazzani, G.M., Di Pasquale, S., Gallerani, V., Morganti, S., Raggi, M. & Viaggi, D., The sustainability of irrigated agricultural systems under the water framework directive: first results. *Environmental Modelling & Software*, **20**(2), pp. 165–175, 2005.
https://doi.org/10.1016/j.envsoft.2003.12.018

[12] Füssel, H.-M., Vulnerability: a generally applicable conceptual framework for climate change research. *Global Environmental Change*, **17**(2), pp. 155–167, 2007.
https://doi.org/10.1016/j.gloenvcha.2006.05.002

[13] Rosenzweig, C., Strzepek, K., Major, D., Iglesias, A., Yates, D., Mccluskey, A. & Hillel, D., Water resources for agriculture in a changing climate: international case studies. *Global Environmental Change*, **14**(4), pp. 345–360, 2004.
https://doi.org/10.1016/s0959-3780(04)00062-7

[14] Arnell, N.W., Climate change and global water resources: SRES emissions and socio-economic scenarios. *Global Environmental Change*, **14**(1), pp. 31–52, 2004.
https://doi.org/10.1016/j.gloenvcha.2003.10.006

[15] Alcamo, J., Flörke, M. & Märker, M., Future long-term changes in global water resources driven by socio-economic and climatic changes. *Hydrological Sciences Journal*, **52**(2), pp. 247–275, 2007.
https://doi.org/10.1623/hysj.52.2.247

[16] Ministry of Agriculture, *Land use land cover map for Lebanon*, Ministry of Agriculture: Beirut, Lebanon, 2002.

[17] Ministry of Environment, *State of the Environment Report*. Beirut, Lebanon, 2001.

[18] Ministry of Energy and Water, *National Water Sector Strategy*. Beirut, Lebanon, 2010.

[19] Mitchell, T.D., et al., *A comprehensive set of high-resolution grids of monthly climate for Europe and the globe: the observed record (1901–2000) and 16 scenarios (2001–2100)*. Tyndall Centre for Climate Change Research Working Paper, **55**: p. 25, 2004.

[20] Fayad, A., *Watershed modeling using integrated hydrologic information system*, in *Department of Civil and Environmental Engineeering*. American University of Beirut: Beirut, Lebanon, 2012.

[21] Kyselý, J. & Beranová, R., Climate-change effects on extreme precipitation in central Europe: uncertainties of scenarios based on regional climate models. *Theoretical and Applied Climatology*, **95**(3–4), pp. 361–374, 2009.
https://doi.org/10.1007/s00704-008-0014-8

[22] Maurer, E., Adam, J. & Wood, A., Climate model based consensus on the hydrologic impacts of climate change to the Rio Lempa basin of Central America. *Hydrology & Earth System Sciences Discussions*, **5**(6), pp. 3099–3128, 2008.
https://doi.org/10.5194/hessd-5-3099-2008

[23] Semenov, M.A. & Stratonovitch, P., Use of multi-model ensembles from global climate models for assessment of climate change impacts. *Climate Research* (Open Access for articles 4 years old and older), **41**(1), p. 1, 2010.
https://doi.org/10.3354/cr00836

[24] Tao, F., Zhang, Z., Liu, J. & Yokozawa, M., Modelling the impacts of weather and climate variability on crop productivity over a large area: A new super-ensemble-based probabilistic projection. *Agricultural and Forest Meteorology*, **149**(8), pp. 1266–1278, 2009.
https://doi.org/10.1016/j.agrformet.2009.02.015

[25] Yates, D., Purkey, D., Sieber, J., Huber-Lee, A. & Galbraith, H., WEAP21—A demand-, priority-, and preference-driven water planning model part 2: aiding freshwater ecosystem service evaluation. *Water International*, **30**(4), pp. 501–512, 2005.
https://doi.org/10.1080/02508060508691894

[26] Yates, D., Sieber, J., Purkey, D. & Huber-Lee, A., WEAP21—A demand-, priority-, and preference-driven water planning model: part 1: model characteristics. *Water International,* **30**(4), pp. 487–500, 2005.
https://doi.org/10.1080/02508060508691893

[27] Hashimoto, T., Stedinger, J.R. & Loucks, D.P., Reliability, resiliency, and vulnerability criteria for water resource system performance evaluation. *Water Resources Research*, **18**(1), pp. 14–20, 1982.
https://doi.org/10.1029/wr018i001p00014

[28] Zongxue, X., Jinno, K., Kawamura, A., Takesaki, S. & Ito, K., Performance risk analysis for Fukuoka water supply system. *Water Resources Management*, **12**(1), pp. 13–30, 1998.
https://doi.org/10.1023/a:1007951806144

[29] Fowler, H., Kilsby, C. & O'Connell, P., Modeling the impacts of climatic change and variability on the reliability, resilience, and vulnerability of a water resource system. *Water Resources Research*, **39**(8), 2003.
https://doi.org/10.1029/2002wr001778

[30] Martin-Carrasco, F.J. & Garrote, L., Drought-induced water scarcity in water resources systems, in *Extreme hydrological events: new concepts for security*. In: O.F. Vasiliev, et al., Editors. 2007, Springer Netherlands. pp. 301-311.

[31] Pulido-Velazquez, D., Garrote, L., Andreu, J., Martin-Carrasco, F-J. & Iglesias, A., A methodology to diagnose the effect of climate change and to identify adaptive strategies to reduce its impacts in conjunctive-use systems at basin scale. *Journal of Hydrology*, **405**(1), pp. 110–122, 2011.
https://doi.org/10.1016/j.jhydrol.2011.05.014

THE ENVIRONMENTAL ADDED VALUE OF WATER: A PROPOSAL

ALFONSO G. BANDERAS TARABAY & REBECA GONZÁLEZ-VILLELA
Mexican Institute of Water Technology (IMTA), Mexico.

ABSTRACT

This work develops a proposal to assign an added value to the water used by man in various productive processes. The cost of energy needed to evaporate a cubic meter under natural conditions is multiplied by the lowest local price per kWh to calculate an 'intrinsic value' of water (IVW). The resulting amount constitutes the monetary unit that is applied to calculate an added value on an entirely environmental basis, regardless of the law of supply and demand and the subjectivity prevailing in the market. Then, it is necessary to calculate the volume required to return to the used water the quantity and quality it had before the production process, or the dilution volume necessary for the wastewater to comply with the corresponding ecological standard for a particular pollutant (indicator). This amount is called the 'restitution volume' (RV), and the product (IVW x RV) constitutes the proposed environmental added value (WEAV). A table showing the evaluation of RV for a set of production processes is included. Several examples are developed to calculate the price of water in this context, including the human right to water and nature conservation, for which it is necessary to include some local costs of the regular water service, such as the federal and municipal costs of construction, operation, maintenance, and administration (COMA) of infrastructure: Water price = IVW + federal COMA + municipal COMA + WEAV, in which IVW represents a payment for the environmental services of the water within the hydrological cycle, such as the support of animal and plant production in the biosphere, climate regulation and shielding against UV rays. A brief discussion on the advantage of this methodology to ensure environmental and human rights to water is included.

Keywords: cost, ecological added value, price, uses, water evaporation heat.

1 INTRODUCTION

The advance of 'civilization' has caused the global deterioration of natural resources and biodiversity, which has led to several international meetings aimed at establishing agreements among countries to take the necessary measures to halt and, as far as possible, to revert natural and social degradation, which is evident and even alarming in many parts of the world Wright and Nebel [1].

Physical causes of deterioration in developing countries are the over-exploitation of natural resources, the production of large quantities of sewage and garbage, and the emission of greenhouse gases into the atmosphere. Behind these physical causes is the economic cause, represented by the gross domestic product of each country, which is an indicator of its economic prosperity and serves as a guarantee for international loans and investments, requested with the aim of promoting 'development' Stern [2].

Water is among the resources affected by deterioration in spite of the fact that it is a fundamental resource for life on the Earth. Over 90% of all visible living matter is plant life. Plants build and renew their tissues—about 25 billion tons annually—through photosynthesis, which combines oxygen obtained from water molecules with environmental carbon dioxide molecules to form carbohydrate molecules and release molecular oxygen to the environment, which is as necessary as water for life Niklas and Spatz [3]; Banderas and Gonzalez-Villela [4].

In addition, water is not an ordinary resource, such as cotton, wood, iron, or petroleum, which can be substituted in many applications by synthetic products, but with all the environmental problems involved in such substitution EEA [5], which affects its value in a free market

DOI: 10.2495/SDP-V13-N3-406-417

economy. Water has an irreplaceable function for life; hence, it should not be subject to a market economy nor be transferred to private companies Barlow and Clark [6–8].

Moreover, the considerations necessary to determine a fair water price are different between northern and southern countries, mainly because in the latter ones water is frequently limited by its heterogeneous distribution, arid conditions and growing competition for it among individuals, communities, and states Anton [9]; Solis [10], resulting in very variable prices that allow for speculation, as will be seen in Table 2.

The values proposed in this paper are taken out from such a commercial context to present an assessment of water and its uses based on purely physical and chemical concepts: 1) the energy required to mobilize it through the hydrological cycle, which enables it to maintain natural phenomena and its use for human consumption; and 2) the relation of concentrations of some chemical indicator of water quality between the influent to the productive process and the effluent with the residual water.

To this end, an intrinsic water value (IVW) Banderas and Gonzalez-Villela [4] will be applied, calculated on the basis of a unique characteristic of water, its latent heat of evaporation (hence its 'intrinsic' character), measured at 27°C, the critical temperature of sea water at which the formation of hurricanes begins Anthes [11], being these last one of the most important water sources for the country Prieto *et al.* [12].

On the other hand, it is common to evaluate the cost-benefit ratio when humans access natural resources, but the cost to nature of human activities is rarely evaluated, and even less the benefit to nature of those activities (if any). In this alternative, an environmental added value (WEAV) is proposed for the water used in several production processes in a purely environmental context, related to the restitution of the quantity and quality of the water before being used, without involving commercial considerations. This WEAV reflects the cost to nature of the degradation of the volume of water used in human productive activities.

With this approach, the authors take the place of the entity affected by the so-called demophoric explosion by Vallentyne [13], which is the same that provides water: nature. At the same time, they help nature impose the price of its environmental and productive services. The importance of IVW is to help nature translate these processes into a monetary base.

The exercise of eliminating value judgments and market laws of the water price is done since they could interfere with the human and environmental rights to this commodity, especially where it is scarce; so that the price only includes service costs, such as construction, operation, administration, and maintenance (COMA) of the infrastructure, whose prices are unitary and therefore auditable, as well as IVW and WEAV, and to show how the human economy is not very different from nature's economy, even when the COMAs, which are usually subject to speculation, are eliminated.

2 BACKGROUND

Mexico receives an annual rainfall volume of 1,489 km^3. Evaporation returns 1,065.994 km^3 (71.6%) to the atmosphere; infiltration accounts for 92.3 km^3 (6.2%); leaving 471 km^3 (31.6%) of available water CONAGUA [14]. To evaporate this rainwater, each year the Sun introduces 3.63 x 10^{12} MJ yr^{-1} or 1.01 x 10^{12} kW h yr^{-1} of energy to Earth's surface. This amount of energy is multiplied by the electric service rate to translate it into money (Solar Financing), resulting in about US$46 billion [3], which serves to give an anthropocentric measure, on physical bases, of what the human species owes nature for water.

The IVW (US $ 0.031) is obtained by dividing the Solar Financing by the total volume of precipitation, and represents the cost of nature to distill a cubic meter of water by means of solar radiation, in the physiographic conditions of the country, and which subsequently precipitates as clean water conserving all its potential uses [4]. From the practical point of view of this paper, the IVW would be representing the human right to satisfy the vital water requirement.

This value is the first investment made by nature for the benefit of life on the planet, including the environmental services of evaporated volume.

The capital gain or added value (in Spanish: plusvalía = Increased value of something for reasons extrinsic to it [15]), varies greatly. For example, the value of a given volume of water may not be the same if used for:

- Washing and personal hygiene than for doing business with water (bottling and use in hotels, public baths, restaurants, and spas)
- Watering crops for self-consumption than for selling them
- Irrigating crop fields than for irrigating a golf course
- Operating a hydroelectric plant than for operating a thermoelectric plant
- Recreational purposes than for fish farming

These cases are presented on a decreasing scale of real basic needs that are compared with commercial and sumptuary uses through value judgments [16], which are accentuated under conditions of scarcity. The authors of the present paper think that human rights to water, food, and a healthy environment must be separated from value judgments and be evaluated objectively on physical bases.

Sumptuary value applications should consider a surcharge to the rent derived from the cost of basic COMA. For example, an aqueduct built to provide drinking water to a city cannot generate the same income as one used for a thermoelectric plant or for oil fracking; furthermore, the latter apparently implies a huge ecological and environmental risk [17].

3 METHODOLOGICAL CONSIDERATIONS

The method described here uses three measurable quantities: 1) the volume used for any purpose; 2) the volume of restitution (VR), where restitution is defined as: 'to set up again; the making good of or giving an equivalent for some injury' [18] and 3) the IVW. Thus, value judgments and speculation are avoided when calculating the surcharge on luxury uses, fines and transactions.

Conservation areas under federal or municipal control fed by rainwater are exempt from environmental added value, as wilderness areas, unless they are under some form of exploitation. Those using surface and groundwater should calculate their volumes and the environmental flows required by law and observe their application by defining the amount, frequency, and timing of environmental flows Richter *et al.* [19]. However, this does not exempt from paying the respective COMAS if they have to use some kind of aqueduct.

Environmental added value for hydroelectric power generation includes two components: the change in water quality of the reservoir due to the damming, which is reflected in its trophic status [20], and the volume evaporated as a result of increased evaporation surface when the river enters the reservoir (ΔVev). The former is measured with an adequate water quality indicator (X_2), such as biochemical oxygen demand (BOD), total nitrogen, or total phosphorus [19], and then divided between the same in the river (X_2/X_1) = k. This quotient represents the number of times that the water in the reservoir must be diluted to return to the quality it has in the river. The restitution volume is equal to k multiplied by the reservoir volume.

Finally, the environmental added value of the hydroelectric generation is obtained by multiplying the IVW by the sum of the restitution volumes resulting from both evaporation and water quality.

This procedure can also be applied to lakes and rivers affected by cultural eutrophication [20], that is, wastewater discharges from communities that inhabit their catchment and river basins.

In other production processes, the appropriate indicator of contamination in the sewage should be sought, and the RV is calculated in relation to the ecological level allowed by the local environmental norm with respect to this contaminant. RV is the amount of water needed to return it to the quality it had before passing through the production process. To do this, a physical, chemical, or biological indicator showing the quality change due to the production must be selected [21]. This procedure would apply in the case of large cities that do not separate gray waters from sewage, as well as for industrial and fracking purposes.

If this indicator is not in the water before passing through the production process, it should be excluded and whoever includes it will be sanctioned until a way is found to remove it from its effluent. These indicators have different weights in different industries, and the most synergistically pernicious, the most toxic to life (polluting type) and that of higher concentration Ostrander [22] should be preferred. There will be the option of investing in treatment to avoid the fine and closure and to minimize payment for environmental gain.

In the case of aquaculture [23]: for self-consumption in river water that does not require infrastructure, the IVW should only be paid by the estimated production volume per unit weight of the exploited species. If stagnant water is used, the IVW should be paid for the used volume; evaporation is not paid, although it can be high; if additional infrastructure is required, the municipal COMA should be paid; commercial aquaculture pays IVW by the volume of water, plus the federal (and/or municipal) COMA and the environmental added value due to changes in volume and water quality.

The environmental added value of water for different types of irrigation contains the same elements; the difference lies in the volume used, the evaporation measured during the agricultural cycle, and the type and quantity of agrochemicals applied. Irrigation by technological means, which uses less water, will be cheaper, which will stimulate their proliferation and facilitate their efficient use. Organic agriculture minimizes WEAV, since it does not apply agrochemicals FAO [24], and therefore does not pollute the receiving water bodies of agricultural return water, rather than with sediment.

Public parks and sports facilities only pay the IVW, the municipal COMA, and evaporation, but sumptuary commercial irrigation, which is used for business—including golf courses and private sports courts—must pay a premium, multiplying the IVW by the number of square (or cubic) meters of development.

In the case of livestock, urban, touristic and industrial uses, the criterion is the same; the only thing that varies is the indicator of water quality before and after the development. In the case of intensive farming, the drains are identifiable. In ranching, the pollution of the environment, surface runoff, groundwater and evaporation in the troughs should be estimated. Small urban settlements without industry will pay an added value relating to BOD, with the option of investing that capital in treatment infrastructure to reduce payment to a minimum. This option is valid for all productive processes.

In the case of fines and penalties for the misuse of water, or for providing defective services, the respective environmental added value must be paid. For example, in the case of the Sonora River [25], data from the Water Quality Monitoring Network [14], obtained on or as close as possible to the site before the event, should be consulted and compared with assessments on the site after the incident, calculating the concentration difference regarding quality at the affected site. This should be done in several places at the edge of the incident, and environmental added value should be calculated, that is, the volume necessary to restore the quality of the aquatic ecosystem before the event. This restitution volume is multiplied by the IVW to calculate the amount of the fine. This penalty shall apply notwithstanding the other impacts to terrestrial ecosystems and human populations.

If federal and/or municipal infrastructure is used, the respective COMA must be charged. In Mexico, there are approximately 4,462 storage dams and levees, 6.5 million hectares of irrigated land, 2.9 million hectares of technified rain-fed land, 631 water treatment plants in operation, 2,029 municipal wastewater treatment plants in operation, 2,186 industrial wastewater treatment plants in operation, and more than 3,000 km of aqueducts [14]. The administration and operation expenses include wages and energy costs.

Fines and penalties for overexploited or contaminated aquifers will directly depend on the depth to which the aquifer is harmed. If the damage is on an intermediate or regional flow [26, 27] it will result in the cancellation of the water allocation and the decommissioning of facilities.

4 RESULTS

The water added value should increase according to the non-exhaustive scale shown in Table 1.

The Vital Human category, which is to meet the daily individual water requirement, is about 4 to 6 L per day per person [28]. Considering 5 L for 120 million people, that is 600 $\times 10^6$ L a day = 0.219 km^3 a year. It is a negligible portion of the annual Vp and 0.046% of the available volume. The water needed to meet other basic human requirements, such as washing and personal hygiene, vary according to personal habits and region. On a base of 100 L individual^{-1} day^{-1}, it would be 4.38 km^3 a year, or 0.93% of the available volume. Mexicans should pay, for the first concept, US$4.03 million a year (Table 4 considers only payment for the IVW), and for the second one US$302 million a year, or US$2.55 per inhabitant a year (the proposed payment is IVW plus federal COMA). This is the current fee paid every two months for minimum domestic consumption (2015 prices).

If municipal COMA is used, with a mean value of US$0.8 per cubic meter [29], the last amount increases to US$32.7 per inhabitant per year. It can clearly be seen that municipal costs increases the price of water. These costs vary greatly in the country due to the commercial added value (Table 2); therefore, the correct COMA should be calculated, which should be close to that of humid areas (US$0.2), where speculation is minimal, because water is abundant. Thus, the payment is reduced to a quarter (US$8.00 per inhabitant per year), which is a reasonable amount considering the country's economy, with an average per capita income of US$ 6,000 INEGI [30].

Roughly, local agriculture produces almost 200,000 T y^{-1} of crops and accounts for 77% of the water consumed in the country: 20.76 groundwater + 41.04 surface water = 61.8 km^3. The evaporation of this volume requires 42 x 10^9 kW h yr^{-1}, which correspond to an IVW of US$ 1.9 billion yr^{-1} for environmental services. However, 50% is lost by evaporation during the production process, so its WEAV is a half of the latter amount: US$ 0.95 billion yr^{-1}, giving a total of US$ 2.85 billion. To get an idea of the benefit provided for this WEAV, the sale of exported agricultural products reached US$ 24.4 billion in 2013 [31, 32], more than 25 times the amount of 0.95 billion. This, supposing we are dealing with organic agriculture that does

Table 1: Prices proposed for water for different uses.

	Use	Rate			
I	**As a human right**	WEAV	IVW	COMA	
				F	M
1	Vital Human	-	x		
2	Basic Human	-	x	x	
II	**Nature Conservation**				
3.1	Municipal, state	-	x	x	
3.2	National	-	x	x	
III	**As a natural resource**				
4	Subsistence fish farming	-	x		
5	Hydropower generation	= f (ΔBOD + ΔVev)	x	x	
6	Thermoelectric generation	= f (ΔT + ΔCOD + ΔVev)	x	x	
7	Commercial fish farming	= f (ΔBOD + ΔVev)	x	x	
8	Subsistence flood irrigation	= f (ΔN,P + ΔCOD + ΔVev)	x		x
9	Subsistence modern irrigation	= f (ΔN,P + ΔCOD + ΔVev)	x		x
10	Commercial rolled irrigation	= f (ΔN,P + ΔCOD + ΔvVev)	x	x	x
11	Commercial modern irrigation	= f (ΔN,P + ΔCOD + ΔVev)	x	x	x
12	Comm sumptuary irrigation*	= f (ΔBOD + ΔCOD + ΔVev)	nx	x	x
13	Public parks and sports facilities	= ΔVev	x		x
14	Subsistence livestock	= f (BOD)	x	x	
15	Commercial livestock	= f (BOD)	x	x	x
16	Urban/municipal service	= f (BOD + COD)	x	x	x
17	Tourism/recreational services	= f (BOD)	x	x	x
18	Industrial services	= f (ΔX,Y + COD + ΔVev)	x	x	x
19	Commercial transactions	= V	x	x	x
20	Fines: waste or pollution	= respective ΔV	x	x	x
21	Groundwater extraction	= f (Vev)	x	x	x

COD, BOD = Chemical and biochemical oxygen demand [19].
Δ(T, N, P, X, Y) = (influent – effluent) temperature difference or the appropriate water quality indicator.
ΔV = Volume.
ΔVev = Evaporated Volume.
ΔCOD = (influent – effluent) COD difference or the appropriate water quality indicator.
ΔBOD = (influent – effluent) BOD difference or the appropriate water quality indicator.
Av = Added value: The volume of water required to achieve the original water quality or the level permitted by the respective ecological standard, multiplied by the IVW.
* =Applies to private property (tennis courts and water and other sports facilities, including golf).
n = Developed area in square meters.

Table 2. Production costs and water rates in some cities randomly selected on the North (Tijuana) – South (Tuxtla Gutierrez) gradient [29].

#	City	Cost MX$	Rate MX$	Difference MX$	Difference average
1	Tijuana	9	21	*12*	
2	Monterrey	9.5	12.5	*3*	
3	Aguascalientes	7	12	*5*	
4	León	10.3	18.6	*8.3*	
5	Pachuca	7.4	11.3	*3.9*	
6	Mexico City	8	15	7	μ= *6.53*/4.83
7	Puebla	8	11.50	3.5	
8	Xalapa	7.32	12	4.7	
9	Cuernavaca	5	8.8	3.8	
10	Mérida	4	6	2	
11	Tuxtla Gutierrez	6	14	8	

not contaminate the water of agricultural return by agrochemicals, nor that the water comes from a dam. If agrochemicals are used and/or there is a dam, the WEAV must be adjusted as we have seen before.

Groundwater agriculture deserves further comment. Crop production began to depend on groundwater until very recently in human history, with a boom in the past century Shah [33]. It is estimated that 97% of the country's water lies underground, but the number of overexploited aquifers is increasing, because arid zones occupy almost 70% of the country [14], Diaz-Padilla *et al.* [34]. To calculate the environmental added value for almost 20.76 km^3 extracted annually, it should be considered that, once on the surface, groundwater is subject to infiltration, runoff, and a more intense evaporation (almost 90%), due to aridity, before and after use, and becomes part of the hydrological cycle. Therefore, this vapor is increasing the potential for rain, which eventually will fall in wet areas prone to more frequent flooding due to global warming [12]. In this case, modernization could apparently increase irrigation efficiency and reduce WEAVE, but the increased frequency and intensity of floods have disastrous effects for mankind.

5 DISCUSSIONS

So far, the ideas developed in this paper draw a picture in which water is no longer a simple natural resource that can be marketed and an excuse to collect money, it becomes a dynamic entity that interacts with the human economy by its irreplaceable vital character and its own dynamics, that man must respect and preserve to ensure water supply in the future.

The IVW is included as a fixed fee per volume unit representing the intrinsic value of water, which is currently immeasurable due to its vital function both inside and outside living things. The volumetric IVW, so-called Solar Financing, together with the proposed environmental added value (WEAV), calculated from the Restitution Volume (RV), allows to estimate nature's investment in environmental services and the cost to nature of human activities in a purely environmental context, related to the restitution of the quantity and quality of water before being used, without involving commercial considerations. This WEAV reflects the cost to nature of the degradation of the volume of water used to support human development.

The WEAVE is a natural added value calculated and applied in this paper to estimate the value of water volumes exchanged between man and nature, but it can also be applied to calculate fines for wasting, misusing, or polluting water, or for having defective infrastructure affecting administration and electricity costs, as well as human and environmental wellness. They could be included in the price given to each unit volume offered to the public. This is:

$$\$ = \text{IVW} + \text{COMA fed.} + \text{COMA municipal} + \text{WEAV.}$$

Excepting the IVW and the WEAV, the components of this equation vary depending on management costs and the type and use of infrastructure, but these are perfectly controllable through unit cost rates and audits, thus avoiding the introduction of charges for spurious items (such as speculation), that are not strictly derived from the environment care.

IVW refers to unpaid ecosystem services, and WEAV is a function of the environmental impact indicators related to use. Therefore, it does not lend itself to speculation, which is very convenient if it is to satisfy the human right to water and sanitation, recognized by the United Nations General Assembly through Resolution 64/292 of 28 July 2010. It 'calls upon States and international organizations to provide financial resources, to promote training and technology transfer to help countries, in particular developing countries, to provide a supply of drinking water and healthy, clean sanitation', and the General Comment No. 15 on the right to water—adopted by the Committee on Economic, Social and Cultural Rights in November 2002, set out in Article I.1—states: 'The human right to water is indispensable for a dignified life', and defines the right of everyone to sufficient, healthy, physically acceptable and affordable water for personal and domestic use. Furthermore, the United Nations Development Program suggests that the cost of water should not exceed 3% of family income.

To meet this objective in a simple and clear way, the price of consumption of a given volume must be the same regardless of the region, which is justified because the money raised should apply to different works and priority actions that depend on the local availability of water. That is, water service comprises: containment, diversion, capture and/or removal, storage, treatment, distribution, use, treatment, disposal, 2nd treatment and recycling. In arid areas, the last three stages are a priority and occur after human use, since the efficiency of a given volume increases due to recycling, while in humid areas there are two stages that occur before use (containment and diversion), especially during the rainy season. In addition, treatment in humid areas becomes more important because environmental conditions are favorable for the development of pests and diseases and requires more attention because local urban infrastructure and water and sanitation systems are inadequate [14].

In this same sense, the above-calculated payment of US$8.00 per inhabitant per year is a reasonable amount considering the country's economy. The authors consider that all people must pay for the water they consume. Giving free water to those who have less is equivalent to re-labeling them as marginalized, which would demonstrate a lesser human quality of those responsible for promoting development, as mentioned in a previous paragraph.

That is, with the exchange rate of 2015, the Mexican people (120 million) should pay USD 305 million per year for their basic and vital water consumption. The income in one year of many professional athletes and actors in the USA and Europe is close to this amount.

Therefore, developed countries would have no problem in giving this money to developing countries, but that is not the solution to marginalization. To many native peoples, water has fallen from the sky for generations. So, the real problems of marginalization are:

1. The historical appearance of fences, walls, alarms, dogs, and even weapons and ideologies that prevent access to areas in which their ancestors found daily sustenance,
2. In developing countries, the diversion of economic resources is more the rule than the exception IT [35], IMCO [36].

The solution to these problems requires, in the first case, education and better legislation. In the second case, greater supervision of the application of national and international economic resources is required, to avoid losses and direct investments to where they are most needed, to create jobs and income for people to have resources to pay for many things besides water. That is, eliminate poverty, and in the case of disabled persons (abandoned elderly, the mentally ill, the homeless), water consumption would be paid by public and private foundations.

6 FINAL CONSIDERATIONS

1. The IVW is a commercial value calculated without prejudice, so it could be a fair, invariable and naturally fair price and a reasonable amount that would guarantee the human right of access to water in sufficient quantity and quality to cover basic needs. Its application would stimulate the optimal functioning of the network of hydrometric stations in the country. The capital raised by this concept would be used for projects of scientific and technological research that would result in works and actions for conserving the volume and quality of water in natural ecosystems —because if the environment falls, there is no water—and to buffer the effects of climate change. The money collected applying the WEAV to fines and penalties would be applied to transparent management and to disseminate updated knowledge on water.
2. If the IVW were to be charged for all precipitated water in the country, a very large amount would be collected, enough to pay (if nature would charge it) for the environmental services provided by the water to maintain an adequate biosphere.
3. In the unlikely event that the government can raise the money for the payment of IVW of available water (31.6% of Vp), which is impossible given its uneven distribution and lack of service coverage and high demand of users (except natural ecosystems), the capital could cover the combined budgets of the CONAGUA and SEMARNAT (Water and Environment and Natural Resources Bureaus) in 2016, both fused and dedicated to caring together, directly and indirectly, technical and naturally, the quantity and quality of territorial water.
4. The proposed WEAV has several advantages if it is applied to water service rates without involving value judgments regarding the necessity and convenience [10]: it can be added to the cost of collection, processing, distribution, maintenance, and expansion of the network; to fix the amount of the transactions or exchanges of water volumes between individuals, communities, areas, regions or basins; to calculate the amounts of fines for waste, pollution or misuse of water; or to sanction any defective infrastructure that does not fulfill its mission.
5. In addition, the IVW allows estimating a monetary value for the so-called water footprint by Hoekstra [37].

7 PROSPECTS FOR RESEARCH

Three issues should be noted:

1. Hydrological data vary over the years due to climate variability. In addition, as global warming promotes evaporation, a greater potential of rain in wetlands and greater aridity in dry areas is expected. It would be necessary to agree on which mean values should be considered in order to obtain precise calculations.
2. It is necessary to complement Table 1 with the greater quantity of productive activities and its WEAV proposed.
3. A study should be conducted to assess the feasibility of this proposal, which would require changing the economic paradigms in a market economy of water for a 'natural economy of water.'

REFERENCES

[1] Wright, R.T. & Nebel, B.J., *Environmental science: toward a sustainable future*. 10th Ed. Pearson/Prentice Hall, 2008.

[2] Stern, N., *The economics of climate change: the stern review*, Cambridge University Press: UK, 2008.

[3] Niklas, Karl J. & Spatz, H-C., *Plant physics*. The University of Chicago Press; USA, 2012.

[4] Banderas, A. & González-Villela, R., The intrinsic value of water: A proposal. *WIT Transactions on Ecology and The Environment*, **200**, pp. 137–145, 2015. https://doi.org/10.2495/WS150121

[5] European Environment Agency. *Environment in the European Union at the turn of the century*. Office for Official Publications of the European Communities, Elsevier Science Ltd, Luxemburg, 2001 (Spanish version).

[6] Barlow, M. & Clark, T., *Blue gold: the fight to stop corporate theft of the world's water*. The New Press, 2002 (Spanish version).

[7] Transnational Institute (https://www.tni.org/en/publication/here-to-staywa ter-remunicipalisation-as-a-global-trend). Consulted 18/01/2017.

[8] Freshwater Organization, available at: http://freshwater.org/wp-content/uploads/ joomla/ PDFs/critical-water/valueofwater.pdf. (accessed 18 January 2017).

[9] Anton, D.J., *Thirsty cities: urban environment and water supply in Latin America*. International Development Research Center, Ottawa: Canada, 1993.

[10] Solís, L., *La Escasez, el Costo y el Precio del Agua en México*. El Colegio de México: Mexico, 2005.

[11] Anthes, R.A., Tropical cyclones: Their evolution structure and effects. *Meteorological Monographs,* **19**(41), 1982.

[12] Prieto, R., Ramírez, R., Cortés, V. E. & Maya, M.E., Impacto del cambio climático en la temporada de lluvias en México. In P. F. Martínez-Austria & C. Patiño–Gomez (Eds.), *Atlas de vulnerabilidad hídrica en México ante el cambio climático*. Semarnat-IMTA, Cap. 3, pp. 65–80. 2010.

[13] Vallentyne, J.R., Freshwater supplies and pollution: Effects of the demophoric explosion on water and man. In N. Polunin (ed.), *The environmental future*. Macmillan Press Ltd, London: UK, pp. 181–211, 1972.

[14] National Water Comission (CONAGUA). Sistema Nacional de Información del Agua (SINA) - *Estadísticas del Agua en México 2011*, México (http://www.cna.gob.mx/ Contenido.aspx?n1=3&n2=60&n3=87&n4=30). (accessed 15 April 2016).

[15] Real Academia de la Lengua Española, *Diccionario de la Lengua Española*. 22ª Edición, Tomo II. Madrid: Esp. 2002.

[16] Raucher, R., The value of water: What it means, Why it's important, and How water utility managers can use it. *Journal AWWA*, **97**(4), pp. 90–98, 2005.

[17] Available at: http://www.greenpeace.org/espana/Global/espana/report/cambio_climatico/Fracking-GP_ESP.pdf

[18] Webster´s Third New International Dictionary. Merriam Webster Inc. USA. V: II. 1986.

[19] Richter, B.D., Warner, A.T., Meñyer, J.L. & Lutz, K., A collaborative and adaptive process for developing environmental flow recommendations. *River Research and Applications*, **22**, pp. 297–318, 2006. https://doi.org/10.1002/rra.892

[20] Straskraba, M., Tundisi, J. & Duncan, A., *Comparative reservoir limnology and water quality management*, Kluwer Academic Publishers, Dordrecht: The Netherlands, 1992.

[21] American Public Health Association (APHA), *Standard methods for the examination of water and wastewater*, APHA, AWWA, WEF, Washington DC, 2012.

[22] Ostrander, G.K., *Tecniques in aquatic toxicology*, CRC Press, Inc. USA, 1996.

[23] Wheaton, F.W., *Aquacultural engineering*. Wiley, USA. 1977. First Spanish Edition AGT Ed. D.F. México, pp. 704, 1982.

[24] FAO, *Organic Agriculture*, Committee on Agriculture, Fifteenth Session, Item 8 of the Provisional Agenda. Rome, Red Room. January 25–29, 1999. Available at: http://www.fao.org/docrep/meeting/X0075e.htm. (Accessed 02 January 2017).

[25] Carretto, B., *A 2 años del derrame contaminante en el Río Sonora, una 'mancha' que no se borra*, available at:http://expansion.mx/?internal_source=MEGAMENU_SECTION

[26] Coloma López, P., Sánchez Navarro, J.A. & Martínez Gíl, F.J., Sistemas de flujo subterráneo regional en el acuífero carbonatado mesozoico de la Sierra de Cameros. Sector oriental. *Estudios Geológicos*, **53**, pp. 159–172, 1997. https://doi.org/10.3989/egeol.97533-4240

[27] Kresic, N., *Hydrogeology and groundwater modeling*. CRC Press, FL USA. 807 pp. 2007.

[28] Gleick, P.H., Basic water requirements for human activities: Meeting basic needs. *Water International*, **21**, pp. 83–92, 1996. https://doi.org/10.1080/02508069608686494

[29] SEMARNAT-IMTA, *Precio del Agua: Subsector agua potable, alcantarillado y saneamiento*, Proyecto DP-1330.1. Instituto Mexicano de Tecnología del Agua, Mexico. 36 pp. 2013.

[30] INEGI. Available at: http://www.inegi.org.mx/sistemas/bie/cuadrosestadisticos/Genera Cuadro.aspx?s=est&nc=583&c=29478.

[31] SAGARPA, *3er Informe de Labores 2014–2015*. Secretaría de Agricultura, Ganadería, Desarrollo Rural, Pesca y Alimentación. México. 148 pp. 2015.

[32] El Financiero. Available at: http://www.elfinanciero.com.mx/economia/un-gigante-expo rtador-de-alimentos.html. (Accessed 30 January, 2017).

[33] Shah, T., Groundwater and human development: challenges and opportunities in livelihoods and environment. *Water Science and Technology*, **51**(8), pp. 27–37, 2005.

[34] Díaz-Padilla, G., Sánchez-Cohen, I., Guajardo-Panes, R.A., Del Ángel-Pérez, A.L., Ruíz-Corral, A., Medina-García, G. & Ibarra-Castillo D., Mapping of the aridity index and its population distribution in Mexico. *Revista Chapingo Serie Ciencias Forestales y del Ambiente*, V. XVII, Ed. Esp.: 267–275, 2011.

[35] IT (International Transparency). Available at: http://www.transparency.org/news/feature/corruption_perceptions_index_2016?gclid=CPj5l_WE9dECFQ8taQodOaQKd
[36] IMCO (Instituto Mexicano para la Competitividad). Available at: http://imco.org.mx/politica_buen_gobierno/mexico-anatomia-de-la-corrupcion/
[37] Hoekstra, A.Y., Methodological advancements in the footprint analysis. *Ecological Economics,* **68**, pp. 1963–1974, 2009.
https://doi.org/10.1016/j.ecolecon.2008.06.021

MOBILE AUGMENTED REALITY FOR FLOOD EVENTS MANAGEMENT

DOMENICA MIRAUDA[1], UGO ERRA[2], ROBERTO AGATIELLO[2] & MARCO CERVERIZZO[3]
[1]School of Engineering, Basilicata University, Italy.
[2]Department of Mathematics, Computer Science and Economics, Basilicata University, Italy.
[3]Environmental Engineering Services srl - academic spin off, Basilicata University, Italy.

ABSTRACT

The frequency of flood events worldwide has increased significantly over the past decades, and with it so has the need to employ information technologies able to help mobile workforces, both technicians and volunteers, during surveys in the emergency phases. In view of this, a client-server framework for the development of a mobile application that uses Augmented Reality (AR) was implemented. This platform, which increases visual perception of the real world merging additional information with the natural scene in real time, allows mobile workforces to more easily reach the most critical areas subject to flooding and rapidly make a decision on the level of flood protection. The performance of the prototype was evaluated on the Bradano river, located in the south-eastern Basilicata region (Italy), both in a real case study and in a simulated one. The obtained results show how the application represents an innovative tool compared to the existing ones, being it able to show, timely and continuously up-to-date, augmented information on various vulnerability scenarios during the emergency phases, helping both technical and non-technical operators to quickly intervene, containing or preventing secondary disasters, thus reducing deaths and injuries, and limiting the resulting economic losses and social disruption.

Keywords: augmented reality, field surveys, flood event, GUI, mobile application, POI

1 INTRODUCTION

Floods are among the most devastating natural hazards in the world, claiming more lives and causing more property damage than any other natural phenomenon. The EMDAT database of the Centre for Research on the Epidemiology of Disasters (CRED) shows that flooding caused the majority of calamities between 1994 and 2013, accounting for 43% of all recorded events and affecting nearly 2.5 billion people (CRED [1]). In the next 15 years flood events in the world are expected to triple due to climate changes, growing of population and development of urban zones. According to the World Resources Institute (Luo *et al.* [2]), the number of people at risk within the flood-prone areas will rise from 21 million today to 54 million in 2030. Despite an increase in flood protection investments during the last two decades, the population tends to grow in socio-economically attractive flood-prone areas, and it is therefore likely that the vulnerability level of such areas will continue to increase in the future.

For these reasons, it is advisable to employ technologically advanced systems, which support emergency and mitigation actions.

Today, a number of applications are available to array and display data for technical experts, to explain programs of flood damage reduction to the decision makers, and to communicate real-time forecasts and warnings to the public. However, the majority of them are generally limited to tabular data, basic 2D plot visualisation, or standard 2D GIS tools designed for the desktop and not suitable for mobile use. Moreover, the existing tools are not interactive enough to be able to update the information easily, nor are they flexible enough to develop scenarios and provide visual and quantitative information regarding the real-time status during the forecasted event, often having restricted connectivity and limited

DOI: 10.2495/SDP-V13-N3-418-424

interoperability. In a more flexible and interactive direction, a client-server framework for the development of a mobile application that uses AR has been implemented by the present research. It could help field workers to quickly move within the critical areas during an extreme event, in order to communicate the state of structures and infrastructures to the operating centre in real time and identify the mitigation measure. AR technology allows the user to see virtual elements generated by the computer overlaying onto the visual perception of physical reality through a camera (Hincapie *et al.* [3]). This virtual-real merge is achieved via the appropriate GUIs on portable devices (smartphones, tablets, etc.) handheld by the user.

Commercially, AR is mainly found in mobile applications and, in recent years, has attracted wider interest from businesses. The fields of application are potentially endless, including the display of information, navigation in real-world environments, advertising, art and games (Geroimenko [4]). In water resources management, the application of AR is still restricted, despite its recognised usefulness during field surveys (Centeno *et al.* [5], Lange & Haynes [6]). In fact, because the present platform is able to leverage the visualisation of geo-referenced sensors measurements and simulation data of hydraulic and vulnerability models (Mirauda & Ostoich [7], Amicarelli *et al.* [8], Albano *et al.* [9], Mirauda *et al.* [10], Greco & Mirauda [11, 12]) in a seamless integrated view of the real-world context, it could be useful if extended out of the field of emergency management into the environmental monitoring in general. In addition, this AR application implemented on mobile devices could support technicians in on-site inspections and in the fault management of gauge stations spread over the basin, as well as in the control and maintenance of works interacting with the flow (e.g., weirs, dams, gabions, piles of bridges, etc.). Finally, users could also benefit from cartographies and guidance materials layered onto the real world in order to understand the phenomenon better and to move around even in less-known areas.

2 APPLICATION ARCHITECTURE

In this work, through the use of AR, an application able to visualise geo-referenced sensors measurements and integrate simulation data of hydraulic and hydrological models on mobile devices was implemented. This application, developed in the Metaio framework, can run on mobile devices equipped with a webcam and sensors such as a gyroscope, a compass and a GPS. It is based on a client-server architecture where: (1) the client (i.e. a smartphone) interacts with a server application (Web Service) through messages; (2) once the request is accepted, the Web Service queries the database and sends the retrieved information to the client; (3) the client interprets the response and renders the information on the screen as virtual objects overlaying onto the real environment being viewed. This architecture is able to manage different information sets associated to some elements, called Points Of Interest (POIs), in the real surroundings. In particular, after capturing a scene on their mobile device, users receive added details for each POI which increase their information content. A POI represents a specific point, a marker detected through geographic coordinates (latitude, longitude, and altitude), corresponding to locations such as a road, a bridge, a church, or a specific address and can be displayed through a graphical symbol on a mobile device. In this paper, the POIs are hydrometric stations or structures/infrastructures within flood-prone areas. They are displayed through billboards overlaid onto a scene captured by the device camera. The billboards are icons that take different shapes and colours based on the POI, each associated with a label that shows different data such as the distance between the mobile device and the POI (Fig. 1).

Figure 1: Some POIs of the presented application shown as billboards.

Furthermore, the platform, together with the POIs, provides images, graphs, videos, notes, tables, and more associated information, as well as POIs on 2D or 3D maps, which allow the user to move easily, to integrate the existing information seamlessly, to import new POIs into the database, and to export data and metadata out to other programs and software.

3 DISCUSSION OF THE RESULTS

A preliminary validation of the prototype performance was carried out on a sample basin represented by the Basento, one of the main rivers in the Basilicata region (Italy), both in a real case study and in a simulated one. The Basento originates in the north-west, flowing to the south-east of Basilicata and into the Ionian sea. It is 149-km long and its catchment is about 1,537 km^2. It is characterised by a seasonal variation of the flows: high levels in winter and low levels in summer. The platform application followed the collection phase of static data (e.g., location of the river gauge stations or the position of the structures/infrastructures at flood risk) and dynamic data (e.g., the water levels of the instrumented river sections or the alert thresholds) acquired both by the regional Civil Protection and through a series of field surveys.

3.1 Testing on a real case

The test includes visiting the Basento basin and locating interesting points where, for instance, gauge stations, buildings, or bridges can be found. Since it is necessary for this application to keep contact with the real environment as well as recognise and reach these points rapidly, it displays the location on the mobile device using virtual elements that enhance the position of the point. As reported in Figure 2, the AR application shows, for example, the gauge stations represented graphically by coloured drops. The colour of the drop can be either yellow, orange, or red if the values of the water depth acquired by the gauge station have reached the ordinary, moderate, or high criticality threshold, respectively. The drop can be oriented according to the user's position and the visualisation perspective through the mobile device screen. Its size increases progressively with the approaching to the POI, showing the

Figure 2: Some gauge stations of the Basento river represented graphically by coloured drops. Labels identify the user-station distance and the hydrometric level in real time.

user-station distance in real time. During the test, all these functions allowed the users not only to clearly visualise the gauge station position, relative distance and direction, but also to know the level of criticality of the gauge station itself. In addition, simply by clicking on the icon of the specific gauge station, the platform is also able to provide the hydrometric level updated on the billboard label and show the trend of the flow depth recorded during the day (Fig. 3a). Thanks to this function, the workforces managed to more easily monitor the river and the flood plains during the surveys. Even the interactiveness was increased by the application association of more information to the station, such as photos, images, videos, animations, and notes carrying indications of the previous field activities (Fig. 3b).

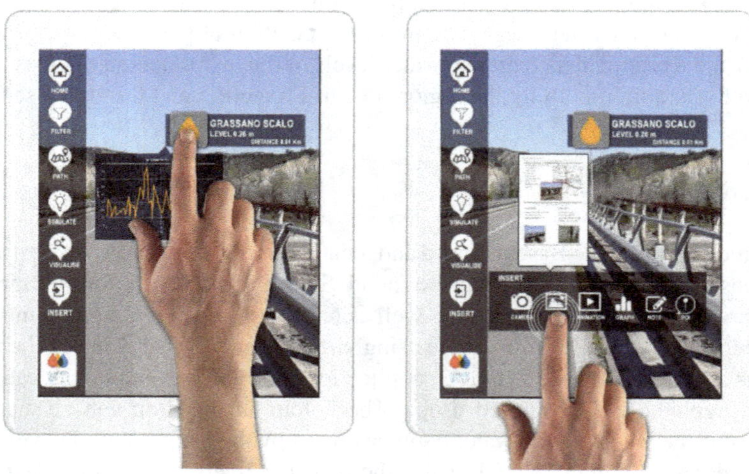

Figure 3: (a) The application shows the trend of the daily flow depths acquired by the gauge station. (b) Example of an image associated to one gauge station in AR mode.

3.2 Testing on a simulated case

The application was also tested simulating a real emergency situation, such as flood event in field, in order to verify the functionality and reliability of the developed program.

For this reason, a sample geographical space, which includes the river and the flood plain, was used to observe different vulnerability scenarios, built on the results of hydraulic models based on the measurements of gauge stations located along the basin. In particular, according to the model output, the platform shows different coloured symbols representing the structures and infrastructures within the flood-prone area (Fig. 4). In the presented survey, for instance, the yellow, orange, and red symbols showed that the POI was located in an area not at flood risk, at moderate or at high flood risk, respectively.

During these preliminary tests, it was noted that the use of such application was relatively easy when the communication between the interface and the models or other applications was available and fast, showing how the modelled situation affected the surrounding environment. The system was perceived as very simple to use, being designed to support total immersion applications when the performance of the mobile device is high. In addition, the workforces found that the integrated visualisation and comparison of different sources of information (libraries of photos, images, animations, videos, etc.) was a major advantage for decision making.

The success of the experiments was sometimes limited because of the hardware being used. Some issues occurred, which were related to the screen visibility under really bright conditions and to the battery life. Such inconvenience was avoided using mobile devices with a superior display resolution, a screen light/dark contrast control and an anti-reflective surface, together with the use of extra and long-lasting batteries.

Figure 4: 2D map visualisation of a sample flood-prone area within the Basento basin. The billboards of the structures and infrastructures display different colours according to the level of criticality simulated by hydraulic models.

CONCLUSIONS

The mobile application here developed, using AR technology, represents an innovative tool able to access a wide range of measurement sensors, deploy various visualisations, consult different data in real time, and integrate simulation results developed specifically to create a shared awareness in the field teams. The user's awareness is enhanced with perception of correspondence between the data and the surroundings in a way that allows the exploration of the relations between them, eliciting understanding and insight through continuous feedback.

For a preliminary validation, a real case study and a simulated one were organised. The results helped to define the benefits of outdoor visualisation, as well as its constraints in some conditions. Thanks to overlaid virtual objects onto real ones, the users are more aware of the potential danger around them in emergency situations, which allows speeding up both the decision-making processes and the crisis responses.

Both in the simulated emergency scenario and in normal environmental conditions, the platform is able to incorporate a variety of models and libraries (of photos, videos, texts, etc.) that drive the field activities, helping to increase knowledge of the phenomenon evolution.

The designed platform has its limitations and is not meant to replace traditional data analysis, but rather to be complementary, supplying localised views on the situation in real time. Further tests on different basins are advisable in order to improve the performance of the application and introduce new functionalities.

With the increasing hardware technology advancement, such platforms could perspectively grow in the environmental monitoring sector and possibly trigger a larger range of opportunities to improve the existing toolset and explore new potential uses.

ACKNOWLEDGEMENTS

The AR application was developed within the project of Innovation and Internationalisation 'Augmented water' – Operative Program ERDF 2007–2013 – Basilicata Region in partnership with the academic spin-off Environmental Engineering Services srl and the regional Civil Protection.

REFERENCES

[1] Centre for Research on the Epidemiology of Disasters (CRED), *The human cost of natural disasters,* Technical report: A global perspective, 2015.

[2] Luo, T., Maddocks, A., Iceland, C., Ward, P. & Winsemius, H., *World's 15 countries with the most people exposed to river floods,* World Resources Institute: Washington, United States, 2015.

[3] Hincapie, M., Caponio, A., Rios, H. & Mendivil, E., An introduction to augmented reality with applications in aeronautical maintenance. *Proceedings of the 13th International Conference on Transparent Optical Networks - ICTON 2011*, pp. 1–4, 2011.

[4] Geroimenko, V., Augmented reality technology and art: The analysis and visualization of evolving conceptual models. *Proceedings of the 16th International Conference on Information Visualisation*, pp. 445–453, 2012.

[5] Centeno, J.A.S., Kishi, R.T. & Mitishita, E.A., Three-dimensional data visualization in water quality studies using augmented reality. *Proceedings of the 6th International Symposium on Mobile Mapping Technology,* 2009.

[6] Lange, E. & Haynes, P., Mobile augmented reality for flood visualisation in urban riverside landscapes. *JoDLA-Journal of Digital Landscape Architecture*, **1**, pp. 254–262, 2016.

[7] Mirauda, D. & Ostoich, M., Surface water vulnerability assessment applying the integrity model as a decision support system for quality improvement. *Environmental Impact Assessment Review*, **31**, pp. 161–171, 2011.
https://doi.org/10.1016/j.eiar.2010.07.003

[8] Amicarelli, A., Albano, R., Mirauda, D., Agate, G., Sole, A. & Guandalini R., A smoothed particle hydrodynamics model 3D solid body transport in free surface flows. *Computers & Fluids,* **116**, pp. 205–228, 2015.
https://doi.org/10.1016/j.compfluid.2015.04.018

[9] Albano, R., Sole, A., Mirauda, D. & Adamowski, J., Modelling large floating bodies in urban area flash-floods via a smoothed particle hydrodynamics model. *Journal of Hydrology,* **541**, pp. 344–358, 2016.
https://doi.org/10.1016/j.jhydrol.2016.02.009

[10] Greco, M., Mirauda, D., & Plantamura Volpe, A., Manning's roughness through the entropy parameter for steady open channel flows in low submergence. *Procedia Engineering, 12th International Conference on Computing and Control for the Water Industry - CCWI2013*, **70**, pp. 773–780, 2014.

[11] Greco, M. & Mirauda, D., An entropy based velocity profile for steady flows with large-scale roughness. *Engineering Geology for Society and Territory - Volume 3: River Basins, Reservoir Sedimentation and Water Resources*, eds. G. Lollino, M. Arattano, M. Rinaldi, O. Giustolisi, J.C. Marechal, G.E. Grant, Springer International Publishing: Switzerland, pp. 641–645, 2015.

[12] Greco, M. & Mirauda, D., Entropy parameter estimation in large-scale roughness open channel. *Journal of Hydrologic Engineering*, **20**(2), pp. 04014047-1–04014047-6, 2015.
https://doi.org/10.1061/(asce)he.1943-5584.0001009

SUSTAINABLE WATER MANAGEMENT IN KUWAIT: CURRENT SITUATION AND POSSIBLE CORRECTIONAL MEASURES

A. MUKHOPADHYAY & A. AKBER
Water Research Center, Kuwait Institute for Scientific Research, Kuwait.

ABSTRACT
Kuwait, an arid country, has no surface source of useable water. It mainly depends on desalination plants for its freshwater needs. Brackish groundwater (salinity <5000 ppm) is used for irrigation and mixing with desalinated water. At the current rate of increase in demand for freshwater, large investment is necessary at close intervals to augment the desalination capacity of the country. With very little natural replenishment, the aquifers yielding the brackish water are also under great stress. Management of both supply and demand is necessary to solve these problems. In addition to the increase in desalination capacity, supply may be augmented by the increased use of renovated wastewater, storage of seasonally higher supply of usable water in the aquifers through artificial recharge, establishment of integrated water transport network through the Gulf Cooperation Council (GCC) countries and exploration of new useable groundwater resources both onshore and offshore. Possible new sources of water are harvesting of rainfall runoff in the wadis and depressions and runoff carried by the storm water network within the urban areas, moisture recovery from the vadose zone, and the paleo-groundwater possibly preserved under the Arabian Gulf. Demand may be curbed through maintaining lower pressure in the water network and some form of rationing, reduction of network loss through adoption of appropriate measures, implementation of updated water code that will help in saving water, control of population through tighter immigration measures and reduction in government subsidy. Finally, water authorities should take steps to increase public awareness about all aspects of water management outlined before in real earnest.
Keywords: artificial recharge, brackish water, desalinated water, groundwater, runoff harvesting, wastewater

1 INTRODUCTION

Kuwait, situated in an arid belt that stretches through the Middle East and North Africa (Fig. 1), has no surface source of useable water. There is only a limited reserve of fresh groundwater in lenses floating over brackish to saline water in north Kuwait. Kuwait depends on desalination plants for the bulk of its freshwater needs. Brackish groundwater (salinity <5000 ppm) extracted in the central and southwestern parts of Kuwait is used for irrigation and for mixing with the desalinated water to make it potable. The demand for fresh water is increasing at rapid rate. If this trend continues unchecked, a large investment will be necessary in the near future to augment the desalination capacity of the country. Aquifers yielding the brackish water are also under great stress, as there is very little natural replenishment for them. Judicious management of the available supplies and demands is the call of the day.

2 SOURCES OF WATER

Rainfall is scarce in the country (110 mm on average annually) and most of it evaporates under high ambient temperature (average annual evaporation more than 3000 mm). Most of the useable groundwater, extracted in the southwestern parts of the country, is brackish (salinity 2500–5000 mg/l). Only in the two large surface depressions of Raudhatain and Umm Al-Aish in the northern part of the country, enough infiltration from the runoff accumulations formed during occasional intense rainstorms occurs to give rise to two near-surface freshwater lenses floating over the brackish to saline groundwater (Fig. 2). No surface water

DOI: 10.2495/SDP-V13-N3-425-435

Figure 1: Location of Kuwait (courtesy of www.worldatlas.com).

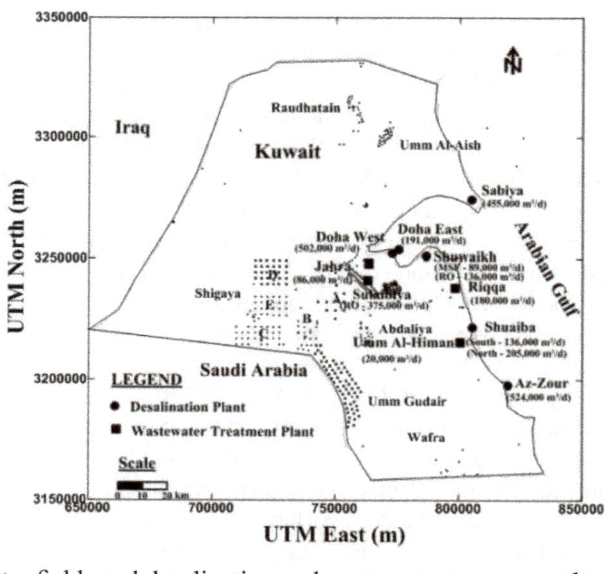

Figure 2: Water fields and desalination and wastewater treatment plants of Kuwait.

(rivers or lakes) exists within the perimeter of the country. For its freshwater needs, Kuwait, therefore, mainly depends on the distillation of seawater.

Municipal wastewater in Kuwait is generally treated to tertiary level. Recently, membrane (reverse osmosis, RO) treatment has been introduced to renovate the municipal wastewater that produces very high-quality water. The treated wastewater is thus a very good additional resource for the country.

3 CURRENT STATUS OF WATER AVAILABILITY AND CONSUMPTION

3.1 Desalinated water

The installed capacity of desalination plants in Kuwait (Fig. 2) in 2014 was 528 MIG/d (2.40 Mm^3/d) [1] (Fig. 3). The average daily demand stood at 398 MIGPD (1.81 Mm^3/d) at the same time, with the maximum consumption during September 2014, hitting 450 MIGPD (2.05 Mm^3/d). The average per capita consumption in that year was 442 l/d/capita (Fig. 3). As noted by Burney et al. [2] compared with industrialized and other developing countries, the average annual freshwater consumption in Kuwait (450 – 500 l/d/capita) is very high (Fig. 3).

The freshwater consumption in Kuwait has a strong seasonal variation. This is illustrated in Fig. 4, which indicates that during the hottest months of the year (June–September), the consumption of freshwater is on average 15–25% higher than the yearly average.

3.2 Groundwater

Brackish groundwater is mainly used for irrigation and landscaping, livestock rearing, construction work, non-potable use in households, and for mixing with desalinated water at ratios of up to 10% of the total volume to make the mixture potable [2]. The production of brackish water reached its peak in 2006 (36,234 MIGPD; 164.7 Mm^3/d), after which there

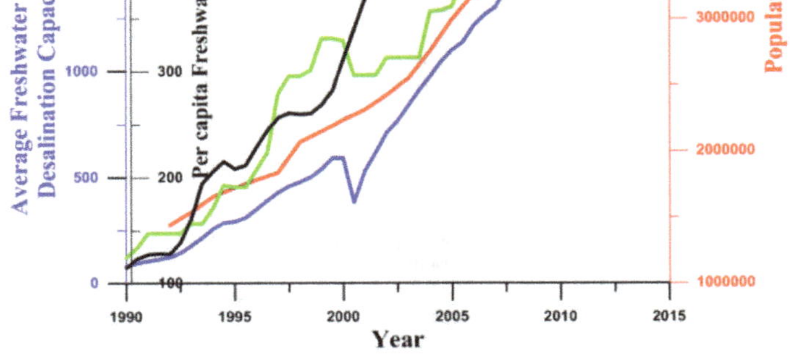

Figure 3: Development of installed desalination capacity, population and water consumption in Kuwait.

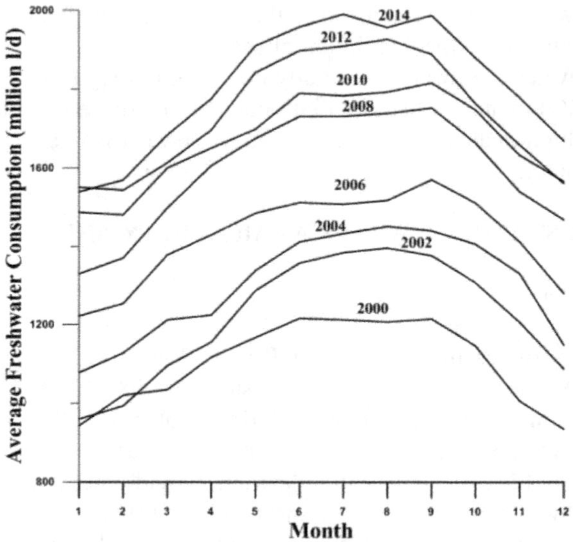

Figure 4: Seasonal variation of water consumption in Kuwait.

was a gradual decline in response to the availability of treated wastewater for irrigation and the awareness within the Ministry of Electricity and Water (MEW) about the need to preserve this strategic resource for posterity (Fig. 5). As of 2014, MEW and the Kuwait Oil Company (KOC) were producing annually 17,996 MIG (81.8 Mm^3) and 1,663 MIG (7.6 Mm^3) of brackish water, respectively (MEW, 2015), giving an average daily production of 53.9 MIGPD (244,849 m^3/d). The maximum daily net consumption of brackish water during the

Figure 5: Development of brackish water consumption in Kuwait.

summer of 2014 reached 73.8 MIGPD (335,495 m^3/d) with a yearly average of 52 MIGPD (236,392 m^3/d) [1]. The average production from the Wafra and Abdali farm wells, as estimated in 1989, was on the order of 200,000 m^3/d from each of the areas [3, 4]. The current withdrawals of groundwater are estimated to be 300,000–400,000 m^3/d in each of the areas. The drop in potentiometric head in the Dammam Formation aquifer due to brackish water withdrawal since 1970s is shown in Figure 6.

In the freshwater fields of Raudhatain, the initial production at a rate slightly above 2 MIGPD (9090 m^3/d) over the period 1963–1967 was reduced to about 1 MIGPD (4545 m^3/d) to control the rising total dissolved solids (TDS) in the produced water. By 1989, total withdrawal from the two freshwater fields during the year declined to 24 MIG (109,105 m^3), approximately 66,000 IG/d (300 m^3/d). During the Iraqi occupation of 1990–1991, the field facilities were destroyed. Subsequently, hydrocarbon pollution of the freshwater lens at the Umm Al-Aish field, caused by the damaged oil wells, was detected. As a result, the Umm Al-Aish field has been abandoned. The Rawdatain Bottling Company is currently producing water at a maximum rate of up to 0.5 MIGPD (2275 m^3/d) from the Raudhatain field with no production by MEW.

3.3 Renovated wastewater

The total average treated effluent of municipal wastewater at the two major conventional treatment plants in operation (Riqqa & Jahra, Fig. 2) was about 226,000 m^3/d in 2005 [5]. About half of it was discharged to the sea and the rest was used for agriculture and landscaping.

Based on the study by Abdel-Jawad *et al.* [6], a municipal wastewater treatment plant using the reverse osmosis (RO) process and constructed by a private consortium on Build – Operate – Transfer (BOT) formula, has come into operation in Sulaibiya in 2004. The plant has a design capacity of 375,000 m^3/d. It accepted an inflow of 355,102 m^3/d and produced 306,218 m^3/d of treated effluent during 2006 [5]. The volume of wastewater generated

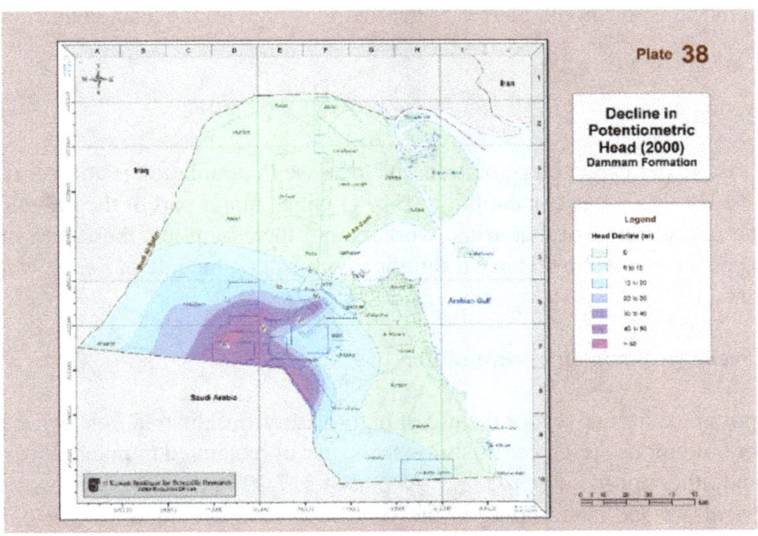

Figure 6: Drawdown in the Dammam Formation aquifer in Kuwait in 2000.

will be increased in future with increases in consumption, and the treatment capacity will also be increased to keep pace [2]. The treated water is planned to be used for irrigation, artificial recharge of the aquifers, and recreational purposes.

4 FUTURE DEMAND FOR WATER

The projections of freshwater demand, carried out by Burney *et al.* [2] have indicated that by 2025, the freshwater demand could be anywhere in the range of 722 Mm^3/y (2 Mm^3/d) to 3036 Mm^3/y (8.3 Mm^3/d). Desalination plant capacity, beyond that available in 2000, would range between 1.2 Mm^3/d and 9.1 Mm^3/d. By 2014, desalination capacity was already augmented to 2.4 Mm^3/d. Further increase in the desalination capacity will depend on the population and economic growth rates in the country.

No detailed study on the historical growth of brackish water demand in Kuwait and its projection for the future has been carried out so far. In view of the availability of treated municipal wastewater for irrigation, production from the brackish water fields has been reduced since 2006 (Fig. 6). These production figures do not, however, take into account the production in the two farm areas (Wafra & Abdali), where current withdrawal is estimated to be in the range of 300,000–400,000 m^3/d [2].

5 AVAILABLE OPTIONS FOR RATIONALIZATION OF WATER DEMAND

Simultaneous adoption of more than one method for the control of water consumption and the preservation of this important resource will be necessary to manage the water resources available in the country optimally [2]. Some of the important steps that may be adopted are summarized subsequently.

5.1 Reduction in subsidy

Rising per capita income together with government-fixed low price have been responsible for a steep rise in average per capita water consumption (from around 113 l/d/capita in 1970 to a maximum around 500 l/d/capita in 2002) [2]. To rein this steep rise in water use, a substantial reduction in the subsidy should be one of the methods high on the agenda and considered in earnest.

5.2 Immigration control

Besides the rise in per capita consumption, the increase in population is another major factor in the rise of Kuwait's water consumption (Fig. 4). Since major part of the growth in population was due to the influx of expatriate workers and their families, the future immigration policy of the government should have a significant impact on the rate of rise of water demand in the country.

5.3 Use of reclaimed municipal wastewater

One of the possible alternatives for the use of high-quality, desalinated freshwater for gardening and other outdoor use (e.g., car washing) is the use of reclaimed municipal wastewater for gardening and other non-potable uses [2]. A total of 317,297 m^3/d of treated wastewater was used for agriculture and landscaping in the year 2005 that amounted to 76% of the treated wastewater generated in that year [5]. This is a welcome sign which will conserve costly freshwater and brackish groundwater to a substantial extent.

5.4 Use of water produced by dewatering wells

The return water from irrigation of private and public gardens, and leakage from septic tanks, the water distribution network and the sewage network have caused rise in groundwater level in the urban areas of Kuwait [7]. A pilot project to study the effectiveness of drainage wells in combating the problem has been completed [8]. With the full-scale implementation of the drainage of subsurface water, a large volume of water will be extracted in the residential areas of Kuwait. Depending on the quality of the produced water, it can be reused for irrigation either directly, or after some processing [2].

5.5 Artificial recharge of aquifers

In 2014, on average, 600,000 m^3/d excess capacity was available in the desalination plants. A part of this idle capacity may be utilized in producing water, especially during the winter months, that could be stored in the aquifers through artificial recharge for later use during a high demand period or in an emergency. The high-quality membrane (RO)-treated wastewater can be another source of water for recharge.

5.6 Exploration and development of new water resources

The following possible sources of additional water may be investigated in Kuwait: a) runoff harvesting; b) usable water in the aquifers under the Arabian Gulf; c) recovery of water from the unsaturated zone; and d) treatment of oilfield produced brine to produce usable water. More elaboration on each of the possibilities follows.

5.6.1 Runoff harvesting

In the urban areas, manmade covered areas (roofs, paved roads and parking lots) prevent rain to infiltrate downwards and the non-evaporated parts of the rainfall generally flows into the stormwater network. This type of runoff collection can be relatively easily implemented in the urban areas of Kuwait.

In the nonurban areas of Kuwait, when the rainfall intensity is relatively high, substantial volumes of runoff can be generated in dry channels (wadis) that concentrate the surface flows from surrounding areas and discharge the generated flows to the depressions where the wadis terminate. The water harvesting can be done from these channels or from the terminal depressions.

5.6.2 Exploration of submarine aquifers

It has been surmised that significant volume of brackish- to freshwater might have recharged the aquifers below the Arabian Gulf during the last pluvial period when these aquifers were exposed to the surface. This water may still be locked up there by the overlying finer grained sediments, deposited during the transgression of the sea at the end of the Ice Age. Besides, recent recharge of some of these aquifers from the Euphrates-Tigris valley in the north and by infiltrating rainfall along the Iranian coast to the northeast is also possible [9, 10]. Seismic interpretation has indicated the presence of a structural high below the Failaka Island [11] that will be potentially suitable for upwelling of freshwater recharged in the Zagros Mountains, if at all it takes place (Fig. 7a). Two paleochannels have been found that run almost West to East across the Kuwait Bay and south of Bubiyan Island, and continue on both sides of Failaka (Fig. 7b) [11]. The existence of usable paleogroundwater bodies in these channels or elsewhere remains possible.

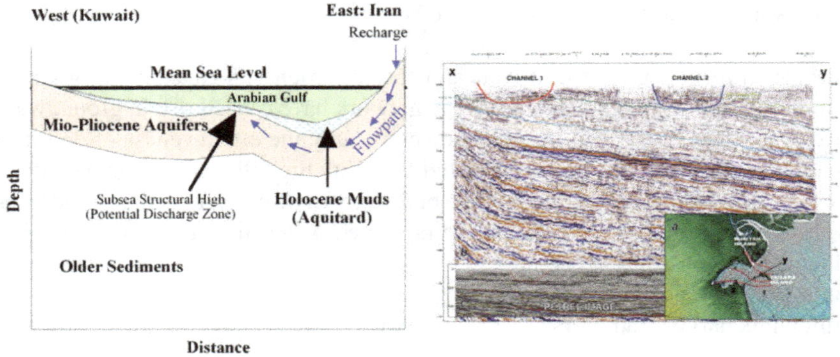

(a) (b)

Figure 7: (a) Conceptual model of upwelling of groundwater in the offshore areas of Kuwait; (b) Seismic section showing paleochannels in the offshore of Kuwait.

5.6.3 Recovery of water from the unsaturated zone

Recovery of water from the unsaturated zone near an urban area utilizing renewable energy like solar power, as proposed by Loo and Luk [12], is an innovative idea and is worth trying in Kuwait.

5.6.4 Treatment of oilfield brine to produce usable water

El-Sayed *et al.* [13] investigated the possibility of treating the huge volume of brine produced in the oilfields of Kuwait to make it usable and at the same time to recover various salts that can be commercially marketed. The aim would be to achieve quality level that will ensure 100% utilization of the recovered water and to have zero liquid discharge (ZLD) to the environment at the end of the treatment.

5.7 Water linkage of the GCC countries

As noted by Burney *et al* [2], by linking the water networks of the Gulf Co-operation Council (GCC) countries, it should be possible to compensate any shortfall (temporary or long-term) in one country by the excess capacities available in others and to rationalize the water use in all of the countries of the region.

5.8 Conservation and protection measures

Burney *et al.* [2] have also advocated for the implementation of measures to protect the available water resources from human activities, environmental degradation and overuse to maximize the utility of the available water resources in the region. As suggested by them, installation of water-saving equipment at the consumer end, introduction of drip irrigation and sprinklers for irrigation of private and public gardens, and limiting the timings of irrigation should save a large volume of water [2]. Protection of groundwater from anthropogenic activities like agricultural development and exploration, production and refining of oil should be pursued with earnest. Renovation of already polluted groundwater should be another goal. The contamination of the desalinated water from the leakage of the distribution pipelines, and intentional or unintentional pollution of the freshwater in the surface reservoirs are also to be guarded against.

5.9 Reduction of network loss

A large volume of water is lost through leakage in the water distribution network and preventive maintenance of the network is one way to minimize such loss [2]. Besides, the adoption of steps like the replacement of asbestos pipes with ductile pipes, application of cathodic protection to the network, reduction of network pressure and installation of automatic leak detection facilities should help reduce network loss substantially [2].

5.10 Legislation

Appeals for voluntary reduction of water use and the implementation of conservation measures may not be very successful. Guidelines and restrictions on water use may need to be imposed on water users through legislation to achieve the fall in demand.

5.11 Institutional measures

Systematic collection and storage of all data and information related to water production and water use, ensuring the accuracy of these data and periodic review of these data and information by the water authorities of the country, should lead to the adjustment in the water management policy that will be beneficial for the country [2]. Motivating the public for implementation of water conservation measures at residences and places of work through media campaigns should also be the responsibility of these authorities [2].

5.12 Adoption of new technologies

Kuwait has shown in the past a high commitment toward the adoption of new technology like multi-stage flash (MSF) and electrodialysis to desalinate seawater and brackish groundwater, respectively, to meet the water requirements of the country. Currently, the application of nanofiltration as a preprocessing method that will make the high-temperature MSF distillation possible is being looked into. The MEW is currently considering the possibility of erecting an RO unit for the desalination of seawater. As mentioned earlier, Kuwait Institute for Scientific Research (KISR) is also investigating the ZLD technology for renovation of oilfield brine. The implementation of these new technologies will certainly make additional resources available for use.

6 CONCLUSIONS

Kuwait depends heavily on seawater desalination for its freshwater needs. Brackish groundwater is exploited for agricultural and greening purposes and for mixing with desalinated water to make it potable. Water demand is escalating at a rapid rate under the increasing population pressure and rise in the standard of living. A substantial investment to augment the desalination capacity will be required to meet the increasing demand [2]. The withdrawal of brackish groundwater is also causing excessive stress in the aquifers. The control of freshwater demand through various means and the replacement of brackish water with renovated wastewater for agricultural and greening purposes are suggested to counter the trend [2]. Strategic storage of water may be created through artificial recharge of aquifers with available excess freshwater that may be recovered during the peak demand period or during an emergency [2]. Exploration and development of new types of water resources should also be pursued to broaden the water resource base of the country. Linkage of the water networks of all the GCC countries, installation of water-saving devices at the consumer end of the water network, prevention of water pollution, reduction of network loss, and adoption of new technologies for the production and treatment of water, as suggested by Burney *et al.* [2], should enhance the available water resources. Rationalization of water use through new legislation and increasing the level of public awareness of the strategic nature of water and need for its sustainable use are expected to yield positive results.

ACKNOWLEDGEMENT

The permission for presentation of the paper from the management of the Kuwait Institute for Scientific Research is gratefully acknowledged (Publication No.: KISR13943).

REFERENCES

[1] MEW, *Statistical year book: water, ministry of electricity and water*. Kuwait, pp. 259, 2015.

[2] Burney, N., Mukhopadhyay, A., Al-Mussallam, N., Akber, A. & Al-Awadi, E., Forecasting of freshwater demand in Kuwait. *The Arabian Journal for Science and Engineering,* **26**(2B), pp. 99–113, 2001.

[3] Al-Sulaimi, J., Viswanathan, M.N., Szekely, F. & Naji, M., *Geohydrological studies of Al-Wafra and Al-Abadally farm areas* (Report No. KISR4404). Kuwait Institute for Scientific Research, Kuwait, 1994.

[4] Al-Sulaimi, J., Viswanathan, M.N., Naji, M. & Sumait, A., Impact of irrigation on brackish groundwater lenses in northern Kuwait. *Agricultural Water Management*, **31**, pp. 75–90, 1996. http://doi.org/10.1016/0378-3774(96)01237-1

[5] Shahalam, A. M., Ahmed, M.E. & Al-Haddad, A., *Wastewater resources in Kuwait: effluent quantity and reuse demand*, Kuwait Institute for Scientific Research and Kuwait Foundation for Advancement of Sciences, Kuwait (*in press*).

[6] Abdel-Jawad, M., Eltony, N., Al-Shammari, S. & Al-Atram, F., *Municipal wastewater desalination by reverse osmosis* (Report No. KISR5224). Kuwait Institute for Scientific Research, Kuwait, 1997.

[7] Hamdan, L. & Mukhopadhyay, A., Numerical simulation of subsurface water rise in Kuwait City. *Groundwater,* **29**(1), pp. 93–104, 1991.

[8] El-Nahhas, F., Sherif, M., Abdullah, W., Hadi, K. & Ghoneim, H., *Long-term operation, monitoring and assessment of the drainage system at Shamiyah and Kaifan, Part 1: Main report* (Report No. 5840), Kuwait Institute for Scientific Research, Kuwait, 2000.

[9] Mukhopadhyay, A. & Fadlelmawla, A., Potential of freshwater reserves under Kuwait territorial water. *34th International Geological Congress (IGC),* Brisbane, Australia, 2012.

[10] Kotwicki, V., Akber, A., Akbar, B., Husain, R., Al-Fares, A., Al-Matar, B., Hussain, T., Singh, P., Hussain, F., van Opstal, R., Groen, J. & Breman, E., *Study of Palaeogroundwater Potential Offshore of Kuwait* (Report No. KISR11954), Kuwait Institute for Scientific Research, Kuwait, 2014.

[11] Mukhopadhyay, A. & Al-Khalid, A., *Initial assessment of paleogroundwater potential in offshore of Kuwait* (Report No. KISR12874), Kuwait Institute for Scientific Research, Kuwait, 2015.

[12] Loo, W.W. & Luk, M.Y., *Economical new water sources for world urban centers*, 2nd International Conference on Sustainable Urbanization, Hong Kong, China, 2015.
[13] El-Sayed, E., Al-Tabtabaei, M., Safar, M., Al-Shammiri, M., Bou-Hamad, S., Ahmed, M., Al-Mesri, A., Al-Jabli, H. & Al-Saffar, A., *Treatment of oil-produced saline water for cost-effective water recovery and its sustainable use* (Phase-I) (Report No.: KISR9385), Kuwait Institute for Scientific Research, Kuwait, 2008.

ANALYSIS FOR WATER MITIGATION OPTIONS USING MCDA: A CASE STUDY IN THE GALAPAGOS ISLANDS

MARIA F. REYES[1], ALEKSANDAR PETRICIC[1], NEMANJA TRIFUNOVIC[1],
SAROJ SHARMA[1] & MARIA D. KENNEDY[1,2]
[1]Envrionmental Engineering and Water Technology Department, UNESCO-IHE (Institute for Water Education).
[2]Faculty of Civil Engineering and Geosciences, Delft University of Technology, Delft.

ABSTRACT
Lately, due to overutilization of scarce natural resources such as water, tourist islands have been severely threatened. This is also the case in Santa Cruz Island, the main inhabited island of the Galápagos Archipelago. Therefore, a Multi-Criteria Decision Analysis (MCDA) was carried out with the aim of mitigating the future water crisis. The proposed solutions were evaluated in terms of environmental, technical, economic and social criteria. The results indicate that the 'best' alternative for municipality representatives is the option including desalination. However, for the other two groups of stakeholders, the chosen option was the one that combined all the proposed alternatives (greywater recycling, specific demand reduction and rainwater harvesting), except desalination. Conclusions show that for most stakeholders the environment is the most important criteria.
Keywords: Galápagos, intervention strategies, multi-criteria decision analysis, tropical islands, water supply indicators

1 INTRODUCTION

Tourism is a major source of income and employment in most small tropical islands [1]. Despite the fact that many islands worldwide have limited water resources, the expansion of tourism over the last 40 years has not stopped [2], affecting natural resources, especially water. As a consequence of tourism and local population growth, water demand has increased dramatically, causing several deficiencies on water supply and sewer services, which is directly related to the provision of facilities and services required by tourists [3–5].

The lack of freshwater resources in Galápagos has been a challenge to residents and visitors since these islands were first discovered. Since 1980, the resident population has quadrupled from 5,500 inhabitants to 26,000 in 2010, while the number of visitors has increased from 17,500 to 200,000 in the same period of time. This increase in the number of residents and visitors has created major challenges for the conveyance of water supply. Local authorities have been unable to keep up with rapid population growth and development, in terms of upgrading and maintaining the water supply infrastructure and service, especially in Puerto Ayora, the main hub of tourism. This is primarily due to: weak and unstable governance, lack of eco-friendly policies, and unplanned urbanization, among others [6, 7].

Based on the study carried out by [8], several population growth rates were proposed, which became the basis for the modelling of future coverage of water demand with supply made by [9]. The later study considered also several alternatives as intervention strategies for solving the future water deficit in Puerto Ayora. Nevertheless, the environmental vulnerability of Galápagos calls for an integrated and more complex analysis, considering several disciplines. According to [10], any issue involving the decision-making process of natural resources assessment, applies for MCDA. This methodology allows the proposed alternatives to be reviewed based on the perspectives of relevant stakeholders. Therefore the alternatives need to be properly weighted in order to evaluate as accurate as possible the potential solutions, and finally decide which is the most feasible and/or sustainable one considering the different conditions.

DOI: 10.2495/SDP-V13-N3-436-444

In this paper, several intervention strategies proposed to solve the future water crisis, are evaluated with the MCDA approach. These strategies were assessed under the feedback of seven groups of stakeholders. Also, an equal weight session was included, in order to analyse equal distribution of the chosen criteria for the analysis.

2 MULTI-CRITERIA DECISION ANALYSIS

The MCDA is an integrated and complete assessment of multiple suggested alternatives, with a set of tools for any decision-making process. This type of analyses comprises applied mathematical algorithms which evaluate a set of measures with different values and factors. The evaluation is usually carried out on problems with conflicting goals, high uncertainty, different forms of data and information, multiple interests and perspectives, and complex and/or evolving biophysical and socio-economic systems [11]. The ultimate goal of this methodology is to define the most feasible and sustainable solution of a certain issue at a low cost and considering all preferences of the participants [12].

Even though this type of analyses has been considered as subjective, the use of MCDA method has increased rapidly since the 1990s, especially with environment-related researches. This has been mainly due to the high complexity of projects, and an increased level of public and stakeholder participation [13]. According to [14], the use of MCDA methodologies provides a reliable method, which allows to rank different proposed alternatives in the presence of numerous objectives and constraints. Moreover, one of the advantages of the MCDA is the capacity to involve several stakeholders' opinions and perspectives. This results in a more thorough understanding of the points of view held by the involved parties [15].

Figure 1 shows the steps that were followed to carry the MCDA of the proposed alternatives for ensuring the future water demand coverage in Puerto Ayora [16].

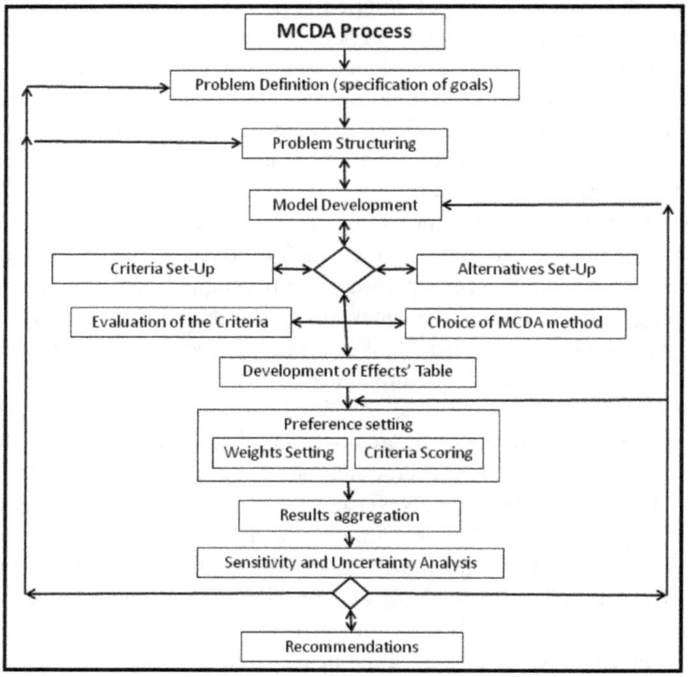

Figure 1: Schematic of the MCDA procedure.

3 RESEARCH METHODOLOGY

First, the problem was defined as the water supply deficit in the town of Puerto Ayora in the year 2044. In order to solve this issue, the study made by [17], developed with the help of the WaterMet^2 model [18] forecasted the water demand coverage over a 30-year planning horizon, proposing five intervention strategies. The results for the end of the forecast period (2045) were the basis for the input data for the MCDA. This analysis was carried out using the DEFINITE software [19], since it is a highly used program for complex-environmental case studies. It was developed to help improve the decision-making process with clear methodologies. Table 1 describes each selected intervention strategy.

Afterwards, four main criteria were selected for this analysis: (1) environmental, (2) technical, (3) economic and (4) social. Each of these was further described with different indicators, which allowed to evaluate the impact and performance of each strategy under the previously defined criteria. The indicators assigned are as follows:

- **ENVIRONMENTAL CRITERIA:** Land use, discharge of wastewater, seawater intrusion, energy consumption, chemical use, impact on endemic species, impact on marine/land ecosystems.
- **TECHNICAL CRITERIA:** Improvement on hours of service, coverage of demand with supply, water losses, robustness of the water supply system, O&M of the water supply system, alternative water sources contribution to overall balance, compatibility with the existing system.
- **ECONOMIC CRITERIA:** Capital cost, O&M cost, non-revenue water income generation, water demand management income generation, employment generation, increase in water tariffs, increase in tourist capacity.
- **SOCIAL CRITERIA:** Social acceptability, willingness to pay, transparency on project implementation process, water quality improvement, annual infection and other water-related diseases risk, compatibility with current legislations.

For this analysis, the DEFINITE software was selected, since it can be used on a wide variety of problems, regarding different disciplines. It is a software developed in order to improve the quality of decision, by a methodical procedure which leads experts through a number of interactive assessment sessions. It uses an optimization approach to integrate all information

Table 1: Intervention strategies selected and modelled with WaterMet^2 software.

Intervention strategy	Description
GALAPAGOS 1	Leakage Reduction + Water Meter Installation
GALAPAGOS 2	Leakage Reduction + Desalination Plant Installation + Water Meter Installation
GALAPAGOS 3	Leakage Reduction + Rainwater Harvesting + Greywater Recycling
GALAPAGOS 4	Water Meter Installation + Rainwater Harvesting + Greywater Recycling
GALAPAGOS 5	Leakage Reduction + Water Meter Installation + Rainwater Harvesting + Greywater Recycling + Water Demand Reduction

Source: [17]

provided by the selected involved stakeholders to a full set of value functions leading to a scientific-based alternative [19].

In the next step, each indicator defined for every criterion was scored. This scoring is expressed in an 'effects table', which refers to the assessment of the performance of each strategy against all the pre-defined indicators in a qualitative or quantitative way. The missing information was taken from the literature, such as potential waste quantities from the different selected strategies, as well as laws and regulations used locally. Another information was also taken from interviews with local experts, including the information for indicators for the social criteria and the technical one. Every indicator was scored using different types of scales/units, either qualitative or quantitative.

Afterwards, the scores assigned to each indicator were standardized. Since the values in the effects table were not in the same units, they were standardized with values between 0 and 1. For this, different methods were used: the maximum method, the goal standardization, the convex function and the yes/no standardization.

Following the standardization, every criterion was allocated with a weight. For this, a questionnaire was distributed to selected stakeholders, clustered into three different categories: municipality representatives, laundries and local community as shown in Figure 1.

4 RESULTS AND DISCUSSION

The MCDA was conducted in a total of four sessions. The first one illustrates the ranking result of the analysis using equal weights of all criteria. The other three analyses provide ranking results according to different weights allocation provided by the stakeholders. The ranking results are illustrated in Figure 3.

As observed in the previous figures, the best overall ranking result is achieved by alternative Galápagos 5. It is ranked on the first place in three different sessions. However, it takes the second place only in the municipality representatives' session. The Galapagos 5 option is the most consistent alternative of all, since the weight score ranges from 0.56 to 0.63, taking always the first or the second place. Even thought the Galápagos 2 alternative is the only one to provide full demand coverage at the end of the project horizon, it only takes the first place in one session (municipality representatives), and then second, but taking the fourth place in

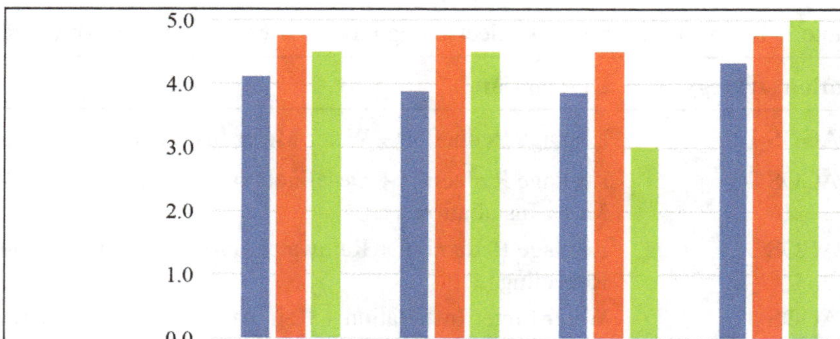

	Environmental	Technical	Economical	Social
Municipality	4.11	3.89	3.86	4.33
Laundries	4.75	4.75	4.50	4.75
Local community	4.50	4.50	3.00	5.00

Figure 2: Ranking preferences of various stakeholders' groups in Puerto Ayora.

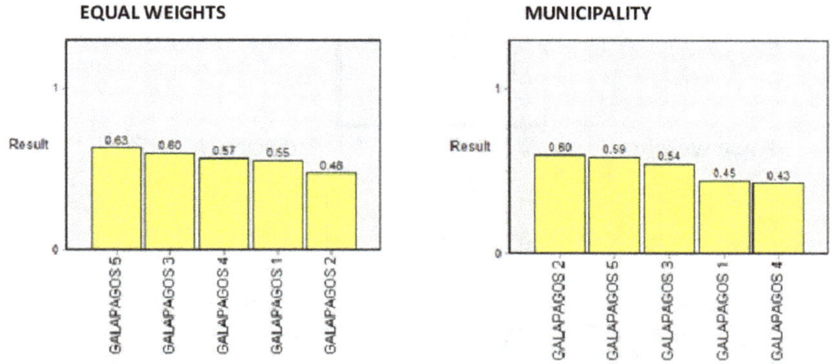

(a)

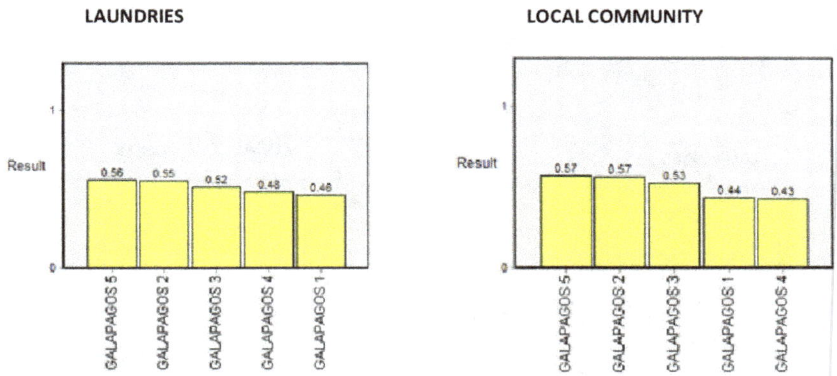

(b)

Figure 3: Ranked results from different stakeholders' perspectives (a) equal weights and municipality representatives, and (b) laundries and local community.

the equal weights' session. This is a consequence of the high costs and environmental threats attributed to this option.

As observed, the dispute is mainly between Galapagos 5 and Galapagos 2 alternatives. This indicates that whenever the environment is given more preference, Galapagos 5 tends to take the lead, while Galapagos 2 takes its place if technical criteria are given preference.

The staked bar charts (Fig. 4) provide more detailed information regarding the contribution level of each criterion on the overall score of the different MCDA sessions. The Galápagos 2 alternative, due to the large contribution from technical and social criteria, is the preferred option for the municipality representatives. This is due to the high scores obtained for the indicators under the social and technical criteria, since it is the alternative that can cover the total demand amount, and improving significantly the water quality.

For the rest of the three sessions, which give priority to alternative Galápagos 5, have assigned a higher weight to the environmental criteria, followed by the technical, and economic criteria respectively. This alternative scores reasonably well in all of the defined criteria, which results in the most consistent distribution of its rank across all of the sessions.

The summary of the ranking results from all MCDA sessions under this analysis is provided in Figure 5. The score allocation, together with the rank positions for each alternative gives a clear picture on resulting values and their frequencies.

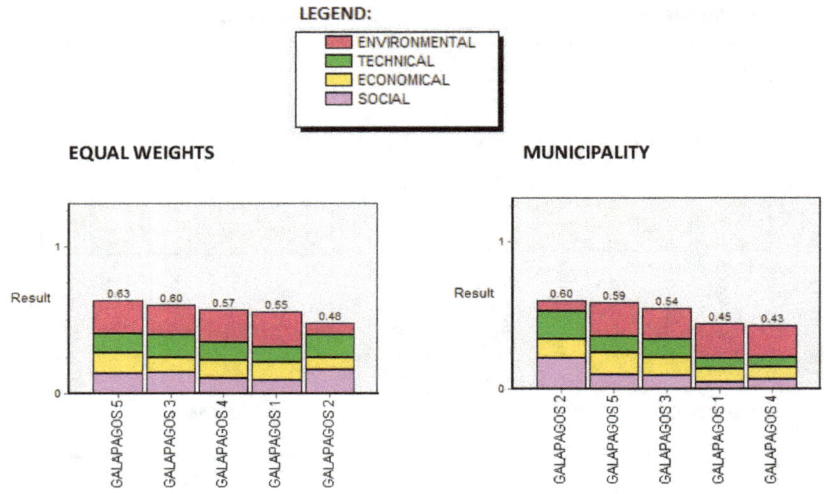

(a)

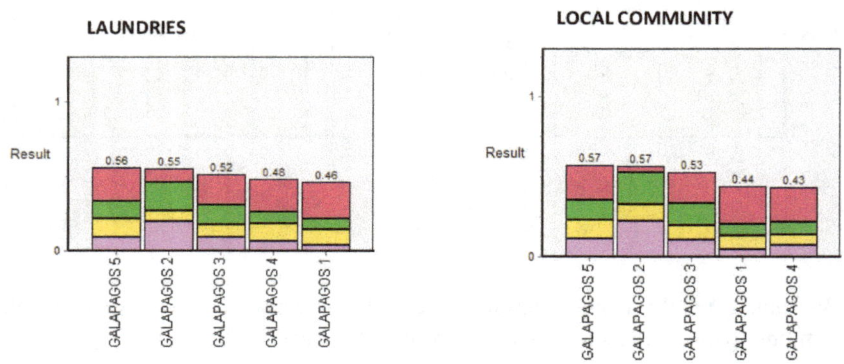

(b)

Figure 4: Stacked bar charts explaining the contribution of each criterion to the total score for each session for (a) equal weights and municipality representatives, and (b) laundries and local community.

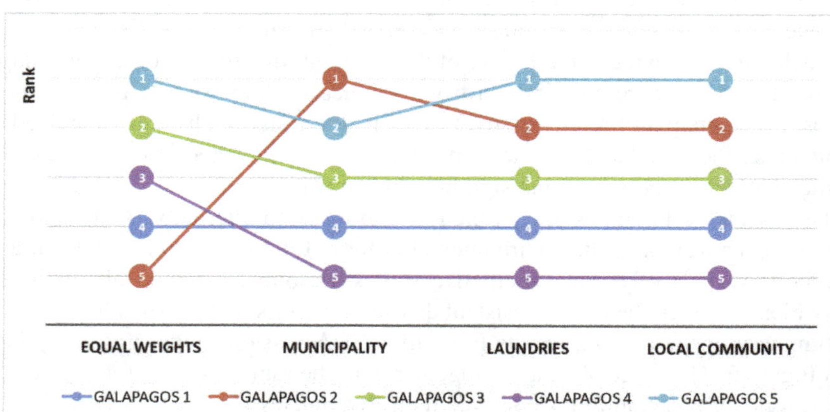

Figure 5: Comparison of MCDA results for Puerto Ayora case study.

As observed in the previous figure, each alternative tends to have a certain ranking, with variations of one step higher or lower, but not drastic changes in the ranking, except Galapagos 2. This alternative, in the equal sessions, takes the last place, suggesting that the technical criteria do not have a significant influence in the final results, due to the importance given to this ecosystem. Therefore, the environment indicators weight more than the technical ones. The most stable alternative is Galapagos 1, which takes the fourth place in all of the sessions. The variation of the rankings, suggests that a sensitivity analysis will help further in the explanation of how the results may vary with variations in weights and input values.

5 CONCLUSIONS

The MCDA methodology has proven to be a suitable decision support tool, which may provide a complete analysis regarding future water supply and demand options. However, the indicators, as well as their original values and ranges, have been proven to be case dependant and case sensitive. In this study, different sets of measures for improving the water supply system of Puerto Ayora were analyzed with four main groups of criteria.

As results showed, the Galapagos 2 and Galapagos 5 alternatives were ranked on the first positions by different stakeholder groups. Galapagos 2, which includes desalination, was preferred by municipality representatives only. On the other hand, Galapagos 5 was preferred by the other three groups of stakeholders, which included all of the options, except desalination. These differences in the results can be attributed to the technical and environmental preferences given. Nevertheless, despite these obvious advantages given by Galapagos 2, due to the higher costs and negative environmental impacts identified, it can easily be replaced by Galapagos 5 in most of the sessions, and other stakeholders seem to prefer less coverage of water demand if it preserves better the ecosystem. This also suggests that many of local population call for more sustainable options must, which replace desalination, in order to avoid possible harmful environmental threats.

Finally, more studies would be needed to arrive at more accurate values of certain quantitative indicators. These studies include long-term modelling (hydraulic, hydro-geological, physical, demographic and economic impact, etc.). More detailed determination of social criteria is needed as well, in order to come up with proper descriptive values for the qualitative indicators.

REFERENCES

[1] Tsiourtis, N.X., Water charge, the Cyprus experience. Water valuation and cost recovery mechanisms in the developing countries of the Mediterranean Region. *CIHEAM-IAMB, Bari*, pp. 91–104, 2002.

[2] Essex, S., Tourism development in Mallorca: is water supply a constraint? *Journal of Sustainable Tourism,* **12**(1), pp. 4–28, 2004. https://doi.org/10.1080/09669580408667222

[3] Charara, N., Cashman, A., Bonnell, R. & Gher, R. Water use efficiency in the hotel sector of Barbados. *Journal of Sustainable Tourism,* **19**(2), pp. 231–245, 2010. https://doi.org/10.1080/09669582.2010.502577

[4] Bramwell, B. Maltese responses to tourism. *Annals of Tourism Research,* **30**(3), pp. 581–605, 2003. https://doi.org/10.1016/s0160-7383(03)00023-9

[5] Gössling, S., Peeters, P, Hall, C., Ceron, J., Dubois, G., Lehmann, L. & Scott, D., Tourism and water use: supply, demand, and security. An international review. *Tourism management,* **33**(1), pp. 1–15, 2012.
https://doi.org/10.1016/j.tourman.2011.03.015

[6] Reyes, M., Trifunovic, N., Sharma, S. & Kennedy, M., Data assessment for water demand and supply balance on the island of Santa Cruz (Galápagos Islands). *Desalination and Water Treatment,* **57**(45), pp. 21335–21349, 2015.
https://doi.org/10.1080/19443994.2015.1119756

[7] González, J. A., Montes, C, Rodrigues, J. & Tapia W., Rethinking the Galapagos Islands as a complex social-ecological system: implications for conservation and management. *Ecology and Society,* **13**(2), pp. 13, 2008.
https://doi.org/10.5751/es-02557-130213

[8] Mena, C., et al., *Determination of social, environmental and economical relations which allow the development based on different processes of modelling, potential scenarios of sustainability of the socio-ecological system of the Galápagos Islands with emphasis on the dynamic of the flux of tourist visitors.* Report for the Ministry of Environment. Unpublished, 2014.

[9] Reyes, M., Trifunović, N., Sharma, S. & Kennedy, M., *Water supply and demand forecasting in Puerto Ayora-Santa Cruz.* Manuscript presented at the 14th Islands of the World Conference, Lesvos, Greece, May 23–27, 2016.

[10] Linkov, I., Satterstrom, F. Kiker, G., Batchelor, C., Bridges T. & Ferguson, E., From comparative risk assessment to multi-criteria decision analysis and adaptive management: recent developments and applications. *Environment International*, **32**, pp. 1072–1093, 2006.
https://doi.org/10.1016/j.envint.2006.06.013

[11] Wang, J.J., Jing, Y.Y., Zhang, C.F. & Zhao, J.H., Review on multi-criteria decision analysis aid in sustainable energy decision-making. *Renewable and Sustainable Energy Reviews*, **13**(9), pp. 2263–2278, 2009.
https://doi.org/10.1016/j.rser.2009.06.021

[12] Linkov, I. & Moberg, E., *Multi-criteria decision analysis: Environmental applications and case studies*. Boca Raton, CRC Press Taylor and Francis Group, 2012.

[13] Salanguste-Anarna, M., *Development of a multi-criteria decision support system for seawater desalination plants,* MSc Thesis, UNESCO-IHE, Institute for water education, The Netherlands, 2009.

[14] Polatidis, H., Haralambopoulos, D.A., Munda, G. & Vreeker, R., Selecting an appropriate multi-criteria decision analysis technique for renewable energy planning. *Energy Sources, Part B*, **1**(2), pp. 181–193, 2006.
https://doi.org/10.1080/009083190881607

[15] Kiker, G.A., Bridges, T.S., Varghese, A., Seager, T.P. & Linkov, I., Application of multicriteria decision analysis in environmental decision making. *Integrated Environmental Assessment and Management Journal*, **1**(2), pp. 95–108, 2005.
https://doi.org/10.1897/ieam_2004a-015.1

[16] Yatsalo, B.I., Didenko, V., Gritsyuk, S. & Sullivan, T., Decerns: a framework for multi criteria decision analysis. *International Journal of Computational Intelligence Systems,* **8**, pp. 467–489, 2015.
https://doi.org/10.1080/18756891.2015.1023586

[17] Behzadian, K., Kapelan, Z., Venkatesh, G., Brattebø, H. & Sægrov, S., WaterMet 2: a tool for integrated analysis of sustainability-based performance of urban water systems. *Drinking Water Engineering and Science Discussions*, **7**(1), pp. 1–26, 2014. https://doi.org/10.5194/dwesd-7-1-2014

[18] Reyes, M., Trifunovic, N., Sharma, S., Behzhadian, K., Kapelan, Z. & Kennedy, M., *Mitigation options for future water scarcity: a case study of Santa Cruz Island (Galápagos Archipelago).* Under Review In Water Journal, 2017.

[19] Janssen, R. & Herwijnen, M., *DEFINITE for Windows: A system to support decisions on a finite set of alternatives (Software packages and user manual) Institute for environmental studies* (IVM). Vrije Universiteit, Amsterdam, 2011.

IMPACT OF GROUNDWATER SALINITY ON AGRICULTURAL PRODUCTIVITY WITH CLIMATE CHANGE IMPLICATIONS

MUTASEM EL-FADEL[1], TANYA DEEB[1], IBRAHIM ALAMEDDINE[1], RAMI ZURAYK[2] & JAD CHAABAN[2]
[1]Department of Civil & Environmental Engineering, Faculty of Engineering & Architecture, American University of Beirut, Beirut, Lebanon.
[2]Department of Landscape and Ecosystem Management, Faculty of Agriculture and Food Sciences, American University of Beirut, Beirut, Lebanon.

ABSTRACT

This study examines the impacts of increased salinity on agricultural productivity and groundwater use for irrigation with the aim to cope with overexploitation associated with potential climate change impacts. For this purpose, a farmers' field survey was conducted at a pilot plain with banana plantations partially irrigated with saline groundwater. The economic burden of increased salinity was examined using a crop-water production function relating water salinity and yield with production cost and selling prices. Current production rates in low salinity plots were greater than those in high salinity plots by an average of 25%, representing the salinity burden incurred by farmers. We close with highlighting mitigation measures and adaptation strategies under potential future climatic changes that are expected to exacerbate irrigation with high salinity groundwater.

Keywords: adaptation, agricultural yield, economic burden, Saltwater intrusion

1 INTRODUCTION

The vulnerability of coastal groundwater resources to saltwater intrusion has evolved to become a challenge of global proportions with the Mediterranean and South-Atlantic coastal aquifers being highly affected [1]. The situation is more critical in arid and semi-arid climatic regions such as the Eastern Mediterranean, where freshwater resources are limited in comparison with continuously increasing water demand [2, 3]. Overexploitation and mismanagement increased the potential of saltwater intrusion, which can negatively affect agricultural yield of coastal crops and plantations through the accumulation of salts causing adverse effects on soils and plants.

Soil salinity reduces water infiltration rates reflected by the breakdown of soil aggregates and the dispersion and swelling of clay minerals that result in poor crop establishment. Moreover, groundwater salinization increases the presence of salts in the root zone, which exerts an osmotic effect on plants forcing them to spend more energy to extract water from the soil and hence consuming this energy instead of using it for growth, flowering or fruiting and thus, the plant's ability to grow is stunted [4]. Increased salinity also reduces plant growth and yield through ion toxicity, defined by an accumulation of Na^+ and Cl^- ions [5, 6], which employ competitive interactions with other nutrient ions (e.g. K^+, NO_3^-, $H_2PO_4^-$) for binding sites and transport proteins in root cells causing nutrient imbalance in plants. Ion toxicity affects crop production directly as diminished nutritional relations (Ca^{2+}/Na^+ and K^+/Na^+) decrease the crop's economic and nutritional value due to reduced fruit size and shelf life, non-uniform fruit shape, and decreased vitamin content [7].

The physiological impacts of salinity on soils and crops/plants, the loss in crop/plant yield and the decreased quality of crops/plants reduce the economic attractiveness of salt-sensitive coastal crops [8–10] that have low tolerance for salts. In this context, this study attempts to assess the impacts of saltwater intrusion on agricultural productivity with associated

DOI: 10.2495/SDP-V13-N3-445-456

economic implications highlighting mitigation measures and adaptation strategies under potential future climatic changes.

2 METHODS

A farmers' field survey, economic assessment, and development of a mitigation / adaptation framework were conducted at a coastal agricultural area along the Eastern Mediterranean in Lebanon (Fig. 1). The pilot area, which relies partially on groundwater resources for crop / plant irrigation, was selected based on field screening surveys involving multiple site visits, interviews with key informants, and water quality data ascertaining the occurrence of saltwater intrusion due to groundwater overexploitation and potential climate change impacts [8]. The main plantation grown in the area consists of banana, which is salt-sensitive and susceptible to the effects of irrigation with saline water [12–14].

2.1 Farmers' Field Survey

The impacts of saltwater intrusion on agricultural production and crop yield were assessed using a field survey questionnaire administered to farmers through a one on one interview process to collect information related to (a) wells (number per plot, elevation and depth, frequency of use, water quality perception), (b) banana production yield and selling prices, (c) irrigation (methods used, frequency, sources, expenditure) and (d) fertilizer (type, quantity, expenditure).

2.2 Economic Assessment

The adverse effect of groundwater salinity on crop yield is translated into an economic burden on farmers, which can be assessed using a crop-water production function that relates water salinity and crop yield coupled with the production cost and selling prices of banana obtained

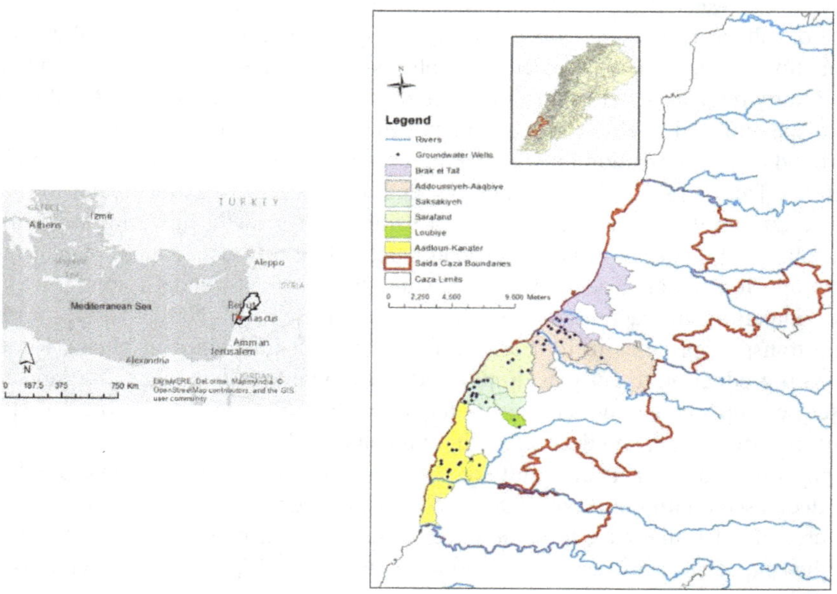

Figure 1: Pilot study area with location of irrigation wells.

through the farmers' questionnaires as expressed in equation (1). Two salinity groups were established based on farmers irrigating with water at $EC_w < 1$ dS/m (low salinity group) and farmers irrigating with water at $EC_w > 1$ dS/m (high salinity group).

$$N_r = T_r - P_c;\ T_r = Y_b x S_p;\ P_c = L_c + GW_c + SW_c + F_c + I_{c,} \tag{1}$$

where N_r = Net revenue ($/m²/yr); T_r = Total revenue ($/m²/yr); P_c = Production cost ($/m²/yr); Y_b = Banana yield (kg/m²/yr); S_p = Selling price of banana ($/kg); L_c = Labor cost ($/m²/yr); GW_c = Groundwater extraction cost ($/m²/yr) (diesel, electricity & maintenance); SW_c = Surface water cost ($/m²/yr) (subscription tariff); F_c = Fertilizer cost ($/m²/yr) (mineral and organic); I_c = Irrigation cost ($/m²/yr) (drip and sprinkler systems).

Additionally, the Ayers & Westcot [15] model (eqn 2) that links banana yields to soil salinity (EC_s) was used whereby EC_s values were calculated from measured groundwater salinity (EC_w) levels as expressed in equation (2a). The model predicts that the maximum banana yield happens when EC_w is < 1 dS/m) with values > 1 dS/m inducing production losses reaching a 100% decrease when EC_w exceeds 6.4 dS/m (Table 1).

$$Y = 100 - b\,(EC_s - a);\ EC_s = 1.5 \text{ x } EC_{w}, \tag{2}$$

where Y = Banana yield (%); a = Salinity threshold value (ds/m) = 0.73 for banana [16]; and b = Yield loss per unit increase in soil salinity (%) = 27.8% for banana [16].

2.3 Mitigation and Adaptation Framework

A framework for mitigation measures and adaptation strategies was developed in an effort to minimize saltwater intrusion and/or cope with increased salinity particularly in the context of future climate change impacts when salinities are expected to increase. The framework discusses the institutions responsible within the water sector, as pertaining to groundwater management and protection. Local and wider-scale measures are outlined with technical, financial and administrative constraints. Adaptation strategies and common irrigation management practices pertaining to agricultural productivity are equally presented under potential future climatic changes.

3 RESULTS AND DISCUSSIONS

3.1 Farmers' Field Survey

A total of 39 questionnaires were administered through the one-on-one interview process with farmers. The land plots in the study area varied in size with an average of ~100,000 m²

Table 1: Salinity impact of banana yield.

Potential Yield (%)	EC_s (dS/m)	EC_w (dS/m)
100	0.7	1.0
90	1.1	1.6
75	1.6	2.4
50	2.5	3.8
0	4.3	6.4

Table 2: Descriptive statistics for plot properties.

Property	Minimum	Maximum	Average
Total plot size (m^2)	1,300	1,000,000	100,000
Total area of banana grown (m^2)	250	350,000	40,030
Number of wells per plot	1	6	1.7
Well elevation (m, above sea level)	1.3	171.6	37.1
Well depth (m)	6	450	123.3

(Table 2). Each plot had at least one well (54.7%), with 45.3% having two or more wells. The wells were located at an elevation ranging between 1.3 and 171.6 m above sea level, averaging at ~37 m. Well depths ranged from 6 m to 450 m with an average of 123 m. While the main source of irrigation for banana is groundwater from wells located at the plots (53.8% of famers), more than 42% of farmers used a regional irrigation project in conjunction with well water for irrigation purposes and only 1.9% relied exclusively on surface water. Average irrigation time was around 4 hours a day with the surface water received through the regional irrigation project being heavily subsidized at a tariff based on the size of the cultivated land (average of 0.03 USD/m^2/year) instead of actual volume used [17,18], which encourages water wastage with less commitment to conservation measures. While no tariff imposed on groundwater with most wells being unlicensed/unregulated, farmers still incur expenses associated with water extraction (pump maintenance and electricity or fuel for operating private generators). Combining the costs pertaining to pump maintenance, electricity and diesel fuel costs, the average cost of providing water from wells over the irrigation season (starting in May and ending with the first rain of the fall-winter season) reaches ~0.38 USD/m^2/year, nearly 13 folds of the governmental surface water tariff.

Banana production in the region is a dominant trade with a strong year-round season and a variety of bananas grown. According to the field survey interviews, 51% of bananas grown are of the Dwarf Cavendish (DC) group with 27.5% of farmers growing both DC and Giant Cavendish (GC) groups simultaneously. The average land size that is grown with banana in the pilot area was around 40,000 m^2, with an average production yield of 7 kg/m^2. The trade season begins in October for a high cost of 1.2 USD/kg that drops by April to reach a low of 0.3 USD/kg because of high production.

3.2 Economic Implications

When revenue was compared across the two salinity groups, the low salinity group (n = 21; $EC_w < 1$ dS/m) had a greater revenue than the high salinity group (n = 18; $EC_w > 1$ dS/m), by ~2.5 USD/m^2/year, representing the monetary burden incurred by farmers due to increased salinity. Soil conditions and the physiological nature of banana are factors that could potentially influence the effect of salinity on its production. Moreover, the use of fresh surface water from the regional irrigation project in conjunction with well water is also expected to reduce the effects of saline groundwater on yields and revenues. As salinity levels continue to rise, their impact on banana production will become more pronounced.

The maximum yield in the study area was set at ~7.4 kg/m^2, which is the average yield reported by farmers cultivating under optimal conditions for banana growth (i.e. yield of farmers with EC_w less than 1 dS/m). Overall, the maximum crop yield in the pilot area is close to the national yield that averages 8 kg/m^2 [16]. On the other hand, the yield at farms with EC_w exceeding 1 dS/m was around 5.6 kg/m^2. This signifies a yield decrease of approximately 25% at a corresponding EC_w of about 2.4 dS/m (Table 1). In comparison to surrounding coastal areas in the MENA region, the reported yields are considered above average when compared to yields reported in dryer surrounding climates such as Egypt (3.9 kg/m^2) [17], Oman (2.3 kg/m^2) [21] and Jordan (1 kg/m^2) [22]. Evidently, current yields will be significantly affected if salinity levels continue to rise. For instance, current EC_w levels along the same coast less than 30 kms to the North, exceeded 16 dS/m as a result of overexploitation and mismanagement of groundwater resources [11]. If such levels reach the pilot area, the plantation of bananas will be economically non-viable. Under such a scenario, farmers will have to switch towards more salt-tolerant crops as was the case in the Al-Batinah region of Oman, whereby banana crops were replaced by salt-tolerant tomato varieties, when EC_s values ranged between 2.5 and 7.5 dS/m [9].

3.3 Mitigation and Adaptation Framework

Within the context of saltwater intrusion, effective mitigation measures to reduce induced effects of groundwater overexploitation are imperative alongside adaptation strategies that involve the recognition of potential climate change impacts that are already occurring and will continue into the future requiring pre-planning to protect ecosystems from potential negative impacts [23]. In fact, the vulnerability of water resources to climate change impacts has been at the forefront of global climate change research. These impacts are reflected in many regions including the study area by an increase in temperature and a decrease in precipitation leading to an overall increase in the net demand for water [24]. Groundwater resources along coastal zones are particularly vulnerable in this context because of decreased aquifer recharge (lower precipitation) and increased water demand (higher temperature), resulting in greater groundwater abstraction and saltwater intrusion [25], potentially exacerbated by sea level rise (SLR) also associated with climate change (Fig. 2).

Naturally, the main apparent measure to control saltwater intrusion is the reduction of abstraction rates from irrigation wells, which can be achieved within an integrated framework under the governance of existing institutions towards protecting groundwater resources (Table 3). The framework encompasses physical control methods coupled with management activities, economic incentives and regulatory enforcement while considering constraints at various levels: technical, financial, as well administrative and institutional.

While mitigation measures have been at the forefront of various policy and management decisions, it has become increasingly evident that mitigation alone cannot be an effective strategy to limit the vulnerability to the causes and consequences of saltwater intrusion. Adaptation strategies, in synergy with mitigation measures, have become an inevitable part of the planning, policies and discussion procedures concerned with salinity management [26]. Such strategies encompass initiatives to reduce the susceptibility of natural and human systems against actual or expected climate change impacts [23]. In the context of adapting to saltwater intrusion, those may include relocating wells inland, changing land use patterns by switching to more salt-tolerant crops, and managing irrigation patterns to improve water use and enhance drainage and leaching of salts [27–29]. Within the study framework, the common agricultural practices that are followed in the pilot area include various on-farm irrigation management that improve crop-water relations, especially with respect to salinity management

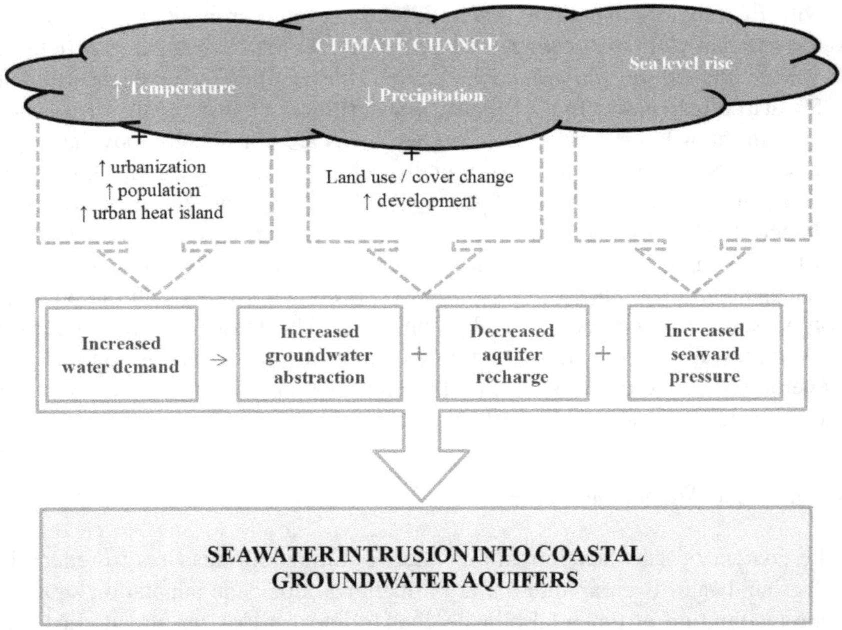

Figure 2: Dynamics of saltwater intrusion with potential climate change impacts.

such as the use of *fertigation via drip irrigation, leaching of salts, and conjunctive use of groundwater and surface water*. These practices are considered as adaptations measures that farmers tend to adopt independently to face aquifer salinization.

Fertigation via drip irrigation systems has proven to be an effective irrigation method to control salinity in soils by applying water precisely and uniformly at high frequencies, potentially increasing the yield and reducing root-zone soil salinity and drainage [38], thus providing improved disease management since only the soil is wetted whereas the leaf surface stays dry [39]. By supplying dissolved fertilizer to crops through an irrigation system, the nutrient-water relationship is more balanced to improve yield, save labor, and reduce compaction in the field resulting in the rapid and uniform uptake of nutrients by plants [40]. However, the initial installation cost for drip irrigation systems is high, with capital investment in the pilot area reaching 1.20 USD/m^2/year. Moreover, if poorly managed, salt accumulation may still occur between drip emitters, therefore it is advisable to design drip irrigation systems with closely spaced emitters and provide adequate *leaching of salts*. The latter is an effective method of adapting to increased soil salinity through excessive application of water to keep salts, which have accumulated in the soil, in soluble form and flush them below the crop root zone to maintain productivity [41]. Mostly, rainfall provides adequate leaching, as is the case in the pilot area. However, farmers tend to practice leaching via their scheduled irrigation patterns. For instance, in the pilot area, farmers have claimed to practice leaching every other time fertigation is performed, meaning that every time fertilizers are passed in the drip irrigation system, the next irrigation is without fertilizers, thus accumulated salts in the soil (from the irrigation water or from fertilizer residues) are flushed. Equally important is the *conjunctive use of groundwater and surface water resources*, which is considered a strategic approach to reduce the susceptibility of crops to salinity (Fig. 3).

Table 3: Framework for controlling saltwater intrusion.

Common physical methods	Constraints [27, 30–33]
Subsurface barriers – Defined as underground semi-impervious or impervious structures constructed in coastal aquifers – Simultaneously impedes inland infiltration of saltwater and increases groundwater storage capacity of an aquifer	– High construction costs – Procedural difficulties of political, social, and economic nature
Artificial recharge through injections wells – Surface water is artificially made to infiltrate into the ground, commonly at rates and quantities greater than natural recharge – Provides artificial recharge of confined aquifers and is effective in the case of highly fractured hard rocks and karstic limestone	– Problematic for arid and semi-arid regions with unavailability of adequate surface water in quantity and quality – High capital, and operation and maintenance costs – Efficacy is minimized for karstic highly fractured aquifers
Management activities	**Constraints [34–37]**
Regulating groundwater exploitation – Reduction of pumping rates through tariff restructuring, metering water wells, and enforcing penalties for over-exploitation *Monitoring groundwater quality* – Establishing monitoring stations with regular testing for water quality parameters relevant to salinity *Regulating well permits and licensing* – Introduction of controls over the import of pumps and drilling equipment *Rehabilitating/modernizing irrigation systems* – Shifting from gravity-fed irrigation to pressurized systems using on-farm irrigation systems (e.g. drip irrigation)	*Technical* – Insufficient number of skilled staff due to lack of up-to-date training and limited equipment for routine maintenance, measurement or monitoring of water supply and quality – Lack of implementation of modern irrigation and water-saving technologies *Financial* – Deficient allocation of resources for proper maintenance and rehabilitation of irrigation water supply and distribution systems – Inadequate collection of fees from consumers leading to poor financial management in water establishments *Administrative & Institutional* – Lack of coordination between regulatory entities (ministries) and water establishments – Lack of regulatory enforcement regarding water use and distribution – Lack of political will in regulating and implementing well permits and payment of fines

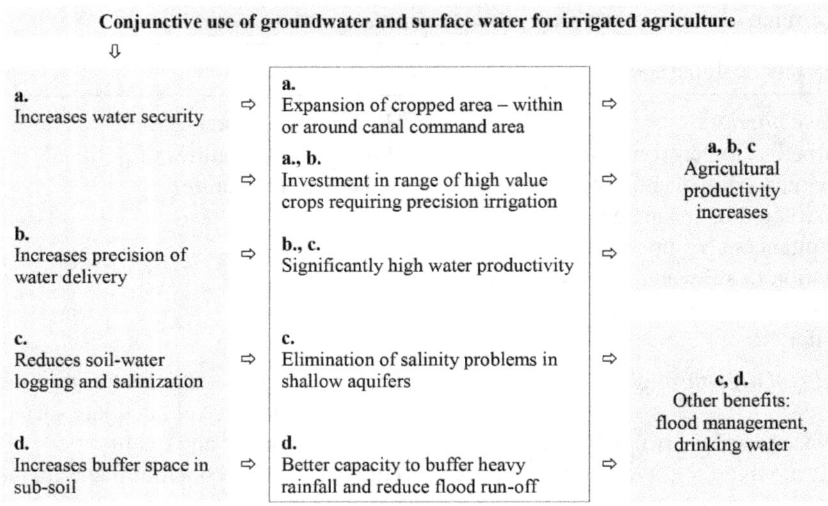

Figure 3: Benefits of conjunctive use for framework implementation.

Conjunctive use of groundwater and surface water (via canals) is critical at stages of crop/plant growth, such as the shooting period for bananas. By increasing the precision of water delivery and reducing soil-water logging and salinization, water productivity is increased significantly and salinity problems in shallow aquifers are reduced, leading to an increase in agricultural productivity [42].

4 CONCLUSION

This study demonstrated through a farmers' field questionnaire (n = 39) and reported groundwater quality that production yield of banana plantation along the Eastern Mediterranean in a low salinity group (n = 21; $EC_w < 1$ dS/m) was higher than in the high salinity group (n=18; $EC_w > 1$ dS/m) by nearly 25% (~2.5 USD/m^2/year), representing the burden of salinity on farmers. Indirect adaptation measures practiced by the farmers encompassing fertigation via drip irrigation, leaching of salts and the conjunctive use of groundwater and surface water, minimized the impact on banana production yield at the encountered level of salinity. Current yields can be significantly affected if salinity levels continue to rise to the extent that the plantation of bananas can become economically non-viable forcing farmers to switch towards more salt-tolerant crops. Saltwater intrusion is expected to increase with climate change due to a projected decrease in precipitation and groundwater recharge, and an increase in water demand leading to further groundwater abstraction during the summer months in particular. The combination of adaptation strategies and methods of banana cultivation (multiple varieties, single and high-density planting techniques) suggest that production can be governed by agricultural and irrigation practices implemented by farmers. The cumulative effects of climate change, particularly decreased precipitation patterns, increase sea level rise and increase in water demand (high abstraction rates), are predicted to aggravate saltwater intrusion with related impacts justifying the adoption of early measures to protect groundwater exploitation. In this context, regulatory tools emphasizing tariff restructuring and well permitting are vital to ensure sustainability.

ACKNOWLEDGEMENTS

The authors would like to acknowledge the support of the US Agency for International Development under the PEER initiative (Grant No. USAID-OAA-A-I1-00012) and the International Development Research Center (IDRC) of Canada (Grant No. 106706-001).

REFERENCES

[1] Stigter, T.Y., Nunes, J.P., Pisani, B., Fakir, Y., Hugman, R., Li, Y., Tomé, S., Ribeiro, L., Samper, J., Oliveira, R., Monteiro, J.P., Silva, A., Tavares, P.C.F., Shapouri, M., Cancela da Fonseca, L. & El Himer,H., Comparative assessment of climate change and its impacts on three coastal aquifers in the Mediterranean. *Regional Environmental Change,* **14**(1), pp. 41–56, 2014.
https://doi.org/10.1007/s10113-012-0377-3

[2] De Montety, V., Radakovitch, O., Vallet-Coulomb, C., Blavoux, B., Hermitte, D., & Valles, V., Origin of groundwater salinity and hydro-geochemical processes in a confined coastal aquifer: Case of the Rhône delta (Southern France). *Applied Geochemistry,* **23**, pp. 2337–2349, 2008.
https://doi.org/10.1016/j.apgeochem.2008.03.011

[3] Masciopinto, C., Management of aquifer recharge in Lebanon by removing seawater intrusion from coastal aquifers. *Journal of Environmental Management,* **130**, pp. 306–312, 2013.
https://doi.org/10.1016/j.jenvman.2013.08.021

[4] Blaylock, A.D., *Soil salinity, salt tolerance, and growth potential of horticultural and landscape plants.* University of Wyoming, College of Agriculture, Larmie, WY, 1994.

[5] Grattan, S.R., *Irrigation water salinity and crop production.* Farm Water Quality Planning, A Water Quality and Technical Assistance Program for California Agriculture (Publication 8066). University of California, 2002.

[6] Rahman, M.H., Lund, T. & Bryceson, I. (2011). Salinity impacts on agro-biodiversity in three coastal, rural villages of Bangladesh. *Ocean and Coastal Management,* **54**(6), pp. 455–468, 2011.
https://doi.org/10.1016/j.ocecoaman.2011.03.003

[7] Ondrasek, G., Rengel, Z. & Veres, S., *Soil salinization and salt stress in crop production.* In Shanker, A. (Ed.). Abiotic stress in plants: Mechanisms and adaptations (pp. 171–191). InTech publishers, Croatia, 2010.

[8] Palacios, M. P., Haman, D.Z., Del-Nero, E., Pardo, A. & Pavon, N., Banana production irrigated with treated effluent in the Canary Islands. *American Society of Agricultural Engineers,* **43**(2), pp. 309–314, 2000.
https://doi.org/10.13031/2013.2706

[9] Naifer, A., Al Rawahy, S.A. & Zekri, S., Economic impact of salinity: The case of Al Batinah in Oman. *International Journal of Agricultural Research,* **6**(2), pp. 134–142, 2011.
https://doi.org/10.3923/ijar.2011.134.142

[10] Mokhtari, R., Economic impact of changes in water supply salinity. In Wilson, E. M. (Ed.). WRPMD'99: Preparing for the 21st Century, *Proceedings of the 26th Annual Water Resources Planning and Management Conference*, Tempe, Arizona, June 6–9, 1999.

[11] El-Fadel M., *Climate change and saltwater intrusion along the eastern Mediterranean: Socio-economic vulnerability and adaptation*. International Development Research Center, Canada. American University of Beirut, 2015.

[12] Arvanitoyannis, I.S. & Mavromati, A., Banana cultivars, cultivation practices, and physicochemical properties. *Critical Reviews in Food Science and Nutrition*, **49**(2), pp. 113–135, 2009.
https://doi.org/10.1080/10408390701764344

[13] Storey, R. & Walker, R., Citrus and salinity. *Scientia Horticulturae,* **78**, pp. 39–81, 1999.
https://doi.org/10.1016/s0304-4238(98)00190-3

[14] Tanji, K. & Kielen, N., *Agricultural Drainage Water Management in Arid and Semi-Arid Areas*. Food and Agricultural Organization of the UN Irrigation and Drainage Papers, 61, 2002

[15] Ayers, R.S., & Westcot, D.W., *Water quality for agriculture*. Food and Agriculture Organization (FAO). FAO Irrigation and Drainage Paper 29, 1985.

[16] California Fertilizer Association., *Western Fertilizer Handbook*. Horticultural edition. Sacramento: Prentice-Hall. pp. 279, 1995.

[17] Litani River Authority, *Qasimiya and Ras Al Ain Irrigation Project*, 2014, available at: http://www.litani.gov.lb/en/?page_id=117. (accessed April 13 2017).

[18] Litani River Authority, *Water tariffs in the three irrigation projects invested,* 2014, available at: http://www.litani.gov.lb/wp/?page_id=1208. (accessed April 13 2017).

[19] Abou Masleh, F., Banana as replacement for citrus in Lebanese coasts. *Al-Akhbar,* (Issue 398), 2007, available at: http://www.al-akhbar.com/node/127740. (accessed April 13 2017).

[20] Khaled, S.M., Shfique, F.A. & Hassan, M.B., Banana-growing tissue and its impact on the economic return per Fedden in Egypt. *Nature and Science*, **8**(10), pp. 267–273, 2010.

[21] Ministry of Agriculture and Fisheries, Sultanate of Oman (MAF) / International Center for Biosaline Agriculture (ICBA), *Oman salinity strategy. Agricultural status and salinity impact (Annex 2)*, 2012.

[22] De Langhe, E., *Banana diversity in the Middle East (Jordan, Egypt, Oman)*. International Network for the Improvement of Banana and Plantain, (INIBAP), Montpellier, France, p. 52, 2002.

[23] International Council for Local Environmental Initiatives (ICLEI), *The mitigation-adaptation connection: Milestones, synergies and contradictions,* 2014, available at: http://itepsrv1.itep.nau.edu/itep_course_downloads/~GeneralAQInfo/Climate%20 Change/ICLEIMitigationAdaptationConnection.pdf. (accessed April 13 2017).

[24] Intergovernmental Panel on Climate Change (IPCC), Coastal systems and low-lying areas. In *Climate Change 2014: Impacts, Adaptation, and Vulnerability. Part A: Global and Sectoral Aspects*. Contribution of Working Group II to the Fifth Assessment IPCC Report. Field et al. (Eds.). Cambridge University Press, pp. 361–409, 2014.

[25] Rasmussen, P., Sonnenborg, T.O., Goncear, G. & Hinsby, K., Assessing impacts of climate change, sea level rise, and drainage canals on saltwater intrusion to coastal aquifer. *Hydrology and Earth Systems Sciences,* **17**, pp. 421–443, 2013.
https://doi.org/10.5194/hess-17-421-2013

[26] Bhamidipati, S., Van der Lei, T. & Herder, P., From mitigation to adaptation in asset management for climate change: A discussion. In Lee, W. B., et al. (Eds.). *Proceedings of the 7th World Congress on Engineering Asset Management* (WCEAM 2012), 2012.

[27] Sherif, M.M. & Hamza, K.I., Mitigation of seawater intrusion by pumping brackish water. *Transport in Porous Media,* **43**, pp. 29–44, 2001. https://doi.org/10.1023/a:1010601208708

[28] Ayers, R.S. & Westcot, D.W., *Water quality for agriculture.* Food and Agriculture Organization (FAO). FAO Irrigation and Drainage Paper 29, 1985.

[29] Kashef, A.I. & Csallany, C.S., Management and control of saltwater intrusion in coastal aquifers. *Critical Reviews in Environmental Control,* **7**(3), pp. 217–275, 1977. https://doi.org/10.1080/10643387709381652

[30] Basri, M.H., *Two new methods for optimal design of subsurface barrier to control seawater intrusion.* PhD dissertation. University of Manitoba, Canada, 2001.

[31] Food and Agriculture Organization of the United Nations (FAO)., Land and water integration and river basin management. *Proceedings of an FAO informal workshop*, Rome, Italy, 31 January – 2 February 1993.

[32] Khomine, A., János, S. & Balázs, K., Potential solutions in prevention of saltwater intrusion: A modelling approach. In Lambrakis, N., et al. (Eds.). *Advances in the Research of Aquatic Environment 1.* Springer. Berlin, Germany, 2011.

[33] Luyun Jr., Momii, K. & Nakagawa, K., Effects of recharge wells and flow barriers on seawater intrusion. *Groundwater,* **49**(2), pp. 239–249, 2011. https://doi.org/10.1111/j.1745-6584.2010.00719.x

[34] El-Fadel, M., Zeinati, M. & Jamali, D.,Water resources in Lebanon: Characterization, water balance and constraints. *International Journal of Water Resources Development,* **16**(4), pp. 615–638, 2000. https://doi.org/10.1080/713672540

[35] Karaa, K., Karam, F. & Raad, R., Moderization of irrigation systems: Measures to reduce pressure on water demand in Lebanon. In El Moujabber M., et al. (Eds.). *Technological perspectives for rational use of water resources in the Mediterranean region.* Bari: CIHEAM, pp. 241–254 (Série A. Séminaires Méditerranéens; n. 88), 2009.

[36] Institutional and Economic Instruments for Sustainable Water Management in the Mediterranean Basin (INECO), *Institutional Framework and Decision Making Practices for Water Management in the Damour River Basin, Lebanon: Towards the development of a strategy for water stress mitigation.* Coordination Action Project supported by the European Commission through the 6th Framework Programme (Contract No: INCO-CT-2006-517673), 2009.

[37] Stephan, R.M., Legal framework of groundwater management in the Middle East (Israel, Jordan, Lebanon, Syria, and the Palestinian Territories). *Water Resources in the Middle East,* **2**, pp. 293–299, 2007. https://doi.org/10.1007/978-3-540-69509-7_30

[38] Hanson, B.R., May, D.E., Simünek, J., Hopmans, J.W. & Hutmacher, R.B., Drip irrigation provides the salinity control needed for profitable irrigation of tomatoes in the San Joaquin Valley. *California Agriculture,* **63**(3), pp. 131–136. https://doi.org/10.3733/ca.v063n03p131

[39] Ahmed, B.A., Mohammed, A.A. & Ihsan, M.I., Effect of drip irrigation system and fertigation on growth, yield and quality of banana cv. Grand Nain (Musa AAA). Research Application Summary. *Second RU FORUM Biennial Meeting*, 20–24 September 2010, Entebbe, Uganda, 2010.

[40] New South Wales Department of Primary Industries, Australia (NSW DPI), *Horticultural fertigation: Techniques, equipment and management*, 2000, available at: http://www.dpi.nsw.gov.au/land-and-water/water/irrigation/crops/fertigation. (accessed April 13 2017).

[41] Fipps, G., *Irrigation water quality standards and salinity management strategies*. Texas A&M Agrilife Extension Service. B-1667 4-03, 2003, available at: http://soiltesting.tamu.edu/publications/B-1667.pdf. (accessed April 13, 2017).

[42] Foster, S., van Steenbergen, F., Zuleta, J. & Garduño, H., *Conjunctive use of groundwater and surface water: From spontaneous coping strategy to adaptive resource management.* GW-MATE Strategic Overview Series 2, World Bank, Washington DC, 2010.

DESIGNING TWO-PART GREYWATER FEES: CONCEPTUAL FRAMEWORK AND EMPIRICAL EVIDENCE FROM GERMANY

BENEDIKT ROTERS & MARK OELMANN
Institute of Economics, Hochschule Ruhr West - University of Applied Sciences, Germany.

ABSTRACT

The wastewater sector in Germany – fragmented local natural monopolies under cost regulation and vulnerable to political intervention – comprises roughly 6.900 fee models. Unlike pricing schemes in other network economies, the prevalent fee models are quite simple as customers pay a single fee per cbm. Such a fee results in high incentives to reduce water consumption and greywater discharge simultaneously, which complies with the Water Framework Directive (WFD). However, the fee lacks financial, economic and social sustainability. Implementing an additional fixed fee seems to be a reasonable measure to remedy these shortcomings, although the fixed fee share in overall revenue remains an important control parameter to guarantee compliance with the WFD. Having argued for a two-part fee, two parameters for the fixed fee component must be defined: First, the fee base and second, the fee rate. Out of a wide range of possible fee bases, the 'Number of dwelling units in a residential building' is not just stable but also corresponds to the costs-by-cause principle and generates manageable transaction costs. However, transaction costs also depend on the extent to which gains and losses among customers due to the fee model switch occur. Testing different fee models with household data in three German cities of different population size, the authors find evidence that the number of dwelling units combined with a digressive rate structure results in less absolute gains/losses for customers. This finding is model-inherent and can be used to rethink fee models just based on the cbm. The authors recommend the fee model switch when (1) water consumption per capita is decreasing, (2) migration of water consumption is high and (3) resource costs are low.

Keywords: fixed fee, greywater fee, sustainability concept, tariff design

1 INTRODUCTION

The wastewater disposal sector in Germany consists out of nearly 6.900 utilities, mostly under the ownership of local municipalities [1]. Unlike in network economies as for example the electricity sector, no incentive-based regulation schemes apply for the wastewater sector, which economists define as 'natural monopoly'. Instead, a form of cost-regulation applies. Cost-regulation does not imply individual monitoring of costs with focus on efficiency, but rather a monitoring of (1) cost accounting principles – among others mandatory full cost-recovery – and (2) tariff design principles, both on request. These principles are encoded in European, national, federal and municipal legal institutions. Cost-regulation schemes are accompanied by voluntary benchmarking rounds, which emphasise efficiency. However, benchmarking reports for public just contain aggregated data and no utility-specific information. Therefore, benchmarking supports dedicated utilities with valuable information about key performance indicators and efficient process design, but limits external pressure to reduce costs. Each utility seems therefore to be well protected. At the same time, each utility is vulnerable to internal pressure via political intervention, which might put pressure on full application of the cost-recovery principle in order to keep the customer fees low. Yearly comparisons of wastewater fees, e.g. from the association of tax payers in North Rhine-Westphalia [2], might explain such behaviour. However, instead of debating if fees are 'too high or too low', different fee models, which evolved in the above-mentioned area of

DOI: 10.2495/SDP-V13-N3-457-468

tension, are discussed. The focus lies solely on **fee models for greywater** [3], ignoring fees for precepitable water, which cover separate costs although technical separation is often missing (e.g. combined sewer systems). Fee models for non-household customers are also excluded, as these need deeper analysis of pollution-based fees [4].

Binding legal rules restrict the design of fee models. These rules are not always very specific, but reasonable regarding their fundamental principles, as several objectives [5] need to be balanced. One major principle is the costs-by-cause principle, which makes customers pay fees according to what extent they contribute to costs. The wastewater disposal sector shows a high percentage of fixed costs, mainly due to the provision of infrastructure. Percentages up to 85% are common, but 95% are possible as well [6]. Such numbers raise the question, what the main service of a wastewater utility is. According to the numbers, it cannot be the disposal and purification of greywater, but rather the provision of sewer system and wastewater treatment plant. Just the former service depends on the quantity of greywater, the latter one is quantity-independent. From an economic point of view, quantity-independent services should be covered by fixed fees – a two-part fee model (volumetric fee + fixed fee) would be reasonable [7]. However, according to an empirical survey, 31 out of 271 cities in North Rhine-Westphalia use two-part fee models. The modified volumetric-only fee (based on metered freshwater consumption ± adjustments, for example, metered freshwater for gardening) is predominant, a statement with can be generalised for whole Germany as spot checks in other federal states show. In Europe, a mixed picture can be found as the following Figure 1 illustrates.

The survey shows the relevance of fixed fees in the ten biggest cities with regard to population size in selected countries. Except Berlin, no other German city uses two-part-tariffs. The cities in the Netherlands show the opposite: All fee models solely rely on fixed fees, based on property value. British and Swiss cities rely heavily on the volumetric fee, as the above calculations for a model household indicate. Though using a fixed fee, the revenue share covered by the fixed fee component is likely to fall short of the fixed costs share. Austrian cities show the widest range of shares covered by the fixed fee with two cities roughly meeting the fixed costs share. Overall, it is obvious that, relative to the high percentage of fixed costs, the volumetric-only fee plays a way too dominant role in fee model design, not just in German cities.

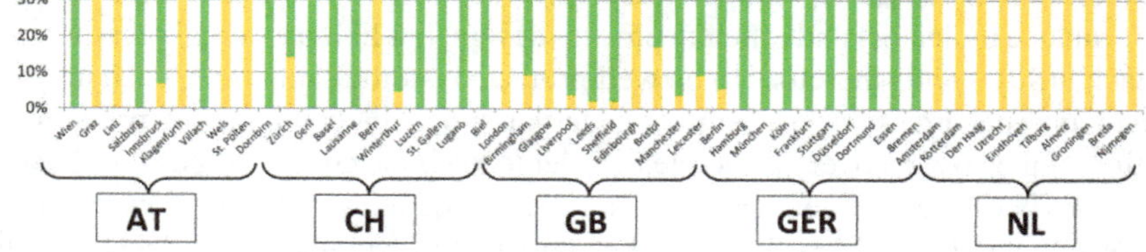

Figure 1: Billing share of fixed fee and volumetric fee for a single-family household (150 cbm) in selected countries 2015.

In this paper, the authors argue that two-part fee models with sufficient revenue share covered by the fixed fee improve sustainability (→ Chapter 2). However, a fee model switch creates distributional effects (→ Chapter 3), which might be the reason why volumetric-only fees prevail. Finally, a different fee model is derived (→ Chapter 4), which is not just sustainable, but also balances the extent of gains/losses due to the fee model switch.

2 JUSTIFICATION OF TWO-PART FEE MODELS

'Sustainability' is a composed criterion, which is here applied to assess the introduction of a fixed fee component into the fee model [8]. Four sustainability dimensions are considered: Economic, ecological, financial and social sustainability [9].

The already mentioned costs-by-cause principle belongs to **economic sustainability**. In case of a volumetric-only fee, customers pay according to their consumption. Households with low consumption, among them holiday homes, pay less than households with high consumption. However, consumption is not the main driver of costs and these customers utilise the provided infrastructure anytime. Therefore, they have to pay for the provision of infrastructure independent of how much greywater these customers dispose into the sewer system. If they do not, customers with the ability to reduce consumption retrieve from funding the infrastructure and the financial burden is redistributed – the financing mechanism does not correspond to the costs-by-cause-principle. A fixed fee component prevents such redistribution and allocates the costs of infrastructurzze provision to those customers who benefit from it. The volumetric fee still allocates variable costs, e.g. energy costs, to customers according to their greywater quantity. However, both fees components should finance different services.

Economic sustainability also requires marginal cost pricing, which does not apply for the volumetric-only fee because of cross-subsidisation mechanisms. With cost-recovering fees, introducing a fixed fee implies a lower volumetric fee. In case the fixed fee fully covers fixed costs, the volumetric fee reflects marginal costs and customers can make efficient consumption and investment decisions. Without fixed fee, customers tend to reduce consumption beyond the economic optimum and invest too much in water-efficient technologies, respectively investments that reduce the disposed/billed greywater quantity (e.g. metering devices for gardening).

Ecologic sustainability requires a tariff design that gives incentives to preserve water resources and prevents pollution. The latter point can be excluded as households tend to have homogenous, manageable dirt freights in their greywater (unlike non-household customers) and comprehensive pollution meters are not an option. However, Art. 9 Water Framework Directive (WFD) requires 'that water pricing policies provide adequate incentives for users to use water resources efficiently'. As volumetric greywater fees exceed in most European countries the volumetric freshwater price [10], incentives from the wastewater disposal side are in comparison stronger. This is also true for countries without organisational separation [11], where greywater fees are included in the water price. At least from consumer perspective, wastewater disposal and water supply sector are anyway part of the same value chain. However, introducing a fixed greywater fees lowers the volumetric fee and incentivises water saving. Therefore, the volumetric-only fee fosters ecological sustainability and every fixed fee decreases the extent to which this dimension is fulfilled.

Financial sustainability requires stable revenue to cover costs occurring in the billing period. The wastewater utility faces price and quantity risks, where the latter one are risks due to volatile consumption patterns. In case of higher than expected consumption among all customers, revenue exceeds costs and vice versa. Both should be avoided as short-term extra revenue has to result in lower fees during the following periods and missing revenue requires intermediate financing measures. The latter one is a concern for municipalities with decreasing water consumption per capita and/or emigration of water consumption. With given costs and volatility of consumption, the volumetric-only fee maximises the difference between revenue and costs. Consequently, the implementation of a fixed fee improves periodic cost-recovery as a stable fee covers fixed costs. The flat rate (NL) is in this context not sustainable as e.g. energy costs vary with greywater quantity and should covered via a volumetric fee.

Social sustainability does not necessarily imply social tariffs but rather questions whether the volumetric-only fee negatively affects low-income households. One circumstance support this hypothesis: Low-income households primarily inhabit houses with multiple dwelling units. Empirical evidence shows that such inhabitants in general react less sensible to fee adjustments. Single-family houses and non-household customers have higher 'price-elasticities' [12] as they perceptibly benefit from quantity adjustments in case of increasing/decreasing prices. As water consumption is declining for several years now and costs increase at the same time not just because of inflation but also because of quality improvements, volumetric only fees rise upwards. Single-family houses and non-household customers reduce water consumption to avoid rising fees. Houses with multiple dwelling units more and more have to bear the financial burden. A self-sustained process emerges where higher fees induce lower consumption and the financial burden for low-income households is likely to rise. The fixed fee slows the 'fee spiral' down and removes social sustainability deficits. This argument applies primarily for municipalities with decreasing total consumption.

Overall, the volumetric-only fee fosters ecological sustainability, but lacks economic, financial and social sustainability. Such imbalance between the four sustainability dimensions can be corrected by introducing a fixed fee. Although this weakens ecological sustainability, the fixed fee strengthens the remaining three dimensions, restores a balance between all four dimensions and improves overall sustainability.

Something that should not be forgotten is that a fee model switch results in gains/losses for customers. If these distributional effects are expected to be substantial, the likelihood of realising the new fee model shrinks. Customers will not accept high losses and obstruct the fee model switch, which in turn makes transaction costs for implementation high. Fee models therefore should carefully balance gains and losses to achieve implementation in the short-run while realising long-term benefits due to improved sustainability. However, designing the fixed fee leaves several options for the wastewater utility and understanding the effects of a model switch is essential. The following chapter provides first insights into such effects.

3 EFFECTS OF A FEE MODEL SWITCH

In Germany, owners of residential buildings pay wastewater fees, as long as the individual dwelling units are not equipped with metering devices. Newer legislation enforces water meters for each dwelling unit, so that households in new constructed buildings pay the fee directly and not via intermediate distribution by the property owner. Given the situation of newer buildings, the individual household i pays a volumetric-only greywater fee p_1^v for each unit of consumption x_i. The fee sums up to

$$p_1^v \times x_i. \tag{1}$$

A two-part fee model with fixed fee component p_i^f results in total fees of

$$p_2^v \times x_i + p_i^f. \tag{2}$$

Assuming that θ is the revenue percentage share covered by the fixed fee component, eqn (2) changes to

$$(1-\theta)p_1^v \times x_i + p_i^f. \tag{3}$$

Assuming that the fixed fee component is the same for each customer ($\rightarrow p_i^f = p^f$), the cost recovery condition equals

$$\sum_{i=1}^{n} p_1^v \times x_i = \sum_{i=1}^{n}\left[(1-\theta)p_1^v \times x_i + p^f\right]. \tag{4}$$

In the equation above, n defines the total number of customers. Simplification and reordering yields

$$p^f = \theta p_1^v \times \bar{x}. \tag{5}$$

Given that each customer has to pay the same fixed fee, the fee itself depends on the percentage share in revenue, the single volumetric rate and the average water consumption (neglecting differences between water consumption and volume of greywater), which is defined as $\frac{1}{n}\sum_{i=1}^{n} x_i$. The individual gain/loss (Δ_i) from the fee model switch is

$$\Delta_i = (1-\theta)\, p_1^v \times x_i + \theta\, p_1^v \times \bar{x} - p_1^v \times x_i. \tag{6}$$

Simplification and reordering shows that an individual gain from the switch occurs when:

$$x_i > \bar{x}. \tag{7}$$

Is the individual water consumption greater than the average water consumption $\bar{x}$, the gain from the reduced volumetric fee compensates for the loss due to the implemented fixed fee. The fixed fee therefore benefits high consumption, as households with less consumption contribute more to the costs of infrastructure provision. The full extent of gains/losses among all customers ($\sum \Delta_i = \Delta$) equals to

$$\sum_{i=1}^{n}\left[\left|(1-\theta)p_1^v \times x_i + \theta p_1^v \times \bar{x} - p_1^v \times x_i\right|\right]. \tag{8}$$

Simplification yields

$$\Delta = n\theta\, p_1^v \times MAD. \tag{9}$$

MAD stands for Mean Average Dispersion and is a common measure of statistical dispersion ($\rightarrow MAD = \frac{1}{n}\sum_{i=1}^{n}\left[\left|\bar{x} - x_i\right|\right]$). As the only **control parameter for fee model design is** θ, a reduction of gains/losses can only be achieved by lowering the percentage share in revenue covered by the fixed fee component. MAD and n are characteristics of the wastewater disposal district and – thought influenced by the single volumetric fee – exogenous parameters at the moment, the fee model switch is conducted. The initial volumetric fee p_1^v is set to recover costs and expected to be higher, when the fee spiral evolves. Wastewater utilities facing such a problem should therefore react in time as gains/losses c.p. increase with increasing volumetric fee.

However, the lump sum fixed fee does not properly correspond to the costs-by-cause principle. Therefore, the fee base complements previous deliberations, mapping the individual costs of infrastructure provision. The fixed fee extends to

$$p_i^f = p_1^f \times b_i. \tag{10}$$

b_i denotes the fee base and is expected to rise as the individually attributable costs of infrastructure provision increase. The 'price' p_1^f of a single fee base unit b_i can be derived via the cost-recovery condition and yields

$$p_1^f = \frac{\theta p_1^v \times \bar{x}}{\bar{b}}. \tag{11}$$

Each single unit of b_i 'pays' the lump sum fixed fee relative to the average value of the fee base among all customers $\bar{b}$. In case all customers show the same characteristic – the wastewater disposal area is completely homogenous (e.g. metering devices are all of the same size) – the fixed fee equals the lump sum fee.

The gains/losses among all customers sum up to

$$\Delta = n\theta p_1^v \times \frac{1}{n}\sum_{i=1}^{n}\left[\left|\bar{\boldsymbol{x}} \times \frac{\boldsymbol{b}_i}{\bar{\boldsymbol{b}}} - \boldsymbol{x}_i\right|\right]. \tag{12}$$

The latter term is not the MAD itself but rather a fee base weighted MAD. However, the equation shows that for minimising gains/losses b_i and x_i must be aligned variables. If a wastewater utility intends to establish a fixed fee to mitigate long-term negative effects due to sustainability deficits, but at the same time worries about high transaction costs due to redistribution effects, it should seek for such a fee base that shows the lowest weighted MAD. Making such a choice is a **second control parameter** to steer gains/losses. However, costs-by-cause considerations should be prioritised.

The derivations above comprise linear rate structures. Without going into detail, the assumption can be modified to allow for also digressive or progressive rate structures. In that case, the fixed fee is calculated as

$$p_i^f = g(b_i). \tag{13}$$

$g(.)$ is a continuous function with the following properties: (1) $\frac{\partial g()}{\partial b_i} > 0$ → the higher the fee base, the higher the fixed fee; (2.1) $\frac{\partial^2 g(b_i)}{\partial b_i^2} > 0$ if rate structure is progressive and (2.2) $\frac{\partial^2 g()}{\partial b_i \partial b_i} < 0$ if rate structure is digressive. Progressive rate structures aren't reasonable in Germany, neither for the volumetric fee (→ no severe water scarcity) nor for the fixed fee (→ no cost progression). However, digressive rate structures seem reasonable as the required infrastructure capacity decreases with regard to the fee base. Assuming that such a connection can be verified by also technical analysis, the wastewater utility can set up a digressive function $g(.)$, which should rely on technical reports to verify the cost digression. The definition of the function is the **third control parameter**. For demonstration purposes, the authors assume a digressive function of the form

$$p_i^f = p_2^f \times \sqrt{b_i}\,. \tag{14}$$

As p_2^f is determined by exogenous parameters, total gains/losses equal

$$\Delta = n\theta\, p_1^v \times \frac{1}{n}\sum_{i=1}^{n}\left[\left\|\bar{x} \times \frac{\sqrt{b_i}}{\frac{1}{n}\sum_{i=1}^{n}\left[\sqrt{b_i}\right]} - x_i\right\|\right]. \tag{15}$$

The latter term is also a weighted MAD, but the function $g(.)$ changes the weights for $\bar{x}$ from function input to function output values. Properly designed functions can substantially reduce the amount of gains/losses. Keep in mind that the cost-recovery-constraint lowers the degrees of freedom by one, so that the functions with more than one free parameter can be selected to be able to reduce the distributional effects. The design of the function – at least in Germany – has to comply with legislation, which constrains free parameterisation in order to keep gains/losses low.

4 DEVELOPING THE FEE MODEL DESIGN

The fee model design requires a parameterisation of the three in chapter 3 derived parameters. The concept of sustainability is applied once more to parameterise θ. The costs-by-cause principle is the leading concept for the choice of the fee base. However, other criteria apply, e.g. data availability. For selected fee bases, a proper rate structure can be derived. The selected fee models have been developed for the city of Leichlingen, Germany.

4.1 Parameterisation of θ

The Parameter θ ranges from zero (→ volumetric-only fee) to 100% (→ flat rate). If the wastewater utility just wants do reduce distribution effects, the value should be low. In following years, it can be raised without too many distortions. However, other utilities want to raise the share immediately to the target level. Both approaches raise the question, what a sustainable and long-term target level should be.

Both, economic and financial sustainability require θ being equal to the share of fixed costs (≈85%). In that case, the volumetric fee equals marginal costs and efficient consumption and investment decisions can be made. It also satisfies potentially the costs-by-cause principle, because volumetric and fixed fee separately cover the costs of the two different services. Cross-subsidisation mechanisms are set off. Any θ above that level diminishes economic and financial sustainability, because of the deviations from the marginal cost pricing principle and non-periodic cost-recovery. Figure 2 illustrates how θ affects the degree of sustainability subdivided by the four dimensions.

Ignoring dirt freights, ecological sustainability in general increases with decreasing θ resp. lower volumetric fee, because of higher incentives to reduce water consumption and preserve water resources. However, disposal areas might not suffer from water scarcity, which implies that resource costs according to the WFD are low and such incentives are dispensable. If

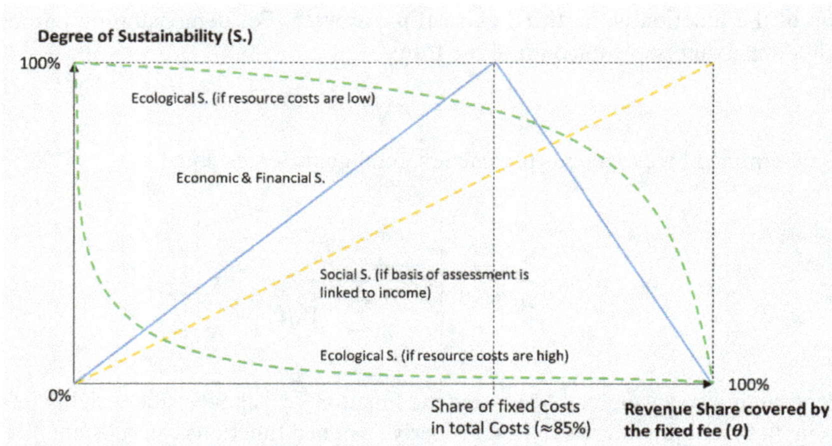

Figure 2: Determinants of θ.

resource costs are low, a higher θ is justified according to ecological sustainability. The flat rate still incentivises wasteful use of water and should be strictly rejected. On the other hand, a lower θ might be sustainable, if the sewer systems need to be flushed with freshwater for hygienic purposes. It is therefore important to consider the whole 'environmental balance sheet' to assess this sustainability dimension.

How the degree of social sustainability depends on θ is uncertain and primarily relies on the fee base. If the base is income-correlated, a θ might lower the greywater fee for low-income households. However, freshwater consumption, the basis of assessment for the volumetric rate already shows (weak) correlation with income [13]. Therefore the fixed fee base must be even closer connected to income, if low-income households should gain from the fee model switch. Not debateable is the fact that a substantial θ helps to stop the fee spiral, which negatively affects low-income-households.

While both, economic and financial sustainability, point towards a θ of roughly 85%, ecological sustainability requires a lower θ. Even when resource costs are low in Leichlingen, there are substantial legal doubts, which is why the authors vote for a compromise to balance sustainability dimensions. A compromise can be 50%, a quite common value also in the water supply sector. A lower θ to minimize distributional effects might be reasonable at first glance, but is not necessary as the results in chapter 4.3 show.

4.2 Selection of the fee base

The main cost drivers of infrastructure provision are the greywater quantity itself (→ wastewater treatment plant) and the network length (→ sewer system). The fee models in selected European cities often apply fee bases related to quantity. Among them are (1) the size of water metering device, (2) the number of dwelling units (DU) in a residential building, (3) the number of toilets and (4) the number of inhabitants. In addition, (5) the property value and (6) the living space partly reflect the network length. Differentiating between (7) different city areas is also a possible fee base, strengthening the costs-by-cause principle if e.g. infrastructure is 'more valuable' in the city center.

For a number of reasons, the majority of fee bases seem not appropriate. The property value for example is not directly linked to the disposed greywater quantity so that it cannot meet the costs-by-cause principle. Other criteria for fee base selection are the costs for gathering resp. maintaining the database and of course data availability in general.

In cooperation with the wastewater utility in Leichlingen the authors developed three two-part fee models. The first model based on the size of the metering device (SMD-Model), the second model based on the number of dwelling units (DU-Model) and a third model with combined fixed fee base (Combi-Model). 'Combined' refers to a fixed fee based on (1) the number of dwelling units and (2) the length of the front of a residential building. The former one covers fixed costs of the wastewater treatment plant, the latter one covers fixed costs of the sewer network. As infrastructure costs of the sewer system depend heavily on the network length, the authors found it appropriate to use a measure, which reflects the individual contribution to the provision of the sewer network length. Some notes on the different fee bases:

The size of the metering device and the number of dwelling units in a residential building are common measures in the water supply sector and also used in the wastewater disposal sector. A serious caveat of the metering size stems from the fact that the standard Q_n 2.5 meter supplies roughly 95% of all residential buildings. Therefore, the capability to segment customers according to their greywater quantity is poorly. According to technical rules [14], the Q_n 2.5 meter can supply buildings up to 30 dwelling units, covering a range of 0 to approximately 2000 cbm. The underlying logic behind the fee base is that even single-family houses could potentially consume the upper cbm limit allowed by the meter size and dispose the freshwater after usage into the sewer system. However, it totally ignores empirical evidence: Single-family households are very unlikely to consume more than 250 cbm. Treating the exception as a rule creates distortions with big residential buildings, in which the meter size is not falsely oversized. For this reason, the weighted MAD for this fee base is quite high and gains/losses expected to be substantial. However, the fee base is popular and widely accepted, which is why the authors decided to include it as benchmark.

The number of dwelling units (DU) in a residential building has more fee base categories (1-DU-Building, 2-DU-Building,...) and can segment customers according to their consumption more precisely. Although the fee base is not able to reflect the network length, it is closely connected to the greywater quantity. The limitation of gains/losses should be easier.

Superior regarding the costs-by-cause principle is the Combi-Model as both cost drivers are covered. However, the front length does not vary with consumption and the distributional effect should be greater. Insofar there is a trade-off: Additional correspondence with the costs-by-cause principle comes at the cost of higher gains/losses.

4.3 Fee structure

While there is no justification for meter size and front length to divert from linear fee structure, a digressive rate structure seems to be appropriate for the number of dwelling units. Three arguments based on empirical evidence in Leichlingen and two other partner municipalities can be urged: First, the disposed quantity per dwelling unit decreases with building size. So even if all inhabitants in a building with multiple dwelling units simultaneously dispose greywater, the quantity per DU would be less compared to the quantity of a single-family house. Second, the stochastic profile of greywater disposal evens out between the

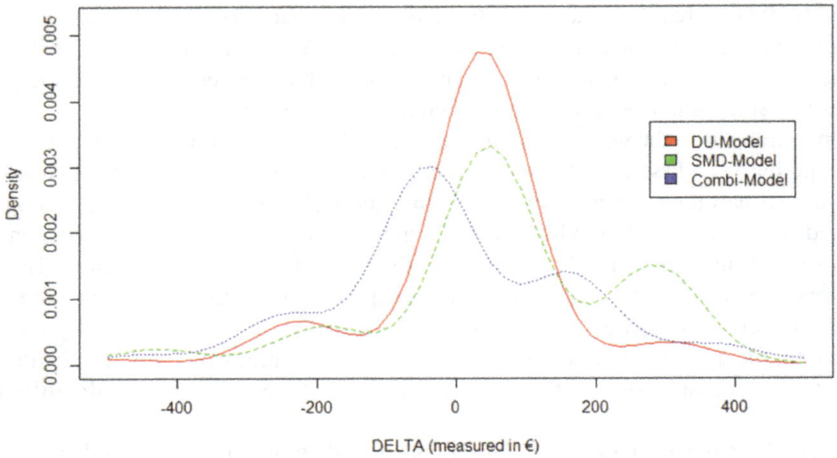

Figure 3: Distribution of gains/losses for customers for different fee models.

different dwelling units in the same building. Third, numbers from Leichlingen indicate that the front length does not increase proportional with increasing number of dwelling units. A 30-DU-Building has less than 30 times the front length of a single-family house. The provided infrastructure is therefore expected to be less for each DU. Cost digression in all three cases justifies a digressive rate structure, which is applied for the number of dwelling units in the DU-Modell and the Combi-Model.

4.4 Distributional effects of selected fee models

Knowing that the digressive rate structure also reflects the consumption patterns in Leichlingen, the weighted MAD and consequently the extent of gains/losses is lowest in the DU-Model. Figure 3 pictures the density functions of gains/losses Δ_i at building level for the three different models. Although Δ_i is no random variable, the figure illustrates the distribution of gains/losses very well. The abscissa is limited to ± 500 €. However, some customers gain/loose well above 10000 €.

Obviously the DU-Model can centralise gains/losses much better around the neutral position (± 0 €) than SMD and Combi-Model. This is due to the segmentation capability of DU. The SMD-Model affects, particularly single-family houses with low consumption. As these customers pay the same fixed fee as e.g. 30-DU-Bildings at lower overall consumption, they lose due to the model switch according to eqn. (6). The fixed fee constitutes losses of 200–400 € for this customer group. The effect can be seen at the right tail of the green dashed line. The Combi-Model seems beneficial to most customers as the "likelihood" of gaining from the two-part model peaks at roughly 50 €. However, a few customers with very high front length bear substantial losses, which would surely provoke resistance.

Both, the SMD-Model and the Combi-Model redistribute more than 30% of the total greywater revenue. The DU-Model redistributes roughly 16%, a number that is confirmed in two other disposal areas with available data for the DU-Model. Households in different buildings with average consumption patterns gain/loose not more than 5%. In a city with favourable settlement structure, consumption patterns and moderate volumetric-only fee, the maximal gain/loss for average customers fell short of 2%. This seems manageable and communicable, so that the authors overall recommend the DU-Model. The model fosters sustainability with given θ and causes only moderate redistribution among customers, making the model well-balanced.

5 CONCLUSIONS

A future-proof greywater fee model can balance different dimensions of sustainability. As the volumetric-only fees lacks economic, financial and social sustainability, two-part fees restore the balance between sustainability dimensions and improve overall sustainability. This is particularly important for municipalities in which water consumption per capita is decreasing and/or water consumption migrates out of the wastewater disposal area. In both cases, a fee spiral starts with negative consequences for sustainability. To properly address ecological sustainability, the share in revenue covered by the fixed fee component should remain below the share of fixed costs. A share of 50% seems to be a compromise, which applies also in other industries as benchmark. However, when resource costs are low, it is reasonable to raise the level above 50% and vice versa. A high share also implies that distribution effects are higher, which makes the fee model switch due to transaction costs less likely. To restrict gains/losses among customers, wastewater utilities can choose the fee base and the fee structure to minimise such distributional effects. Both parameters – this is the restriction – have to comply with the costs-by-cause principle to guarantee sustainability. Testing three different fee models, the 'number of dwelling units in a residential building' is capable to meet the costs-by-cause principle and reduce gains/losses properly, at least in case of the three cities. To account for reduced consumption per dwelling unit with increasing building size, a digressive rate structure was modelled, which is both, justified by the costs-of-cause principle and favourable in terms of gains/losses. A fee model of this form performs better than two benchmark models. Overall, the authors recommend the number of dwelling units as sustainable fee base, which creates moderate gains/losses and makes the fee model easy to implement.

REFERENCES

[1] Profile of the German Water sector 2015; ATT et al., available at: https://www.bdew.de/internet.nsf/id/profile-of-the-german-water-sector-2015-de. (accessed 9 March 2017).

[2] Comparison of wastewater fees 2016; Association of tax payers NRW, available at: http://www.steuerzahler-nrw.de/files/75620/8_Abwasser_Gesamttabelle.pdf. (accessed 9 March 2017).

[3] Oelmann, M. et al., *Nachhaltige Bestandserhaltung öffentlicher Abwasseranlagen und ihre Refinanzierung*, available at: https://www.lanuv.nrw.de/landesamt/forschungsvorhaben/details/?tx_mmkresearchprojects_pi1[uid]=448. (accessed 13 March 2017).

[4] Mogden-Formula; Ofwat, available at: http://www.ofwat.gov.uk/nonhousehold/yourwaterbill/hownonhousehold/trade/mogden. (accessed 21 July 2015).

[5] American Water Works Association (AWWA), *Principles of water rates, fees, and charges – Manual of water supply practices*, 6th Ed., Denver, 2012.

[6] Leptien, C. et al., Wirtschaftsdaten der Abwasserbeseitigung 2014, Bad Honnef, 2014.

[7] Coase, R., The marginal cost controversy. *Economica*, *13*, pp. 169–182, 1946. https://doi.org/10.2307/2549764

[8] Oelmann, M., Roters, B. & Gawel, E., Sustainable fee models in the wastewater sector, *KA Korrespondenz Abwasser*, to be published.
[9] Organisation for Economic Co-operation and Development (OECD), *Pricing water resources and water and sanitation services*, Paris, 2010.
[10] Bundesverband der Energie- und Wasserwirtschaft e.V. (BDEW), *Comparison of European water and wastewater prices*, VEWA survey, Berlin, 2015.
[11] European Environmental Agency (EAA), *Assessment of cost recovery through water pricing*, EAA technical report, No 16/2013, Luxembourg.
[12] Oelmann, M. & Gendries, S., Auf dem Weg zu einem neuen Tarifmodell in der deutschen Wasser-versorgung. Teil 1: Anforderungen aus Sicht eines Wasserversorgers, Prozessgestaltung und Datengenerierung, in: gwf-Wasser | Abwasser, pp. 820–827, 2012.
[13] Schleich, J. & Hillenbrand, T., Determinants of Residential Water Demand in Germany, Working Paper Sustainability and Innovation, No. S3, 2007.
[14] Deutsche Vereinigung für Wasserwirtschaft, Abwasser und Abfall e.V. (DWA), Worksheet W 406.

EFFECTS OF CLIMATE CHANGE ON WATER AVAILABILITY FOR THE USUMACINTA RIVER ENVIRONMETAL FLOW (MEXICO)

REBECA GONZÁLEZ VILLELA & MARTÍN J. MONTERO MARTÍNEZ
Instituto Mexicano de Tecnología del Agua, Mexico.

ABSTRACT

The water availability in the Usumacinta River sub-basin is determined through the analysis of the amount and frequency of precipitation, associated with the quantity, frequency and magnitude of flow regimes (environmental flows). The river is located in south-eastern Mexico. The precipitation of pre-impact (1960–1983) and post-impact (1984–2008) periods was analysed using a climatological mesh database created by CICESE (Center of Scientific Research and Higher Education of Ensenada) covering the period 1960–2008. For the analysis of flows, the hydrometric information the IHA V7 program was used to define the main trends in the temporal variation of the daily flows of the pre and post-impact periods. A number of several factors such as: the increase in monthly precipitation in the rainy season and a decrease of the precipitation in the dry season in the post-impact period, the significant increase in the number of days with zero precipitation, the increase in the number of days a year with a greater amplitude in the maximum rainfall, a positive tendency of precipitation in rainy season and a significant decrease in the dry season; implies that now in the wet period it rains more and in dry season it rains less, indicating that the climate is more extreme, aspects that can be associated with the effects of climate change. Also, torrential rains that can be associated to the changes in the precipitation rates are due to the effects of climatic change. The natural flow shows a slight decrease in the flow during the rainy season, or a significant decrease in the flow rates for some months in the post-impact period. This condition does not coincide with the increase in precipitation for this period and in this part of the basin, a situation that may be related to the anthropic use of the resource. The Usumacinta basin is relevant for the environmental services that it provides.

Keywords: climate change, Environmental Flows, water management.

1 INTRODUCTION

Climate change impacts unpredictably the lake systems; therefore, it is necessary to generate models and analytical methodologies to predict their trends and possible changes, in order to carry out a sustainable management of the associated resources [1, 2]. For this, it is necessary to generate standardized tools that assess the biophysical impacts of climate change where socio-economic elements of ecosystems are included, to generate the strategies that help to face the effect of climate change in a harmonized way with sustainable development [3–5]. For this reason, studies of climate impact assessment through environmental flows are important for species conservation, for assessing the role and resilience of aquatic ecosystems, and for welfare of the people who depend on them. Because, they support and provide goods and services to the population, such as: drinking water supply, fishing, recreation, agriculture and electricity [4, 5]. However, the limnological information of the country is insufficient. On the other hand, the difficulty in quantifying the impact that climate change is exerting on water quality and its vulnerability in surface water bodies. In the present study, water availability in the Usumacinta River sub-basins was determined through the analysis of the amount and frequency of precipitation, associated with the quantity, frequency and magnitude of flow regimes (environmental flows), to determine how much of the flow in the river has been altered by the effects of climate change or the number of changes that have been introduced in the regimes of flow by human influence.

DOI: 10.2495/SDP-V13-N3-469-481

2 STUDY AREA

The Usumacinta sub-basin, stretching from north-western Guatemala to the states of Chiapas and Tabasco in Mexico, covers an area of 122,000 km^2 [6]. The basin is located in south-eastern Mexico, at the geographic coordinates 18.71 °–15.22 ° N and 94.25 °–90.38 ° W. The Grijalva and Usumacinta rivers, which drain into the Gulf of Mexico through a rectangular fluvial system, are one of the most important basins in Mexico and North America. This sub-basin covers 87,686 km^2, and in volume the Usumacinta River is the most important current in the Gulf of Mexico after the Mississippi and the seventh worldwide (Fig. 1). It is one of the main regions with the highest precipitation in Mexico, especially in the transition between the coastal plain and the central saws where the average annual precipitation can reach more than 4,500 mm, with an average annual flow of 3,727 m^3/s [7]. The annual catchment water in this region, equivalent to 30% of the total surface runoff of the country, according to data from Ref. [8].

The rainforest is found in the humid coastal plain and in the lower parts of the east, while the Central Depression is dominated by dry tropical forest due to a double rain shadow. In the mountains of northern Chiapas (1400–1800 m), the central plateau or Altos de Chiapas (1200–2800 m) and the mountains of Chiapas (1500–3050 m), you can find the different types of temperate and cloudy forest. The huge amounts of surface water feed a large number of inland and coastal bodies, flood plains and wetlands on the coastal plain with an area of more than 400,000 ha. The lowlands of the sub-basin create a large and complex system of productive tropical canals and wetlands [7, 9].

3 METHODS

3.1 Base of climatological data

For the analysis of precipitation variation, a climatological mesh database created by CICESE (Center of Scientific Research and Higher Education of Ensenada), will be used. For the comparative analysis of this work in the Usumacinta sub-basin, the periods were defined: pre-impact (1961–1984) and post-impact (1985–2008) with the idea of comparing potential changes in the environmental flows due to changes in the climate. The data was obtained from the official climatological database of the SMN (National Meteorological Service of Mexico), which comprises more than 5000 stations and is stored in the CLICOM

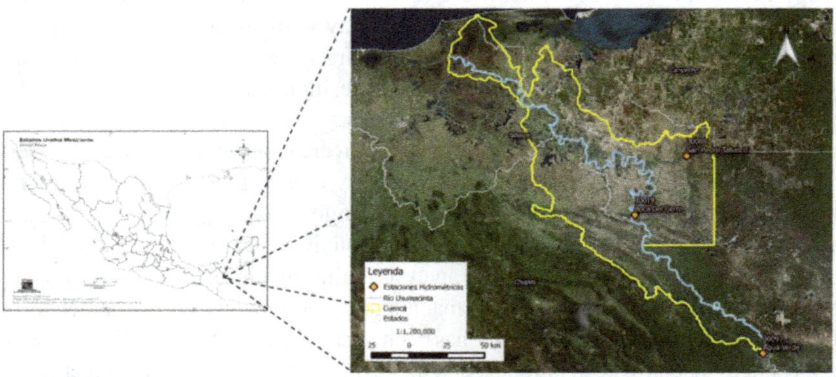

Figure 1: Location of the Usumacinta River and the Weather Stations.

(CLImatological COMputing) system. Some quality tests were applied to this database, such as: 1) removing negative rainfall; 2) check the maximum temperatures less than or equal to the minimum temperature of the same day (in such cases, both temperatures were placed as missing values); 3) review days with the same value repeated 10 or more consecutive times (except for zero precipitation); for these cases, all values were placed as missing values; daily values that significantly exceeded climatological values were also revised. In such cases, the data was removed and replaced with a lost value. The above data were used to calculate the Standard Precipitation Index (SPI), created by Mc Kee [10, 11], which is a widely used index for calculating potential drought or intense rainfall events in a given location.

The station data already filtered with the above-mentioned data quality analysis were interpolated to a regular mesh using the Synaptic Mapping System (SYMAP) [12]. This mesh uses the weighted average (based on the inverse of the square of the distance to the mesh cell in question) of all the records in the vicinity of a mesh cell, to produce a daily database for precipitation and temperature (maximum and minimum), of surface, in a mesh of spatial resolution of 1/8° for all Mexico [13, 14]. The graphic platform was developed by CICESE (http://clicom-mex.cicese.mx/malla). The SPI ranges recommended by the World Meteorological Organization [15] were used (Table 1).

The data filtered of precipitation were analysed using graphs and statistics through the IHA V7 (Indicators of Hydrological Alteration) [16], program to define the main trends in the temporal variation of daily precipitation. The hydrological alteration indicators (IHA) and the Range of Variability Approach (RVA) of intra and inter-year time series were taken into account.

3.2 Environmental Flows (IHA, Indicators of Hydrologic Alteration)

For the analysis of the hydrometric information of the pre-impact and post-impact periods in the environmental flows of the Usumacinta River, we considered the Boca del Cerro station (30019), located in the upper part of the river basin, with the amount of available data (lat: 17.43° N, lon: -91.48° W). The data were analysed using graphs and statistics using the IHA V7 program, to define the main trends in the temporal variation of daily expenses and the RVA indicators of the intra and inter-annual time series of 32 hydrological parameters from the average daily flows. The parameters were classified into six groups: Low Extreme Flows, Low Flows, High Flows, High Flow Pulses, Small Floods and Large Floods of 2 and 10 years of return, respectively. For each indicator, parametric and non-parametric analysis was

Table 1: Values of the normalized precipitation index.

VALUE	PRECIPITATION INDEX
2.0 and more	Extremely wet
1.5 a 1.99	Very wet
1.0 a 1.49	Moderately moist
-0.99 a 0.99	Normal or about normal
-1.0 a -1.49	Moderately dry
-1.5 a -1.99	Severely dry
-2 and less	Extremely dry

performed, estimating the means or medians and their dispersion, by standard deviations or percentiles for the years considered. The alterations imposed by flow regulation and/or by changes in the climate (post-impact) were compared with the time series of the natural system (pre-impact) using the range of variability of each parameter given by ± the standard deviation or between the percentiles 25% and 75%. This range of variability was used to assess the impact of climatic or anthropic change on environmental flows in the Usumacinta River, and possible scenarios were established for species of flora and fauna.

4 RESULTS

4.1 Variation of precipitation in the high basin of the Usumacinta River

The monthly precipitation obtained from the analysis of the 1/8° spatial resolution mesh in the Usumacinta River for the upper basin (Boca del Cerro) through the IHA Software, indicates significant differences considering the ranges of variation of the values to 75% with respect to the median for June and September, months of higher rainfall and intra-drought, between (1961–1984) and the post-impact period (1985–2008). The significant variation occurs in the rainy season, implying that it rains more in the wet season in the post-impact period (Fig. 2). The historical analysis of precipitation during the post-impact period (1985–2008) shows the high intensity of a storm in 2003 (350 mm) and the high frequency of small storms with a precipitation of 105 mm in the years 1977, 1978, 1987, 1988 and 1997. Likewise, the predominance of the high pulses of precipitation (60 mm) in the upper part of the Usumacinta basin is common.

In the post-impact period, the number of days with zero precipitation decreases significantly from 1989 to 1997, an aspect that can be associated with rainy years in this region, and on the contrary, from 1999 to 2008 the number of days with zero precipitation it is incremented, which can be associated with dry years. The weather was more extreme.

The regression analysis of precipitation in the Usumacinta high basin (1960–2008) indicates a significant negative trend in the dry season and a significant increase in rainfall (July and October), which indicates a trend towards a climate more extreme (Table 2). As for the analysis of the standardized precipitation index, was calculated the SPI-12 months for the upper basin (Fig. 3), which emphasizes the wet and dry periods of long duration (about 1

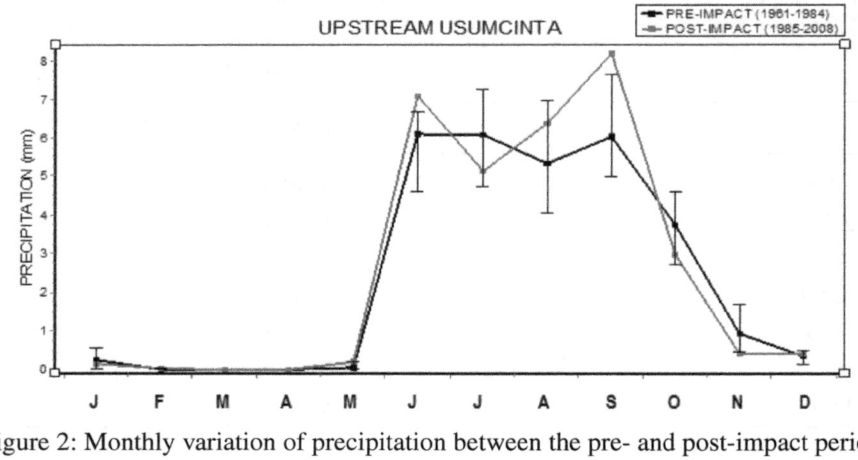

Figure 2: Monthly variation of precipitation between the pre- and post-impact period.

Table 2: Regression and correlation analysis of average monthly trend of PP.

Month	Regresión Equation	R^2
J	Y = - 0.020 X + 41.41	0.042
F	Y = - 0.006 X + 13.71	0.004
M	Y = 0.007 X - 12.21	0.012
A	Y = - 0.024 X + 49.43	0.042
M	Y = 0.022 X – 38.57	0.014
J	Y = 0.076 X– 139.1	0.089
J	Y = 0.028 X - 47.18	0.023
A	Y = 0.056 X - 102.5	0.061
S	Y = 0.100 X – 187.7	0.101
O	Y = 0.023 X 36.63	0.010
N	Y =- 0.001 X + 3.29	0.000
D	Y = - 0.13 X + 29.55	0.012

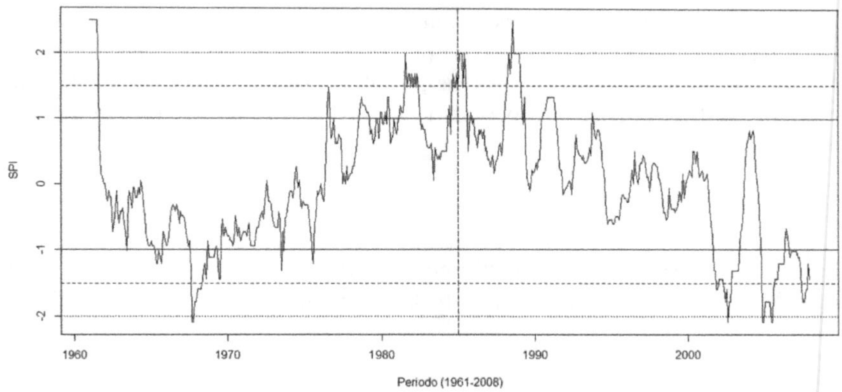

Figure 3: Standard SPI-12 precipitation index in the Upper Usumacinta River basin (lat: 16.1875 ° N, lon: 91.3125 ° W). The vertical line marks the division of the period before impact (left) and after impact (right).

year). For the pre-impact period, an extreme wet period for the first few months of 1961 can be noted at the outset, hence a gradual decrease towards another dry and long period by the late 1960s.

Hence a slow rise in the SPI with a tendency to periods moderate and severe, and long-lasting wet periods peaking around the mid-1980s, which coincides with the division of the pre-impact and post-impact periods. For the post-impact period, there is an almost monotonous decrease in SPI until it reaches again moderate and severe drought periods at the end of this period.

4.2 Variation of precipitation in the lower basin of the Usumacinta River

The monthly rainfall in the Usumacinta River for the low basin, considering the ranges of variation of the values to 75% with respect to the median, indicates that the dry period comprises the months of November to May. There was no significant variation between the pre-impact period (1961–1984) and the post-impact period (1985–2008), except for November showing an increase in precipitation, in contrast to the upper basin that indicates the month of September with greater precipitation (Fig. 4).

The high intensity of the 2003 storm (180 mm) is noticeable, coinciding with the high basin but of smaller magnitude and the high frequency of small storms with a precipitation of up to 140 mm. Likewise, the high precipitation pulses of 90 mm are abundant but of smaller magnitude with respect to the high basin, situation that indicates for this zone a similar behaviour with the high basin but with less precipitation. In the post-impact, the number of days with zero precipitation decreases since 1985. This aspect can be associated with wet years in this region. On the contrary, in the pre-impact-period, days with zero precipitation are more frequent, coinciding with the upper basin.

The regression and correlation analysis applied to the study period (1961–2008) for the rainfall in the lower basin of the Usumacinta River shows a non-significant decrease in all months of the year (except for May and November), unlike high basin that only shows a decrease in dry season (Table 3). Aspect that shows a regional variation of the distribution and tendency of the precipitation in the Usumacinta sub-basin. On the other hand, the normalized precipitation index SPI-12 months for the low basin show a very different behaviour to that of the high basin (Fig. 5). Here, it is observed, in general, a more variable behaviour throughout the period. Among the most prominent are the dominant drought periods in the 1970s, as well as the wet periods in the latter part of the pre-impact period (1978–1985). The first part of the post-impact period shows the prevalence of droughts (1985–1995), followed by two extreme and long-range wet periods (1997–2003) to finally close with two dry and short periods at the end of the period.

4.3 Environmental flows

The results of the analysis of the environmental flows through the IHA (Indicators of Hydrologic Alteration) in the upper part of the Usumacinta river basin show that the day with

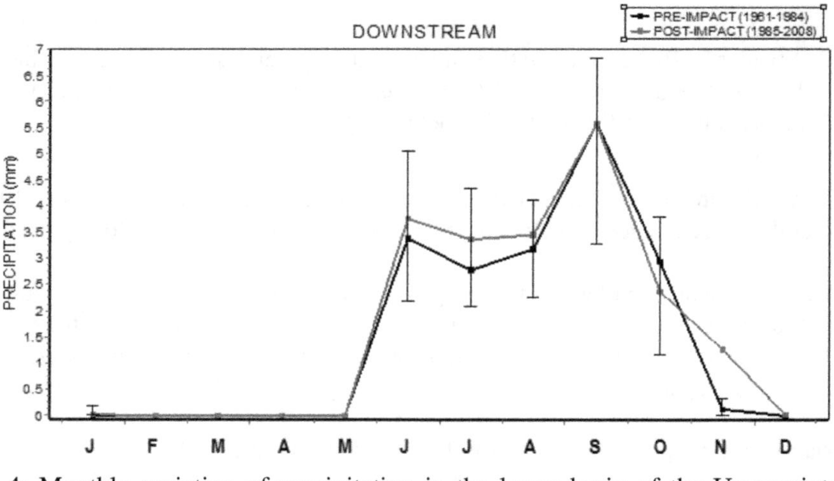

Figure 4: Monthly variation of precipitation in the lower basin of the Usumacinta River between the pre- and post-impact period and historical behaviour of precipitation.

Table 3: Regression and Correlation analysis of average monthly PP of lower basin.

Month	Regression Equation	R^2
J	Y = - 0.022X + 47.49	0.035
F	Y = -0.0003X + 3.18	0.000
M	Y = - 0.012X + 25.05	0.016
A	Y = - 0.023X + 48.18	0.050
M	Y = 0.007X - 10.62	0.000
J	Y =-0.007X - 5.23	0.001
J	Y = - 0.005 + 16.53	0.001
A	Y = - 0.008X + 22.37	0.001
S	Y = - 0.029X + 6879	0.009
O	Y = - 0.018X + 44.32	0.006
N	Y = 0.020X 35.26	0.008
D	Y = -0.017X + 37.73	0.017

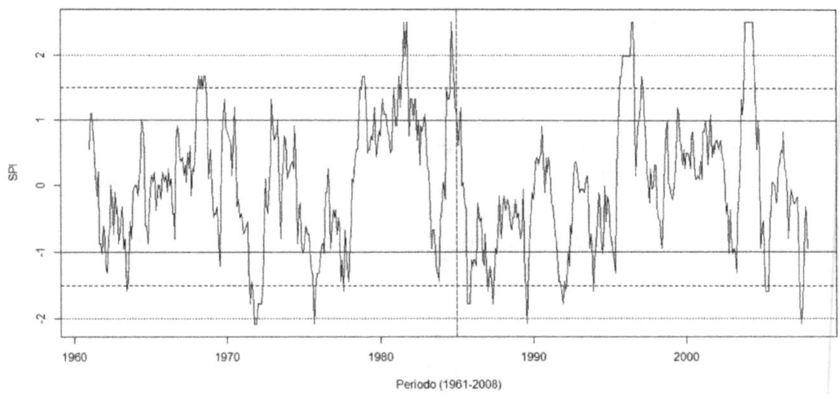

Figure 5: Standard SPI-12 precipitation index by months in the lower basin of the Usumacinta River (lat: 17.4375 ° N, lon: 91.3125 ° W). The vertical line marks the division of the pre-impact period (left) and post-impact (right) period.

the Low Extreme Flow was at the beginning of April, with 138.8 (m^3/s), with a duration of 50 days. The day with the highest flow was on August 11, with an average flow threshold of 2,293 and a duration of 11 days. The base flow was 259.9 m^3/s and the high threshold flow was 2,538 m^3/s. The High Flow Peaks with 3328.0 m^3/s, with a duration of 11 days in early September and a frequency of 5 times. The peaks of the Small Floods with a flow of 5271.0 (m^3/s), with a duration of 63 days, at the end of September. Peaks of Large Floods with 6571.0 m^3/s, lasting 42 days and until the 3rd, October week. Flows that provide the information to establish the proposals and strategies for the rational management of aquatic resources and the assessment of the effect of climate change on the rivers between the pre and post impact period in the environmental flows.

4.3.1 Monthly average flows Upper River basin Usumacinta

The comparative analysis of monthly average flows in the upper Usumacinta River basin, between the pre-impact (1961–1984) and post-impact (1985–2008) periods, indicates that flows were 567.2 and 367.3 (May) and from 3048 to 1920 m^3/s in September, respectively. This indicates the significant decrement of the flows for the rainy months in the post-impact and that does not coincide with the increase of precipitation for this period and in this part of the basin. Likewise this could be related to the increase of the drought in this period. The base flow index of 0.25–0.06 (m^3/s) indicates a monthly depletion of significant natural flow in the rainy season in the post-impact period (June to September, Fig. 6).

4.3.2 Components of the environmental flow of the upper Usumacinta River basin

Environmental flow components in the upper Usumacinta River basin in this period show the greatest variations between pre- and post-impact for Extremely Low Flows and Low Flows, these are most frequent in post-impact (1980 to 2000), as well as, the Small Floods, that coincide with a period of drought. This implies that the natural flow rate shows a small decrease in the dry periods and an increase in the rainfall with torrential rains, making evident the change towards more extreme conditions and coincident with the precipitation changes.

4.3.3 Components of hydrological alteration between the pre and post-impact periods in the upper Usumacinta river basin

The components of hydrological alteration for upper Usumacinta River basin indicates alterations between the pre-impact and post-impact period corresponding to: 1) increase of the average monthly flows of all the months of the year, increase in the quantity pulses of Low Flows and High Flows, as well as in the number of investments of the flow and rate of change. On the contrary an alteration of the date in which the maximum flow is given, in the decrease in the duration of the Low Flow Pulses and in the High Flow Pulses, in the increase of the duration of the Low Flow Pulses; in the contrary it is observed an increase in the change rate, factors that indicate that rains are more torrential and that there are longer periods of drought during the year for the post-impact period.

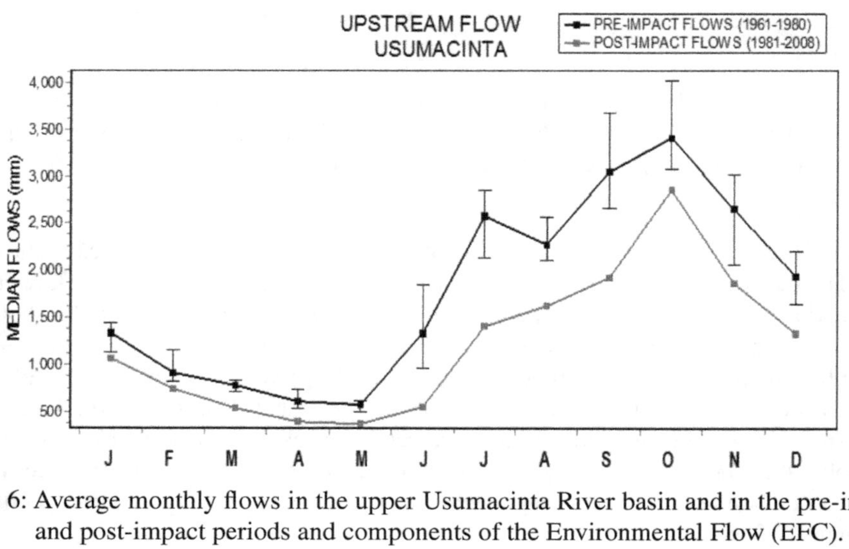

Figure 6: Average monthly flows in the upper Usumacinta River basin and in the pre-impact and post-impact periods and components of the Environmental Flow (EFC).

Table 4: Regression analysis and correlation of the monthly flows of the upper Usumacinta river basin in the study period (1961–2008) in the Boca del Cerro station (30019).

Month	Regression Equation	R^2
J	Y = 4.535X 7736	0.035
F	Y = 2.717X - 4414	0.017
M	Y = 0.687X - 616.1	0.001
A	Y = 2.717X - 2561	0.006
M	Y = 0.0193X - 558.2	0.000
J	Y = 5.97X + 12980	0.001
J	Y = - 21.8X + 45230	0.087
A	Y = - 19.72X + 41140	0.080
S	Y = - 26.69X + 55560	0.096
O	Y = - 19.13X + 41010	0.057
N	Y = -16.27X + 34580	0.047
D	Y = -2.302X 6290	0.001

4.3.4 Regression Analysis and Correlation to Flows for the Upper Usumacinta River Basin in the period of study

The regression analysis and linear correlation of the flows in the study period 1961–2008, for the upper basin of the Usumacinta River, show to the rainy season with a negative trend in the flows, with significant decreases in the months of October, November and December (Table 4).

On the contrary, the dry season shows an increase in non-significant flows. However, the smallest amount of flow in the rainy season, may be due to defacing in the quantity and magnitude of precipitation (droughts and steeper rains) due to the effects of climate change that have an impact on flow rates and the use by human influence resource.

5 DISCUSSION

The increase in monthly precipitation in the high basin in the rainy season and a decrease in the dry season in the post-impact period (a significant decrease in the number of days with zero precipitation from 1989 to 1997) can be associated with years rainy. The increase in the number of days with zero precipitation from 1999 to 2008 may be associated with dry years with a climate more extreme. Likewise, the regression analysis of rainfall in the Usumacinta high basin (1960–2008) indicates a significant negative trend in the dry season and a significant increase in rainfall (July and October). The greater intensity of precipitation (an increase in the number of days a year with a greater amplitude in precipitation) also shows a tendency toward extreme climates and the effects of climate change, a situation that coincides with those pointed out by Kobashi *et al.* [17], where greenhouse gases in the world can lead to floods and droughts, and depending on the season and latitude these can be more severe and frequent, an aspect that generates hydrological changes in the rivers and therefore a change in the availability of water. Wetherald [18], also point out a precipitation increase in tropical regions, as well as to higher latitudes in almost all parts of the world and an overall decrease in precipitation. The regression and correlation analysis in the lower basin of the Usumacinta river shows a significant decrease in all months of the year (except for May and November), showing a negative trend of precipitation throughout the year, unlike the upper basin which only decreases in the dry season. Aspect indicating a regional variation of the distribution and trend of precipitation in the Usumacinta sub-basin. Also, the high increase in days with zero precipitation can be associated to a dry period in this area and with it the regional variation in the availability of water in the basin.

The Usumacinta basin is very important for ecological or environmental services of direct economic importance. These include volumes of carbon captured of the forest masses, the collection and supply of water and nutrients of the lands of northern Chiapas and Tabasco (low basin), as well as estuarine areas of importance to fisheries in the Coast of the Gulf of Mexico. The forests and rainy forests of the Usumacinta basin are part of the most important carbon sinks in Mesoamerica [19–21], systems that can be disturbed by changes in the availability of water in the area due to the effects of climate change.

The comparative analysis of the average monthly flows in the upper Usumacinta river basin, between the pre-impact (1961–1984) and post-impact (1985–2008) periods, shows that flows were 567.2–367.3 m^3/s and from 3048 to 1920 m^3/s by September, respectively. This indicates the significant depletion of the flow rates for the rainy months in the post-impact and that does not coincide with the increase of the precipitation for this period and in this part of the basin and therefore, this decrease can be associated to the utilization of the resource by the population. Also, the ecological flow components show the greatest variations for Low and Low Extreme Flows, which were more frequent in post-impact (1980–2000), and in Small Floods, which coincide with a period of drought. This implies that the natural flow shows a small decrease in the dry periods and an increase in rain season, making evident the change towards more extreme conditions and coincident with the precipitation changes. The regression analysis and linear correlation indicated the rainy season with a negative trend in flows, with significant decreases in the months of October, November and December observed, point out, the least amount of rainfall can be due to changes in the quantity and magnitude of precipitation (droughts and steeper rains) due to the effects of climate change that have an impact on the decreasing flows, aspect that also is combined with the use of water by human influence. De la Maza & Carabias [22] indicate that effects of climate change have an impact on the flows decreasing. Also, mention that the main tails of deterioration in the Usumacinta basin are due to the accelerated change in land use, unplanned colonization, the irrational exploitation of forest resources, the indiscriminate use of fire in agricultural activities, the disorderly development induced by oil exploitation, the increase of forest fires in the dry season, the magnitude of floods, and the effects of climate change on rainfall regimes.

The essential ecological processes of the basin in the Usumacinta depend on the high zone located in Guatemala and the Selva Lacandona de Chiapas [23]. The ecology of lowland marshes is of great importance for migratory bird populations that take refuge there as well as for fisheries in the Gulf of Mexico, that depend on the nutrients provided by the upper and middle basin ecosystems that are transported through the hydrological network. A number of threatened or endangered species require this connectivity to maintain the health of their populations, including the manatee *Trichechus manatus*, crocodile *Crocodylus acutus* and *C. morelettii*, and the white turtle *Dermatemys mawii* [12]. For this reason, this basin can be converted into a space to capture CO_2 emissions through the conservation of natural spaces, eliminating the use of agricultural fire and optimum management of livestock to reduce methane emissions. However, current trends in land use change, overexploitation of wild resources, alterations by the energy sector, pollution, population growth and dispersion, expansion of the agricultural frontier, lack of governance and climate change can lead to degradation social, economic and political area [9].

Climate impact assessment studies on ecological and environmental flows are important for species conservation, for assessing the role and resilience of aquatic ecosystems, as well as for the well-being of people dependent on the rivers. It is therefore necessary to determine the quantity and frequency of flow regimes in rivers. The advantage of determining environmental and ecological flows is that decision-makers can know how much of the flow can be used for the purposes of the population and how many changes have been introduced into river flow patterns by human consumption or climate change [22]. Adaptation to climate change requires determining the flow regime in rivers, such as: change in water consumption to compensate precipitation rates, relocation of industries to regions of higher humidity, or change in the morphology of cities to compensate for floods. Ecological flows (Low Extreme Flow, Low Flow, High Flows, High Flow Pulses, Small Floods and Large Floods) provide the information to establish the proposals and strategies for the rational management of aquatic resources and the valuation of the effect of the climatic change in the availability of water in the flows of the rivers.

ACKNOWLEDGMENT

To the financing granted by the Sectorial Fund for Environmental Research SEMARNAT-CONACYT, Call S0010-2014-1, through project 249435.

REFERENCES

[1] Jones, Ph. D. & Wigley, T.M.L., Estimation of global temperature trends: what's buckets important and what isn't. *Climatic Change,* **100**, pp. 59–69, 2010. https://doi.org/10.1007/s10584-010-9836-3

[2] Swart, R., Bernstein, L., Ha-Duong, M. & Petersen. A., Agreeing to disagree: uncertainty management in assessing climate change, impacts and responses by the IPCC. *Climatic Change,* **92**, pp. 1–29, 2010. https://doi.org/10.1007/s10584-008-9444-7

[3] Arnell, N.W., Adapting to climate change: an evolving research programme. *Climatic Change,* **100**, pp. 107–111, 2010. https://doi.org/10.1007/s10584-010-9839-0

[4] Cohen, S.J., From observer to extension agent—using research experiences to enable proactive response to climate change. *Climatic Change*, **100**, pp. 131–135, 2010. https://doi.org/10.1007/s10584-010-9811-z

[5] Yarime, M., Takeda, Y. & Kajikawa, Y., Towards institutional analysis of sustainability science: a quantitative examination of the patterns of research collaboration. *Sustainability Science,* **5**, pp. 115–125, 2010. https://doi.org/10.1007/s11625-009-0090-4

[6] Komar, P.D., *Beach processes and sedimentation.* Englewood Cliff, NJ: Prentice Hall, 1976.

[7] Fischer, S. & Kummer, H., Effects of residual flow and habitat fragmentation on distribution and movement of bullhead (Cottus gobio L.) in an alpine stream. *Hydrobiologia,* **422/423**, pp. 305–317, 2000. https://doi.org/10.1007/978-94-011-4164-2_25

[8] INEGI, *Síntesis geográfica y anexo cartográfico del Estado de Tabasco: México D.F*, 1986.
[9] Kolb, M. & Galicia, L., Challenging the linear forestation narrative in the Neo-tropic: regional patterns and processes of deforestation and regeneration in southern Mexico. *The Geographical Journal,* **178**(2), pp. 147–161, 2012. https://doi.org/10.1111/j.1475-4959.2011.00431.x
[10] National Water Commission, *Statistics on water in Mexico*. Conagua, 2014. Available at: www.conagua.gob.mx.
[11] March, M.I., & Castro, M., *La Cuenca del Río Usumacinta: Perfil y Perspectivas para su Conservación y Desarrollo Sustentable*. INECC (Instituto Nacional de Ecología y Cambio Climático, 2016. Available at: www2.inecc.gob.mx/publicaciones/libros/639/rusumacinta.pdf. Consultado el 26/enero/2016.
[12] McKee, T.B., Doesken, N.J., & Kleist, J., The relationship of drought frequency and duration to time scales. In: Proceedings of the 8th Conference on Applied Climatology. *American Meteorological Society,* **17**(22), pp. 179–183, 1993.
[13] McKee, T.B, Doeskin, N.J. & Kieist, J., Drought monitoring with multiple time scales. *Proceedings of the 9th Conference on Applied Climatology,* January 15–20, 1995, American Meteorological Society, Boston, Massachusetts, pp. 233–236. 1995.
[14] Shepard, D.S., Computer mapping: The SYMAP interpolation algorithm. *Spatial Statistics and Models* (G. L. Gaile and C. J. Willmott, Eds., D. Reidel,), pp. 133–145, 1984.
[15] Zhu, C., Dennis, P., & Lettenmaier, D.P., Long-term climate and derived surface hydrology and energy flux data for Mexico: 1925–2004. *Journal of Climate,* **20**, pp. 1936–1946, 2007. https://doi.org/10.1175/jcli4086.1
[16] Muñoz-Arriola, F., Avissar, R., Zhu, C., & Lettenmaier, D.P., Sensitivity of the water resources of Rio Yaqui Basin, Mexico, to agriculture extensification under multiscale climate conditions. *Water Resources Research*, **45**(11), 2009. https://doi.org/10.1029/2007wr006783
[17] OMM (2012). Índice normalizado de precipitación. *Guía de Usuario*. OMM-No 1090. ISBN 978-92-63-31090-3.
[18] TNC (The Nature Conservancy), Indicators of hydrologic alteration, Version 7. User Manual. USA, 2006.
[19] Kobashi, T., Severinghaus, J.P., Barnola, J.M., Kawamura, K., Carter, T. & Nakaegawa, T., Persistent multi-decadal Greenland temperature fluctuation through the last millennium. *Climatic Change*, **100**, pp. 733–756, 2009, available at: Springerlink.com; doi:10.1007/s10584-009-9689-9. (accessed 20 March 2016). https://doi.org/10.1007/s10584-009-9689-9
[20] Wetherald, R.T., Changes of time mean state and variability of hydrology in response to a doubling and quadrupling of CO_2 Changes. *Climatic Change*, 102, pp. 651–670, 2009. available at: http://www.springer.com/earth+sciences+and+geography/atmospheric+sciences/journal/10584.html. (accessed 19 March 2011). https://doi.org/10.1007/s10584-009-9701-4
[21] Kolb, M. & Galicia, L., Challenging the linear forestation narrative in the Neo-tropic: regional patterns and processes of deforestation and regeneration in southern Mexico. *The Geographical Journal,* **178**(2), pp. 147–161, 2012. https://doi.org/10.1111/j.1475-4959.2011.00431.x

[22] Rodiles-Hernández R., González-Díaz, A. A. & Pérez-Mora, E. *Inventario ictiofaunístico en tres regiones hidrológicas prioritarias de la Cuenca del Grijalva-Usumacinta en el Estado de Chiapas.* El Colegio de la Frontera Sur (ECOSUR-Unidad San Cristóbal). Informe final SNIB-CONABIO proyecto No. FM020. México D. F. 2011.

[23] De la Maza, J. & Carabias, J. (2011). *Usumacinta. Bases para una política de sustentabilidad ambiental.* Instituto Mexicano de Tecnología del Agua-Natura y Ecosistemas Mexicanos A.C. México.

[24] King, J., Brown, C. & Sabet, H., A scenario-based holistic approach to environmental flow assessments for rivers. *River Research and Applications,* **19**, pp. 619–639, 2003. https://doi.org/10.1002/rra.709

WATER RETENTION MEASURES – CASE STUDY FROM SLOVAKIA

MARTINA ZELEŇÁKOVÁ[1], MARIA MANUELA PORTELA[2] & PETR HLUŠTÍK[3]
[1]Institute of Environmental Engineering, Faculty of Civil Engineering, Technical University of Košice, Košice, Slovakia
[2]Department of Civil Engineering, National Technical University of Lisbon, Lisbon, Portugal
[3]Institute of Municipal Water Management, Faculty of Civil Engineering, Brno University of Technology, AdMaS centre, Brno, Czech Republic

ABSTRACT
Water retention measures are multi-functional measures that aim to protect water resources using natural means and processes. These measures can contribute to reducing the risk of floods and water scarcity and drought while also improving the status of surface and ground water bodies. Water retention measures can support the achievement of the goals of a range of EU policies, including those for surface water, groundwater and coastal management, nature conservation, agriculture, forestry, urban water management, disaster risk management, green growth and climate change mitigation and adaptation. This article presents water retention measures suitable for application in urban areas. Infiltration of rainwater, taking into account the suitability of local conditions, appears to be the most environmentally suitable solution of rainwater drainage. The connection of conventional method of draining rainwater through sewer systems and infiltration of rainwater appears to be the most suitable drainage concept for sustainable cities.
Keywords: infiltration, rainwater management, urban areas

1 INTRODUCTION

Water retention measures are multi-functional measures that aim to protect and manage water resources using natural means and processes, thereby building up Green Infrastructure, for example, by restoring ecosystems and changing land use (European Union, NWRM project [1, 2]). These measures have the potential to provide multiple benefits, including flood risk reduction, water quality improvement, groundwater recharge and habitat improvement. As such, they can help achieve the goals of key EU policies such as the Water Framework Directive 2000/60/EC [3] and the Floods Directive 2000/60/EC [4]. Figure 1 illustrates selected EU policy initiatives recognized in the potential role of water retention measures in contributing to the achievement of their objectives.

Water retention measures can contribute to both Water Framework and Flood Directive goals and can enhance synergies between the implementation of both directives and support the coordination between the River Basin Management Plans and Flood Risk Management Plans. The uptake of these measures in management plans is triggered through the potential to improve or preserve hydro-morphological conditions, as well as the water quantity and quality.

New approach towards urban water management includes a number of tools which enable to reduce environmental effects of urbanization and maintain purpose of the area. In recent years, the implementation of several different types of object infiltration in Europe and overseas has led to errors in their planning and operation caused by neglected scientific claims. However, in an area where poorly permeable rocks appear, subsurface water infiltration must be carried out very cautiously. Meeting the requirements for collecting rain water for the land property must proceed with the professional care so that it does not damage the objects (European Union, NWRM project [1, 2]). Nowadays, runoff process and rainwater management

DOI: 10.2495/EI-V1-N3-213-220

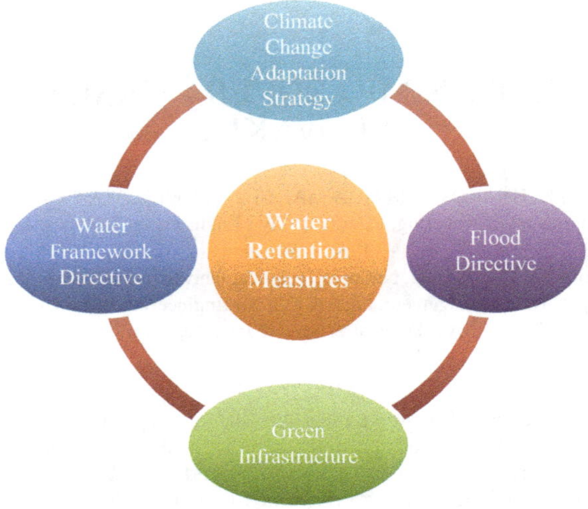

Figure 1: Water retention measures and selected EU policy initiatives.

are inevitable topics of research in the time of climatic variability. Rainwater management should be considered as a sustainable strategy for reconstruction of rural and urban settlements from the aspects of environmental management and social criteria (Markovič et al. [5]). There are many possibilities which can reduce the risk of flooding and regulate the quantity and quality of groundwater. Most of these possibilities are proposed only by theoretical calculation based on standards (Standard DWA-A 138E [6]). Best management practices reflect the natural process of rainwater management that can be found in non-urban basins and their transformation to urban areas.

The following sections present selected water retention measures applicable in urban areas – a case study from Košice City in Slovakia.

2 STUDY AREA

Košice is the second largest city in the Slovak Republic. It consists of 22 city boroughs, among which is the new project Nová Terasa located on the border of the city borough Staré Mesto in the near vicinity of Košice's historical centre. The project brings modern architecture, a characteristic feature of which is the exclusiveness of mutually linked buildings, which together form an enclosed interior courtyard. The Nová Terasa project (Fig. 2) offers a wide selection of different types of flats. In the first phase a total of 199 one- to four-room flats were built with floor space from 31 to 110 m². The flats are orientated directly on the centre of Košice, in close vicinity of the hospital, or on the slope of the hills.

A system for drawing away rainwater from the roofs, car parks and local paths and roads has been designed for the site. Rainwater from roads and car parks from all stages will be harvested using rainwater traps. The water, including rainwater from roofs, will subsequently be carried into a retention tank with throttled discharge using a whirling flow regulator at 25 L/s per one stage. From this it follows that from all four stages the permitted amount of rainwater (100 l/s) will be released into the sewerage system. Rainwater will be taken from the retention tank to an oil–water separator with a guaranteed discharge up to 1.0 mg NEL/l Macurová [7].

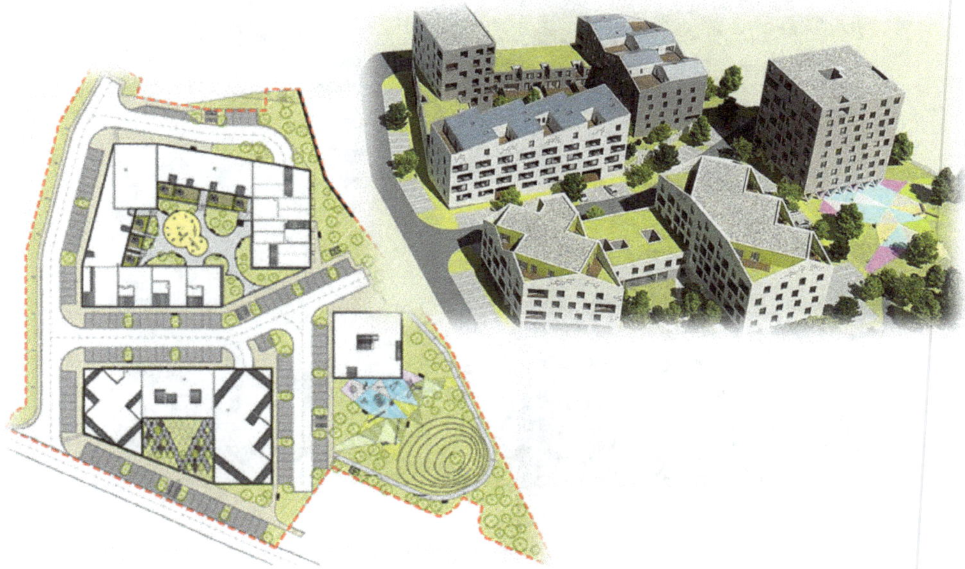

Figure 2: Nová Terasa project.

The original design resolves the possibility of taking rainwater away from the residential area and into the public sewerage system. The pipes will be connected to the existing sewerage network, terminated by a sewerage shaft in the pavement. The route will consist of ribbed sewerage pipes and sewerage shafts of ø 1.0 m will be placed along it. On the request of Eastern Slovakia Water Supply Company, to limit the discharge of rainwater to 100 l/s, a throttling section with a 0.6% gradient will be established on the main path, to which a pipe connected to the existing sewerage system will be led. Slowing the discharge of rainwater will be resolved in the scope of drainage of roads and paved surfaces by establishing a retention tank with a throttled discharge to 25 l/s.

3 RESULTS

Alternative option for rainwater management, which is proposed for this new urban settlement, is presented in the following.

From a geological survey of the site where construction of the first stage of the project is currently taking place, it follows that the territory is not suitable for absorption facilities (Zelenakova and Hudakova [8]). Therefore, a retention tank was considered in the original proposal. In the subsequent proposal, alternative facilities are proposed for the trapping and accumulating of rainwater, namely

- Absorption blocks
- Absorption wells
- Infiltration pond

These alternative possibilities for trapping rainwater are proposed for the nearby street, to where the rainwater would be taken. Based on the geological survey performed, conditions on nearby street are evidently suitable for absorption.

Figure 3: Absorption blocks.

3.1 Absorption blocks

Rainwater is led underground into hollow absorption boxes (Fig. 3), where it is temporarily accumulated, and based on the permeability of the soil, infiltrates little by little into the ground. The blocks are also suitable where there is a smaller amount of space for absorption. They are expanded to all sides and combined into optionally large facilities. They also contain a transverse inspection tunnel, which enables inspection by camera or the flushing of the entire system.

Rainwater from car parks will be taken to an oil–water separator, from which, after removal of hazardous substances, it will be fed together with rainwater from the roof into the underground absorption space located near the Elementary School on a street in the vicinity of Kuzmány Street in Košice. The amount of water q which is absorbed, with a value of 0.78 $l/s/m^{-2}$, is calculated on the basis of the coefficient of filtration known from a hydrogeological survey in the given area. This means that for every 1 m^2 of surface, 0.78 l of rainwater is absorbed into the subsoil. The total amount of rainwater carried for absorption from all stages is known: 695.5 l/s. The absorption area is the area which is formed for the most part by the base on which the absorption blocks are placed of 480 m^2 – area available for placement of absorption blocks.

3.2 Absorption wells

Absorption wells (Fig. 4) collect rainwater from the roofs of houses, car parks, surfaces, etc. and return it to the groundwater. They have a large spatial volume, i.e. such that each well may absorb a great amount of water. They are usually made from perforated pipes through which rainwater is absorbed into the subsoil. Drilled wells compared with a dug well absorb deeper groundwater, which is not directly dependent on precipitation and its amount remains more consistent over the course of the whole year. The drawn water also wields better quality. In the second alternative solution for trapping rainwater, water from roofs and car parks is taken to a retention container with a throttled discharge. From there it passes through an

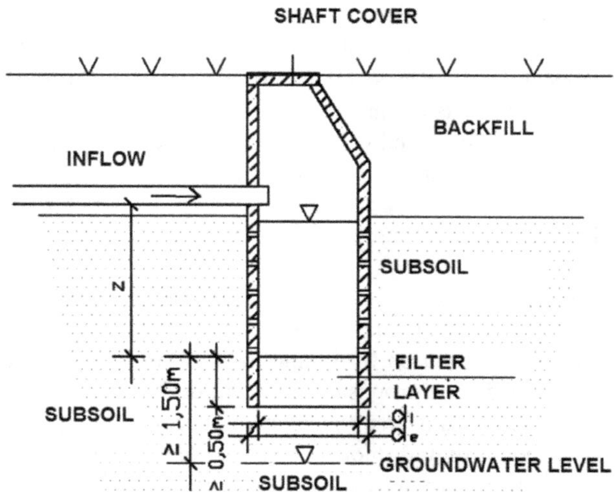

Figure 4: Absorption wells.

oil–water separator and is then taken to drilled absorption wells. Only sinkwater is released into the sewerage system.

In the scope of the second alternative solution for trapping rainwater, 10 absorption wells with a separation of 2 m will be proposed for the selected site. The wells will be mutually interconnected by pipes of diameter DN 150. The dimensions of absorption wells will be as follows: the depth is 15 m, diameter 0.35 m.

3.3 Infiltration pond

The already created shallow depression in the lawn enables the larger part of the water to remain on the site. If this depression is put in a place with permeable soil, it can serve for infiltration (Fig. 5).

Large buildings (like the newly designed Nová Terasa site) ensure a sufficient amount of water for establishing a closed infiltration pond, which enables the permanent planting of banks with wetland vegetation. An accumulation space for the trapping of rainwater is created over the permanent level of the water. The banks above the plain of the permanent level are made from permeable rock, and the held water gradually soaks into the surrounding soil. The third alternative solution is designed using an infiltration pond. The water from roofs

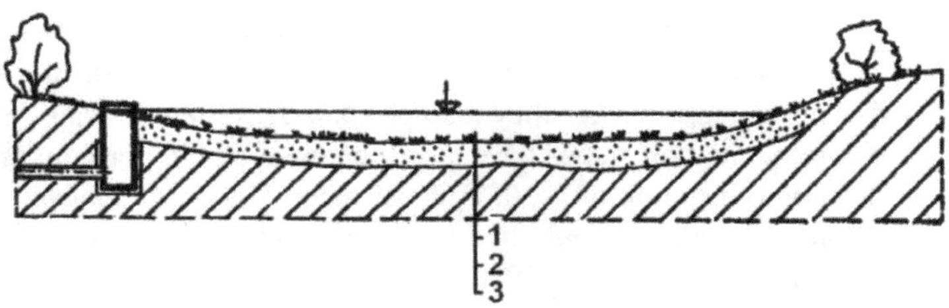

Figure 5: Infiltration ponds, 1-infiltration basin, 2-permeable ground, 3-sand.

together with water from car parks flows into a retention tank with a throttled discharge, from where it is taken for absorption to the infiltration pond. Sinkwater is taken independently to the public sewerage system. The proposed pond absorbs rainwater from all constructions. It has the shape of a basin, i.e. it has the greatest depth in the centre and towards the edges its depth declines. Area of infiltration pond will be 18 × 25 m. Average depth of the pond will be 1.5 m. Finally the total volume of the pond will be 675 m^3.

3.4 Evaluation of the alternative proposals

This work presents a proposal of alternative possibilities for the trapping of rainwater for the newly built project Nová Terasa on Ondavská Street in Košice. This proposal consists in the selection of the most suitable solution from among absorption blocks, absorption (drilled) wells and an infiltration pond. The proposal of absorption blocks is considered as the most suitable alternative solution for rainwater management, because though they are more expensive than an infiltration pond, they do not represent a danger to the children living in the vicinity, and an additional playground can be built over them. All alternative methods of trapping rainwater ensure an easing on the public sewerage system, which was also an aim of this work.

Effective devices in urban areas designed for fluent infiltration of rainwater from the roofs of buildings and paved surfaces are infiltration facilities (Slys [9]). The basic principle and function of all types of infiltration facilities is to divert rainwater as quickly as possible to infiltration zone, and there it infiltrates into the surrounding soil (Markovič et al. [3]). Design and use of infiltration facilities as a sustainable method of rainwater runoff disposal become an integral part of the drainage management and projects of sewerage system of buildings or other paved surfaces. Disposal, especially safe drainage of rainwater, is a problem of almost every new building in an urban area and in area with undersized sewage systems.

Wastewater disposal systems, sewage treatment plants and recipients of water flows are overloaded and the risk of floods grows. This represents another reason for waterworks companies to have the possibility of forbidding the disposal of rainfall water from surface detention into the wastewater disposal system already present during the creation and fulfillment of the development plan. This leads to the necessity of using rainwater infiltration systems or capturing rainwater for further reuse. Infiltration systems not only solve a technical problem that partially relieve a sewer network but also contribute to the ecological stability of ecosystems; in addition, they reduce the expense for disposal of rainwater. Although infiltration systems are not new in urban drainage, they were seldom used in the past. Nowadays its utilization is increasing, which presents the novelty of urban drainage solution. These systems contribute to refurbishing the aquifer and river basins generally.

4 CONCLUSION

Many cities and urban areas are located in flood plains because the land is fertile and flat which is suitable for agriculture and urban development. Rivers provide water supply for domestic, industrial and irrigation uses; they also provide convenient means for navigation, transportation and communication. Cities have large percentage of impervious areas that prevent effective infiltration of rainfall into soil. To have successful flood control and flood risk management, we should consider not only hydraulic and engineering aspects but also socio-economic and environment aspects (Slys [9]). Flood management should have involvement of various stakeholders, including concerned authorities such as urban planners, civil and water resources engineers, civil disaster defense authorities and health and social services. The best flood mitigation measures from all main points of view – social, economic and environmental – are natural water retention measures.

Water retention measures cover a diversity of measures that are implemented by different sectors or considered in different planning processes dealing with water, food risk management, biodiversity protection, climate change adaptation or urban planning. Some of these measures aim to directly modify the ecosystem, while others focus on changes of practice of economic operators.

The concept of drainage in cities, which aims to mitigate the impact of urbanization on the hydrological regime of the country and on aquatic ecosystems, come from our experiences and knowledge of current method of sewerage. New concept of capture and use water from surface run-off provides a platform for a new technical and non-technical measures, both that is drained on each property and in the public part of urban drainage area.

Suitability of choice of type of infiltration facility is dependent on local conditions. In each case, it is to be considered carefully, which drainage concept in combination with the percolation of precipitation is ecologically sensible, technically possible and economically justifiable (Slys et al. [10]).

This work deals with the issue of rainwater management. This article presents water retention measures suitable for application in urban areas. The practical part presents a proposal for trapping rainwater on the selected territory of the newly designed project Nová Terasa in Košice.

This proposal resolves alternative possibilities for taking away rainwater rather than sending it to the public sewerage system. In the original design, rainwater is taken for accumulation into a retention tank with a safety overflow and throttled discharge. After passing through an oil–water separator, it is taken along with sinkwater into the public sewerage system. For alleviating the sewerage network, a system for taking off rainwater for absorption was proposed for the selected territory, which due to the poor soil permeability at the selected locality will be located on Kuzmány Street in Košice.

These methods for trapping and accumulating rainwater reduce the burden on sewerage networks and thus also relieve the public sewerage network from rainwater. Precipitation water is an essential part of the natural water cycle, leading to water falling on the earth's surface in the form of rain. After impact on the earth's surface, it is taken, together with sinkwater, into sewerage systems, which are today at full capacity. In order to prevent this, it is advantageous to instead take rainwater to localities with permeable soils for absorption. With suitably selected systems for trapping rainwater, the negative overburdening of sewerage systems can be avoided.

ACKNOWLEDGEMENT

The contribution is written thanks to support of project VEGA 1/0609/14.

REFERENCES

[1] European Union, *EU Policy Document on Natural Water Retention Measures*. By the drafting team of the WFD CIS Working Group Programme of Measures (WG PoM), Luxembourg: Office for Official Publications of the European Communities, 2014.

[2] NWRM project, *Concept Note: Natural Water Retention Measures (NWRM) and the WFD and Other Daughter Directives*, available at http://www.nwrm.eu, January 2014 (accessed 14 November 2017).

[3] European Union, *Directive 2000/60/EC* of the European Parliament and of the Council of 23 October 2000 establishing a framework for Community action in the field of water policy. Official Journal of the European Communities L 327/1.
[4] European Union, *Directive 2007/60/EC* of the European Parliament and of the Council of 23 October 2007 on the assessment and management of flood risks. Official Journal of the European Communities L 327/1.
[5] Markovič, G., Zeleňáková, M., Káposztásová, D. & Hudáková, G., Rainwater infiltration in the urban areas. *Environmental Impact 2*, ed. G. Passerini & C.A. Brebbia, WITT Press: Southampton, Vol. 181, pp. 313–320, 2014.
[6] Standard DWA-A 138E, *Planning, Construction and Operation of Facilities for the Percolation of Precipitation Water*, 2005.
[7] Macurová, M., *Rainwater Management in Urban Areas* (in Slovak). Thesis. Technical University Kosice, Bratislava, 2012.
[8] Zeleňáková, M. & Hudáková, G., The concept of rainwater management in area of Košice region. *Procedia Engineering: 16th Conference on Water Distribution System Analysis*, WDSA2014 – Urban Water Hydroinformatics and Strategic Planning, Bari, Italy, Vol. 89, pp. 1529–1536, 2014. DOI: 10.1111/j.1747-6593.2008.00159.x.
[9] Slys, D., Potential of rainwater utilization in residential housing in Poland. *Water and Environment Journal*, **4**, pp. 318–325, 2009.
[10] Slys, D., Stec, A. & Zelenakova, M., A LCC analysis of rainwater management variants. *Ecological Chemistry and Engineering S-Chemia i Inzynieria Ekologiczna S*, **19(3)**, pp. 359–372, 2012. DOI: https://doi.org/10.2478/v10216-011-0026-7.

INTEGRATED WATERSHED MANAGEMENT AND FLOODPLAIN PROTECTION ON THE CARSON RIVER IN THE WESTERN USA

JOHN COBOURN & STEVEN R. LEWIS
University of Nevada Cooperative Extension, USA

ABSTRACT
The Carson River in western Nevada, USA, flows approximately 200 miles (320 km) from its headwaters in the central Sierra Nevada range to its terminal wetlands in Nevada's Great Basin Desert. The nearly 4,000 square mile (10,400 km^2) watershed is home to about 156,000 people. A year after a major flood event in 1997, local citizens and agency staff came together to form an Integrated Watershed Management group called the Carson River Coalition (CRC). Rather than focusing solely on flood problems, the group pledged to address all problems of flooding, water quantity, water quality and wildlife habitat in an integrated fashion through improved communication and collaboration. In 2003, the group decided to determine its 'main message', the single most important message the public needs to understand about the watershed. Over 50 group members participated, and the message was defined as, 'Protect the floodplain from future development'. Since 2003, CRC members, working with the staff of a regional water management agency, have written the Regional Floodplain Management Plan of 2008 (RFMP). It explains the functions, values and ecosystem services of natural floodplains. Much of the floodplain near the channel is used for privately owned farms and pasturelands. This land use is compatible with the RFMP's goals. However, the population of the watershed is projected to grow, and the floodplain is beginning to be developed for urban infrastructure. To help focus public attention on the value of protecting floodplains, CRC members engaged a County geographic information systems specialist to map the location of all properties in the riverine floodplain protected by conservation easements or public ownership. This map was introduced in a University of Nevada Cooperative Extension Special Publication in 2015. It serves as a call to action and baseline inventory for measuring progress towards the community goal of protecting the Carson River's natural floodplains.
Keywords: Carson River, floodplain management, GIS, integrated watershed management, non-structural, protected floodplain, riverine, storm events, watershed

1 INTRODUCTION

Most of the headwaters of the Carson River arise from the snowpack of the central Sierra Nevada range in California. The high drainages spring from peaks of 10,000 to 11,000 feet (3,000–3,300 m) in elevation, and they flow east and north into the Great Basin Desert of Nevada, the driest state in the USA. The river itself is about 200 miles (320 km) long, and the lower reaches lie in the rain shadow of the Sierra Nevada. The watershed is home to about 156,000 people, but approximately 90% of surface and groundwater use is for irrigated agriculture [1, 2].

European-American settlement began at the time of the California Gold Rush in 1849 and accelerated with the boom in gold and silver mining in the famous Comstock Lode in Virginia City, Nevada, in the 1860s and 1870s. The boom ended by 1900, and for the first half of the 20th century, the economy of the watershed was based mostly on ranching and alfalfa farming. The population surged again after the end of the Second World War, when the state Capital of Carson City and the nearby towns of Minden, Gardnerville, Dayton and Fallon began to grow (Fig. 1).

Because many of the forests in the watershed were logged during the mining boom, and because thousands of logs were rafted down the Carson River for transport to the mines, the

DOI: 10.2495/EI-V1-N3-221-231

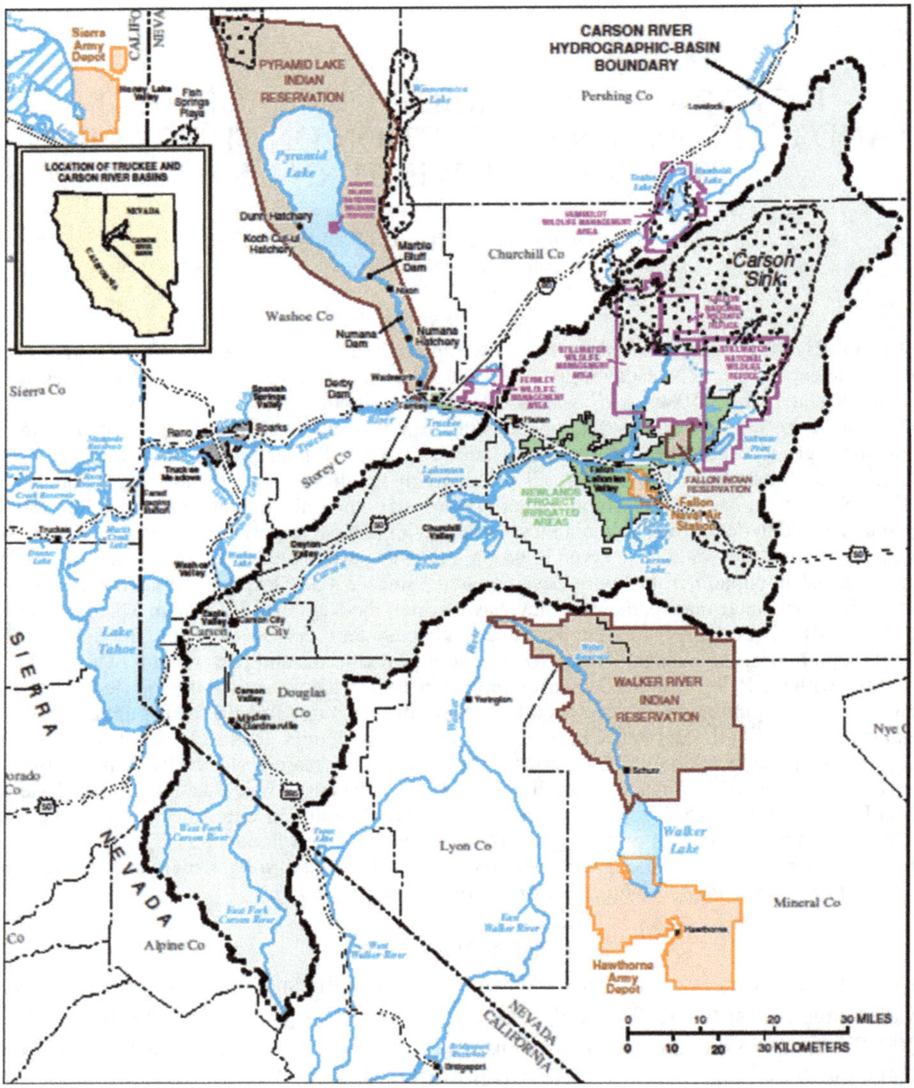

Figure 1: USGS hydrologic features map of the Carson River watershed and surroundings [2].

river channel became incised in valley floors, and water quality suffered. By the 1990s, local communities teamed up with state and federal agencies to stabilize raw, eroding banks and attempt to meet the standards of the Clean Water Act [1].

Between 30 December 1996 and 3 January 1997, a large, warm atmospheric river storm unleashed up to 17 inches of rain in the headwaters areas of the watershed. Known as the New Year's Flood of 1997, this 'rain-on-snow' storm event produced severe flooding on many large watersheds originating in the Sierra Nevada, including the Carson River and the Truckee River just to the north. These floods crippled downstream communities for days. Hundreds of millions of dollars of flood damages occurred in the city of Reno, Nevada, on the Truckee River, just 30 miles north of Carson City. In the Carson Valley, near Minden and Gardnerville, agricultural fields flooded and created a shallow lake over 10 miles long and 3–4 miles wide in places [2, 3]. See Fig. 1.

While many Nevada residents were shocked by this severe flood in the western side of the Great Basin Desert, this flood was not unique. Over the past 160 years, large river floods have occurred on western Nevada rivers about every 10–20 years. Floods in 1950 and 1955 were comparable in size to the New Year's flood of 1997 [2].

In the year following the 1997 flood, citizens, agency staff and University of Nevada Cooperative Extension educators came together to organize the first major watershed conference on the Carson River. Many believed that the Carson River was so important to the community that it deserved to be cared for and managed using the practices of integrated watershed management (IWM). Two weeks after the watershed conference, which was attended by 250 participants, the organizers held a follow-up workshop where they asked participants to consider the pros and cons of establishing a watershed management group based on the principles of IWM. At the end of the workshop, participants overwhelmingly voted to support such a process, and the Carson River Coalition (CRC) was born. A regional water management agency known as the Carson Water Subconservancy District (CWSD) was selected to house and support this new citizen and inter-agency watershed group.

2 DEFINING IWM

The concept of watershed management has been developed by scientists, government officials and educators over the past century. In the early 1990s, the Universities Council on Water Resources devoted the autumn 1993 and the summer 1995 issues of its journal, *Water Resources Update*, to articles on the concept of IWM. Cobourn mined these articles for key ideas and created a list of the 'defining characteristics of IWM' [4]. These were used to evaluate the degree to which the Truckee River was planning and managing the watershed according to IWM [4]. Some of the most important characteristics are the following:

1. It is holistic. It looks at many issues and problems and their interrelationships, not single issues or 'projects' [5, 6].
2. It addresses issues of water quality, water quantity, habitat and floodplain management in an integrated fashion [5].
3. It is interdisciplinary. It addresses the whole ecosystem, including the human dimension. Property owners, developers, engineers, resource managers, agricultural producers, attorneys, etc., need to join the team [7, 8].
4. It aims at long-term sustainability, considering current and future problems, solutions and consequences [9, 10].
5. Since watersheds cross jurisdictions, collaboration is needed between all levels of government. This requires communicating, joining forces and sharing 'territory' [7, 11, 12].

More recent definitions echo many of these characteristics. In Canada, the group Conservation Ontario stresses the interaction of human and natural processes in its definition: 'Integrated Watershed Management (IWM) is the process of managing human activities and natural resources on a watershed basis. This approach allows us to protect important water resources, while at the same time addressing critical issues such as the current and future impacts of rapid growth and climate change' [13].

University of Arizona professor J.E. de Steiguer [14] emphasizes the importance of stakeholder input to the success of this kind of planning. 'Integrated Watershed Management (IWM) has emerged worldwide as the preferred model for watershed planning. IWM uses the watershed as the basic geographic planning unit while integrating social, economic, ecological and policy concerns with science to develop the best plan. Stakeholder input is key to successful IWM' [14].

3 GETTING STARTED WITH IWM PLANNING

The Carson River watershed conference organizers were aware of the principles of IWM in 1998 and wanted to give this ambitious concept a try. One of the organizers was an engineer who was on contract to manage the CWSD. This regional body had been formed by the Nevada Legislature in the 1950s to plan for possible storage reservoirs on the upper Carson River (which were never built).

Today, the CWSD is a six-county, bistate regional watershed agency. Its 14 member board of directors consists of elected representatives from each county in the watershed as well as several representatives from the agricultural community [15]. The fact that this board is composed of elected officials from several local governments and by agricultural stakeholders is crucial to the success of this agency and its watershed management efforts.

CWSD's mission is to work within existing governmental frameworks to promote cooperative action for the watershed that crosses both agency and political boundaries. CWSD acts as lead agency for integrated watershed planning and funds the CRC. The CRC is a watershed-wide stakeholder group that serves as the steering committee for the integrated watershed planning process. CWSD strives to involve all counties and communities within the watershed in the effort to meet future water needs, enhance the health of the river system, protect the floodplains and provide outreach and information to the community [15].

In 1998, CWSD hired a full-time general manager and soon after obtained funding from the Nevada Division of Environmental Protection to hire a watershed coordinator for the CRC. Attendance at monthly CRC meetings was regularly 30 or more people, and the group soon hammered out a list of 'Watershed Management Guiding Principles.' These principles became the basis for the Carson River Watershed Adaptive Stewardship Plan, prepared by CWSD in 2006 [15]. These principles are shown in Table 1.

In 2003, University of Nevada Cooperative Extension faculty asked the members of the CRC to conduct a group process to determine the single most important message to convey to the general public about the watershed. Over 50 members participated, and they reached consensus that this 'main message' was

> Protect the floodplain from future development. Once the floodplain and especially the river's meander belt corridor are impacted by development, the river loses the ability to reestablish its natural functions. Agricultural fields near the channel are critical for floodwater attenuation, ground water recharge, non-point source pollution buffering and providing habitat for wildlife [16].

4 FLOODPLAIN MANAGEMENT IN THE CONTEXT OF IWM

As noted above, there was a major river flood in 1997, a little over a year before the Carson River Conference was held in 1998 and the CRC was formally established. The conference organizers did not see the conference and the CRC simply as responses to the flooding. Note

Table 1: Carson River watershed management guiding principles [15].

1) Manage the water's resources for economic sustainability, quality of life and protection of private and public property rights.
2) Acknowledge and respect the watershed's natural processes in land use decisions.
3) Maintain or improve the quality of the water to support a variety of beneficial uses.
4) Protect the headwaters region as the system's principal water source.
5) Recognize and respect the interests of all stakeholders upstream and downstream by fostering collaborative and mutually respectful relationships.
6) Maintain the riverine and alluvial fan floodplains of the Carson River watershed to accommodate flood events.
7) Protect and manage uplands, mountain ranges, wetlands and riparian areas to enhance the quality of surface flow, groundwater recharge and wildlife habitat.
8) Promote conservation of water from all sectors of the community's water users for the benefit of municipal, industrial, agricultural domestic, recreational and natural resources.
9) Encourage management of growth that considers water quality and quantity, open space preservation and maintenance of agriculture in floodplains.
10) Protect and support opportunities for public recreational access to natural areas throughout the watershed – including the river corridor – where appropriate.
11) Promote understanding and awareness of watershed resources and issues through cooperative education efforts throughout the watershed.

that the guiding principles in Table 1 call for protection of the riverine floodplain not just to attenuate floodwaters but also to maintain the river's 'natural processes,' recharging groundwater and maintaining surface water quality and wildlife habitat. CRC members certainly agreed that the magnitude of large river floods such as the 1997 New Year's Flood demanded preparedness and floodplain management [17]. But they insisted that flood control structures were not the best answers to the problem.

One of the keynote speakers at the 1998 conference was Dr. Jeff Mount, at that time, the chair of the Geology Department at University of California at Davis. Dr. Mount, author of *California Rivers and Streams* [18], spoke about what he referred to as the vicious cycle of 'serial engineering' that often occurs on riverine floodplains. Echoing many of the conclusions in the US Army Corps of Engineer's 1994 'Galloway Report,' [19], Dr. Mount pointed out that once structures such as homes or businesses are built in a floodplain and are flooded, property owners call for flood control structures to protect them. After levees are built, people feel safe, and more structures are built there. Then a larger flood occurs, and this cycle of flooding and levee construction continues until the river is put into an extremely expensive concrete channel, such as the channel of the Los Angeles River. When that happens, the fish, wildlife and water quality values are lost. Furthermore, preventing the storage of flood waters on floodplains increases peak flows during both rain-on-snow and spring snowmelt flood events. These increased flows increase flood hazards for all downstream communities.

After the CRC agreed on its 'main message [16]', its members and coordinator began work on the Carson River Watershed Regional Floodplain Management Plan (RFMP) of 2008 [20]. This plan was grounded on guiding principles 2, 6 and 9 (from Table 1) and on the 'main message.' It also developed an important economic argument that spending money now on floodplain protection will save much larger sums of money that would be spent on damages in the future if the floodplain turns into subdivisions and industrial parks. One key element in this argument was the fact that flood damages in the built-up floodplains of the Truckee River in Reno and Sparks, about 30 miles to the north, were 30 times more expensive than flood damages in the relatively undeveloped floodplains of the Carson River watershed.

The RFMP analyses the hazards of riverine flooding and presents a long list of 'suggested actions' for 'flood risk reduction and floodplain protection' [20]. The central thrust for these actions is: 'Protect Natural Floodplain Functions and Values' [20]. This goal is described in terms of what is called the 'Living River Approach' [20]. The goal of this approach is to maintain a river environment which:

- provides un-impeded flow conditions and connection of the river to its floodplain;
- minimizes disruption and alteration of river and riparian habitat;
- conveys variable flows and restores habitat in floodplain;
- balances sediment input with sediment transport;
- provides fish and wildlife habitat;
- enhances water quality and supply;
- maintains aesthetic and recreational qualities [20].

Though the Carson River's channel is incised in many places, and many riparian areas are denuded because of steep eroding banks, at least the floodplain is still used mainly for agriculture. The RFMP recognizes that if riverfront homes and urban infrastructure can be kept to a minimum, the cycle of 'serial engineering' can be prevented. If the above characteristics of a living river can be restored, they will provide flood water attenuation and other ecosystem services for the community in the future. Non-structural floodplain management methods are recommended over structural flood control measures. For example, the suggested actions of the RFMP urge river communities to protect the natural functions and benefits of floodplain lands, to consider flood hazards when prioritizing acquisition targets for open space programs and to 'retain lands that provide floodplain storage and maintain or restore connection of the river with its floodplain through land acquisition, conservation easements, local open space programs, transfer of development rights (TDR) … and other protection methods' [20].

5 THE CHALLENGE OF PROTECTING FLOODPLAIN LANDS

Most of the floodplain land along the channel of the Carson River is private property, owned by cattle ranchers and alfalfa farmers. However, the watershed and particularly the Carson and Dayton Valleys are near enough to Reno and Carson City to experience growth pressure. As ranchers near retirement age, they sometimes find that their children do not want to take on the hard work of running the ranch. During times of economic growth, land developers often look for opportunities to divide ranches into small-acreage ranches ('ranchettes'). Many people are still drawn to the lure of a riverfront home.

County planners in Douglas County have zoned much of the riverine floodplain for a maximum of one house per 19 acres (7 ha). They have also created a Transfer of Development Rights (TDR) programme which allows property owners to sell development rights of floodplain properties to developers in 'receiving areas' which are not in flood hazard zones [21]. In addition, NGOs such as The Nature Conservancy have raised funds to purchase and retire the development rights on some ranches, placing a permanent conservation easement on the deed.

Members of the CRC reaffirmed that protection of floodplain lands from development was still their 'main message' in 2012. They wondered at that time if the message was actually getting out to the public and, more importantly, if floodplain lands were actually being protected. CRC members felt that an inventory of protected floodplain lands would be helpful to act as a 'report card' for showing progress towards this community goal. In 2013, two Cooperative Extension faculty members formed a small group to research this question and to develop a map and inventory.

6 CREATING A FLOODPLAIN PROTECTION INVENTORY

The group who worked on the inventory included the floodplain managers and engineers from the three counties in the watershed with the most floodplain lands at risk from development. It also included local open space managers, consultants and staff of local NGOs working on conservation easement opportunities, plus the two cooperative extension faculty. Douglas County contributed the time and support of its geographic information systems (GIS) supervisor.

The group decided to inventory riverine floodplains only. Alluvial fan flood zones would be the subject of a future study. They chose to define the floodplain as the Federal Emergency Management Agency's (FEMA) 'one-percent-chance' flood zone, which has a 1-in-100 chance of flooding in any given year, combined with the boundaries of the area actually flooded in January 1997, as mapped in the Nevada Bureau of Mines and Geology's 'Flood Extent Map' [2]. These two areas are similar in size and shape. The FEMA one-percent-chance flood zone is also known as the 100-year flood zone or Zone A, and it defines zones of risk called Special Flood Hazard Areas. Since the FEMA flood zone is to some degree hypothetical, we included additional areas that flooded in the large (80+-year) 1997 flood as a way of ensuring a conservative picture of lands known to be at relatively high risk of flooding. This combined floodplain is shown in blue on our maps.

For the purpose of this inventory, 'protected floodplain' refers to land that is either publically owned or privately owned and not likely to be developed. Briefly, 'protected' parcels include the following:

- Open space and parks owned by federal, state, county, city or general improvement district agencies.
- Privately owned land with a conservation easement and/or development restrictions (including those small parcels in Douglas County's Transfer of Development Rights Program).
- Privately owned open space – by homeowner associations (common areas) [22].

Group members were asked to forward legal descriptions of protected floodplain lands to the GIS mapmaker. The results of this inventory are illustrated in e-maps (Figs 2 and 3), with the unprotected floodplain shown in blue and with parcels of protected land shown as green polygons.

We tallied the total number of acres of floodplain lands and protected parcels in each county in Table 2. The total area of the combined riverine floodplains amounts to 39,879 acres (16,138 ha) within Douglas County, Carson City and Lyon County. Within this floodplain area, 31% have been identified as protected and 69% as unprotected. Specific acreages and percentages for each county are shown in Table 2 [22].

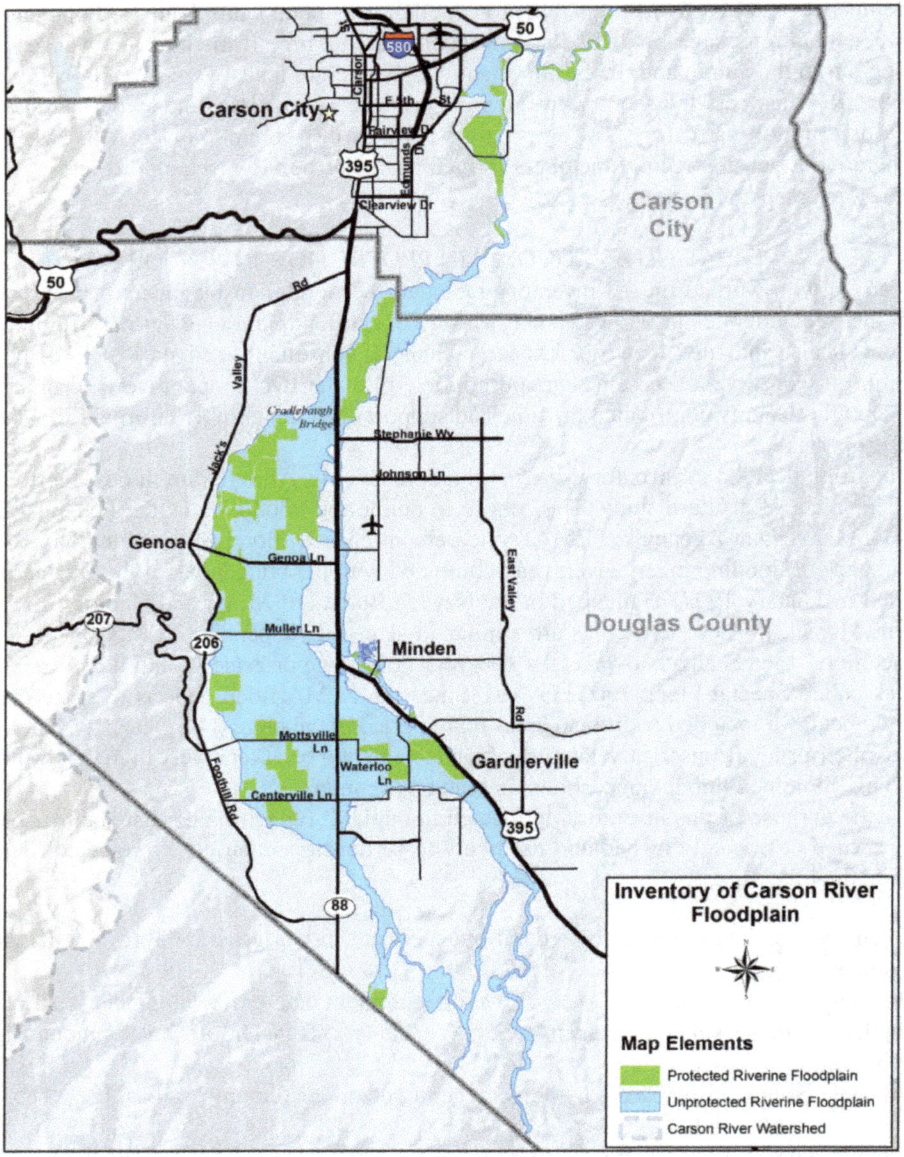

Figure 2: Map of protected and unprotected riverine floodplain lands within Douglas County and Carson City in the Carson River watershed as of 2015 [22].

7 USING THE INVENTORY FOR PUBLIC OUTREACH EDUCATION

The CRC members who called for this inventory wished to get the information out to the residents of the watershed, so the data and maps were compiled into a readable 'Special Publication' and published by Cooperative Extension. The 6-page publication explains the functions and values of undeveloped floodplains, explores why and how they could be protected and presents the maps and data in clear terms. It also poses the choice that the people and local

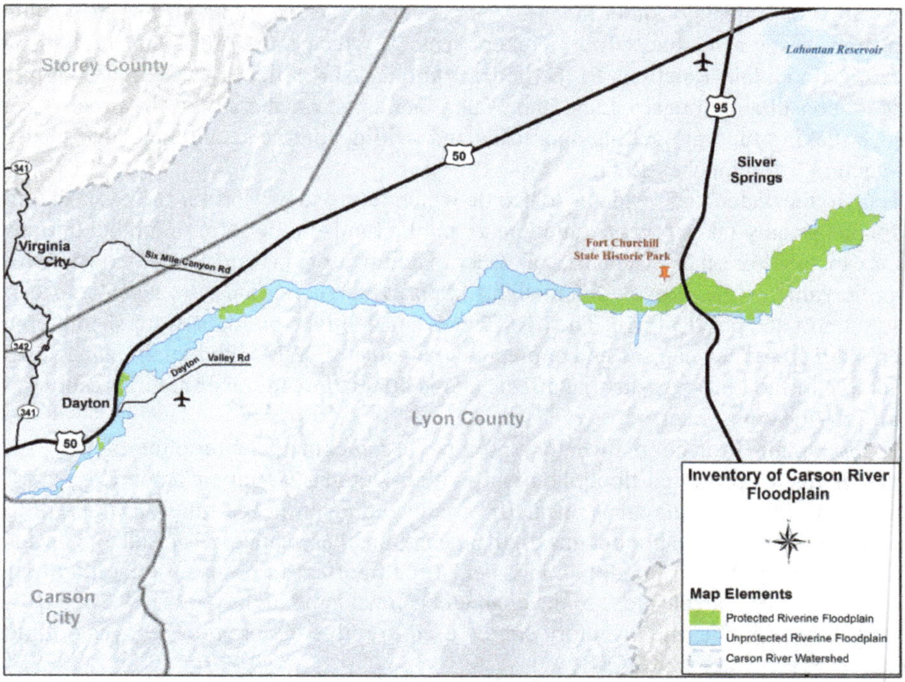

Figure 3: Map of protected and unprotected riverine floodplain lands within Lyon County in the Carson River watershed as of 2015 [22].

Table 2: Acres and percent of protected and unprotected riverine floodplain in Douglas County, Carson City and Lyon County within the Carson River Watershed as of 2015 [22].

County	Protected floodplain acreage (ha)	% Protected floodplain	Unprotected floodplain acreage (ha)	% Unprotected floodplain	Total riverine floodplain acreage (ha)
Douglas	6,009/2,431	23	20,182/8,113	77	26,191/10,544
Carson	903/365	36	1,588/638	64	2,491/1,003
Lyon	5,403/2,186	48	5,794/2,329	52	11,197/4,515
Total	12,315/4,983	31	27,564/11,081	69	39,879/16,064

governments in the watershed need to make. It is a choice between protecting undeveloped floodplains and letting them be gradually developed for subdivisions, business parks, resorts and small ranches. The latter choice could lead to the eventual construction of an expensive concrete-lined flood channel to replace the river and its aquatic and riparian ecosystems. This could increase flood hazards to downstream communities, including Fallon, Nevada, below the Lahontan Reservoir.

In order to achieve the goal of protecting riverine floodplains, millions of dollars of funding is needed to compensate private landowners for the cost of retiring development rights using conservation easements or other means. Many local ranchers would consider such solutions if the money was available. CRC members seek to spread the word about the benefits of protecting floodplain functions to the general public. NGOs like The Nature Conservancy and local consultants, Legacy Land and Water, can assist ranchers with the process, which includes seeking funding, getting appraisals and writing contracts, and they have succeeded in protecting a few ranches so far.

It is up to the watershed residents to decide which future they want for the river. Protecting floodplains usually takes private donations or public funds in the form of a bond initiative, a sales tax increase or other mechanism in order to compensate private property owners for the economic value of their development rights. Approving such a measure takes considerable public support and political will. The CRC works constantly to build the watershed literacy of local residents, so they can see and appreciate the value of IWM. Floodplain management is an example of the kind of issue that affects and is affected by all other natural resources and human activities in a watershed.

The Floodplain Protection Inventory provides a benchmark or baseline of the location and percentage of protected floodplain land within Douglas County, Carson City and Lyon County, Nevada. The same analysis can be conducted again in the future to track progress toward protecting riverine floodplains. In the meantime, these data can be used to help answer questions such as: Are all floodplain lands equally important to protect, or should floodway and other high-risk parcels close to the channel be considered higher priority? So in addition to educating the public, the inventory can be used to refine the broad objectives outlined in the Carson River watershed RFMP.

REFERENCES

[1] Azad, G., Cobourn, J., Kaffer, D., Lewis, S.R., Riedl, M.K. & Skelly, J.A., *The Carson River Watershed Map: Our Lifeline in the Desert*, University of Nevada Cooperative Extension, Special Publication-03-02, available at www.unce.unr.edu/publications/files/ag/2003/sp0302.pdf, 2003 (accessed 7 March 2017).

[2] Trionfante, J.V. & Peltz, L.A., *Hydrologic Features of the Truckee and Carson River Basins and Adjacent Areas, Western Nevada and Eastern California*. USGS Report 93-368, available at https://pubs.usgs.gov/of/1993/0368/report.pdf, 1994 (accessed 7 March 2017).

[3] Rigby, J.G., Crompton, E., Berry, K., Yildirim, U., Hickman, S.F. & Davis, D.A., *The 1997 New Year's Floods in Western Nevada*, Nevada Bureau of Mines and Geology, Special Publication 23, Reno, NV, 1998.

[4] Cobourn, J., Integrated Watershed Management on the Truckee River in Nevada. *Journal of American Water Resources Association*, **35**(3), pp. 623–632, 1999. DOI: 10.1111/j.1752-1688.1999.tb03618.x.

[5] Nelson, D., Water education in Integrated Watershed Management. *Water Resources Update*, **93**, pp. 34–37, available at http://ucowr.org/files/Achieved_Journal_Issues/V93Water%20Education%20Integrated%20Watershed%20Management.pdf, 1993 (accessed 20 Sept 2017).

[6] Ballweber, J.A., Prospects for comprehensive, Integrated Watershed Management under existing law. *Water Resources Update*, **100**, pp. 19–23, available at http://opensiuc.lib.siu.edu/cgi/viewcontent.cgi?article=1353&context=jcwre, 1995 (accessed 9 September 2017).

[7] Wayland III, R.H., Comprehensive watershed management: A view from EPA. *Water Resources Update*, **93**, pp. 23–26, 1993.
[8] Burt, J.P., New water resource challenges. *Water Resources Update*, **93**, pp. 27–28, 1993.
[9] Cairns Jr., J., The economic basis for a partnership between human society and natural ecosystems. *Water Resources Update*, **93**, pp. 18–22, available at http://ucowr.org/files/Achieved_Journal_Issues/V93The_Economic_Basis_for_a_Partnership(1).pdf, 1993 (accessed 20 September 2017).
[10] Heaney, J.P., New directions in water resources planning and management. *Water Resources Update*, **93**, pp. 3–8, available at http://opensiuc.lib.siu.edu/cgi/viewcontent.cgi?article=1412&context=jcwre, 1993 (accessed 20 September 2017).
[11] Funk, W.H., A water problem that continues to be rediscovered: nonpoint source pollution. *Water Resources Update*, **93**, pp. 9–12, available at http://opensiuc.lib.siu.edu/cgi/viewcontent.cgi?article=1413&context=jcwre, 1993 (accessed 20 September 2017).
[12] Margerum, R.D., Integrated Watershed Management: Comparing selected experiences in the U.S. and Australia. *Water Resources Update*, **100**, pp. 36–47, available at http://opensiuc.lib.siu.edu/cgi/viewcontent.cgi?article=1356&context=jcwre available at: http://opensiuc.lib.siu.edu/cgi/viewcontent.cgi? article=1356&context=jcwre, 1995 (accessed 20 September 2017).
[13] Conservation Ontario, *Integrated Watershed Management*, available at conservationontario.ca/what-we-do/what-is-watershed-management/integrated-watershed-management (accessed 7 March 2017).
[14] de Steiguer, J.E., Duberstein, J. & Lopes, V., The analytic hierarchy process as a means for Integrated Watershed Management. Presented at First Interagency Conference on Research in the Watersheds, Benson, AZ, 2003.
[15] Carson Water Subconservancy District (CWSD), *Carson River Watershed Adaptive Stewardship Plan*, available at www.cwsd.org/carson-river-watershed-adaptive-stewardship-plan/ (accessed on 7 March 2017).
[16] Lewis, S.R. & Cobourn, J., *River Corridor Protection: Carson River Coalition's Main Message*, University of Nevada Cooperative Extension, Fact Sheet 04-71, available at unce.unr.edu/publications/files/nr/2004/fs0471.pdf (accessed 7 March 2017).
[17] Nevada Division of Water Planning, *The Flood of 1997*, Final Report, available at water.nv.gov/home/pdfs/the%20flood%20of%201997%20final%20report.pdf (accessed 7 March 2017).
[18] Mount, J.F., *California Rivers and Streams: The Conflict Between Fluvial Process and Land Use,* University of California Press: Berkeley, CA, pp. 292–312, 1995.
[19] U.S. Army Corps of Engineers, *Sharing the Challenge: Floodplain Management into the 21st Century*, available at www.floods.org/PDF/Sharing_the_Challenge.pdf, 1994 (accessed 7 March 2017).
[20] Carson Water Subconservancy District, *Carson River Watershed Regional Floodplain Management Plan*, available at cwsd.org/library/Final%20floodplain%20plan%209-08.pdf (accessed 7 March 2017).
[21] Douglas County, Nevada, Chapter 6: Growth management. *Douglas County Master Plan of 2011*, Vol. II, available at www.douglascountynv.gov/DocumentCenter/Home/View/936 (accessed 7 March 2017).
[22] Cobourn, J. & Lewis, S.R., *Floodplain Protection Inventory for the Carson River*, University of Nevada Cooperative Extension, Special Publication 15-05, available at unce.unr.edu/publications/files/nr/2015/fs1505.pdf (accessed 7 March 2017).

ARE THE BENEFITS OF INTEGRATED CATCHMENT MODELLING BEING REALIZED IN THE UNITED KINGDOM?

WILLIAM RUST & PHOEBE VENN
Infrastructure, Atkins, United Kingdom

ABSTRACT

Historically, hydrological systems have been modelled separately to investigate flood risk. This is due to computational limits and the devolved responsibilities set out in the existing legislation for risk management authorities. This method may result in the true impact of flooding being misrepresented as the interaction between hydrological systems and drainage infrastructure unaccounted for. In contrast, integrated catchment modelling (ICM) is a methodology in which various hydrological systems are explicitly represented in a single flood model. This enables more realistic assessments of flooding sources and mechanisms, and allows for improved communication of risk. This article provides an overview of the use of ICM in the UK; discusses the benefits of and barriers to the use of ICM; and provides case studies that demonstrate the benefits of ICM for multiple end users. Cost–benefit analyses are traditionally carried out using results from separate models, and alleviation schemes are developed and funded based on the outcome. The economic benefits may be underestimated and the scheme under-designed due to inaccuracies inherent in separate modelling approaches. The use of ICM allows for the development of flood alleviation schemes (FASs) that provide multiple benefits by protecting people, properties and infrastructure at risk from combined sources of flooding. As set out in the Flood and Water Management Act 2010, Lead Local Flood Authorities (LLFA) are responsible for strategic flood risk management and as a result are moving towards the use of ICM to open more funding streams for FAS development. ICM provides a tool that demonstrates to multiple stakeholders how they can benefit from a single alleviation scheme. This increases the chances of a scheme reaching construction due to greater funding potential. While there are still barriers to the use of ICM, the benefits are beginning to be realized by LLFAs and other regulatory bodies.

Keywords: flood economics, flood risk management, integrated catchment modelling, hydraulic modelling, partnership funding

1 INTRODUCTION

A key issue that faces flood risk managers in the UK is the procurement of funding for the construction of flood alleviation schemes (FASs). Funding is dependent on two factors: a favourable cost–benefit ratio (BCR) for the scheme construction, and sufficient partners or stakeholders willing to contribute capital and maintenance costs towards a scheme. Given the complex interactions between flooding and the ever developing urban environment, establishing a robust BCR can be problematic without a comprehensive view of integrated risk mechanisms. Additionally, an incomplete view of all flood risk sources can limit the number of partners who may be shown to be benefiting from, and therefore willing to contribute towards, a scheme. A sufficient view of integrated flooding mechanisms is therefore crucial for successful scheme implementation in many urban areas.

Partnership funding (PF), where money for a scheme is accrued from multiple stakeholders, provides a mechanism for stakeholders to financially contribute towards a FAS. This is usually used in conjunction with government funding from the Department for Environment, Food and Rural Affairs. The Environment Agency (EA) has a national target to obtain 15% of funding for FASs, from non-governmental money throughout the period 2015–2021 [1].

As set out in the Flood and Water Management Act 2010 [2], the responsibility for managing the different sources of flood risk is divided between separate regulatory bodies.

DOI: 10.2495/EI-V1-N3-232-239

This devolution of responsibilities means that hydrological systems are often assessed in isolation to predict flood risk. Given the integrated nature of flooding between natural drainage and man-made infrastructure in the built environment, a siloed approach to assessing flood risk can result in an inaccurate understanding of flood risk in many urban areas. Historically, models have been developed to assess single-source risk for use by individual regulatory bodies (e.g. surface water drainage or river networks). In addition to an incomplete view of flood risk sources and mechanisms, this siloed approach means funding efficiencies are often lost where combined-system flooding prevails. With the growing demand for new development in the UK, flood mechanisms are becoming increasingly integrated, and the need to understand combined flooding is becoming more important, which cannot be easily achieved with the current approach.

In addition to legislative restrictions, hydraulic models used to understand flood risk and develop alleviation schemes have historically been simplified due to computational limitations. However, the increase in computational ability in the past two decades now allows hydraulic models to estimate larger and more complex, integrated flooding issues. Integrated catchment modelling (ICM) is a concept in which multiple connected hydrological systems across entire catchments are explicitly represented in a single flood model. This enables more realistic assessments of flooding sources and mechanisms, therefore, allowing an improved understanding and communication of risk. ICM provides a tool which demonstrates to multiple stakeholders how they can benefit from a single alleviation scheme. Using the results from integrated models to develop alleviation schemes can increase the BCR as it allows for a single scheme to be developed to protect people and property from combined sources of flooding. Consequently, the increased number of stakeholders benefitting from a scheme opens more funding streams for PF. Despite these developments in modelling ability, many FASs undertaken in the UK today still use siloed modelling approaches for combined urban flooding issues. This means fewer schemes are able to progress to design and construction as the number of stakeholders involved is limited.

This article will review three recent case studies where ICM has been used to inform a more complete view of combined flood risk and open more funding streams from stakeholders. These case studies will then be discussed within the wider context of adoption of ICM approaches, and the current barriers to the use of ICM in the UK.

2 CASE STUDIES

2.1 M1 junction 4–5 flood study

The M1 carriageway between junctions 4 and 5 is prone to flooding with a history of standing water on the carriageway, which has caused significant disruption to vehicles. The M25 Design, Build, Finance and Operate study identified this section of the M1 as the number one flood hotspot within the study area. Drainage improvements, such as recessed gullies and new drainage conduits, have been made in an attempt to reduce the impact of flooding. However, these improvements, coupled with an enhanced maintenance regime, have provided no reductions in flooding.

Atkins was commissioned by Highways England (HE) and Connect Plus Services to assess the flood risk areas, sources and mechanisms on the M1 between junctions 4 and 5 and to develop conceptual alleviation scheme designs. It was decided to use ICM to determine the mechanisms of flooding on the carriageway. This was because the flooding was not represented in existing separate hydrological system models, such as fluvial or highways drainage, and an integrated catchment model is able to represent the combined systems, allowing for a more efficient alleviation scheme to be designed.

The M1 is located in a cutting through the fluvial catchment of the Hilfield Brook. The study area primarily focused on the carriageway between junctions 4 and 5 on the M1. The Hilfield Brook and Tykes Water catchments were also included to take into account overland flow and the fluvial component of the Hilfield Brook catchment. The catchment is largely agricultural fields with some residential areas.

Traditionally, highways drainage systems are designed using hydraulic models that apply rainfall to the road and allow for a small percentage increase in the rainfall to runoff from the wider catchment. In the case of the M1, it was clear that, due to the location in a cutting, the surface water runoff would be a key component of the flood risk. The Atkins team determined that an integrated catchment model with the rainfall, fluvial network and drainage infrastructure represented would better assess the flood risk, sources and mechanisms. The model boundaries were set using the Hilfield Brook and Tykes Water catchment boundaries as illustrated in Fig. 1, and LiDAR was used for the ground model to simulate runoff over

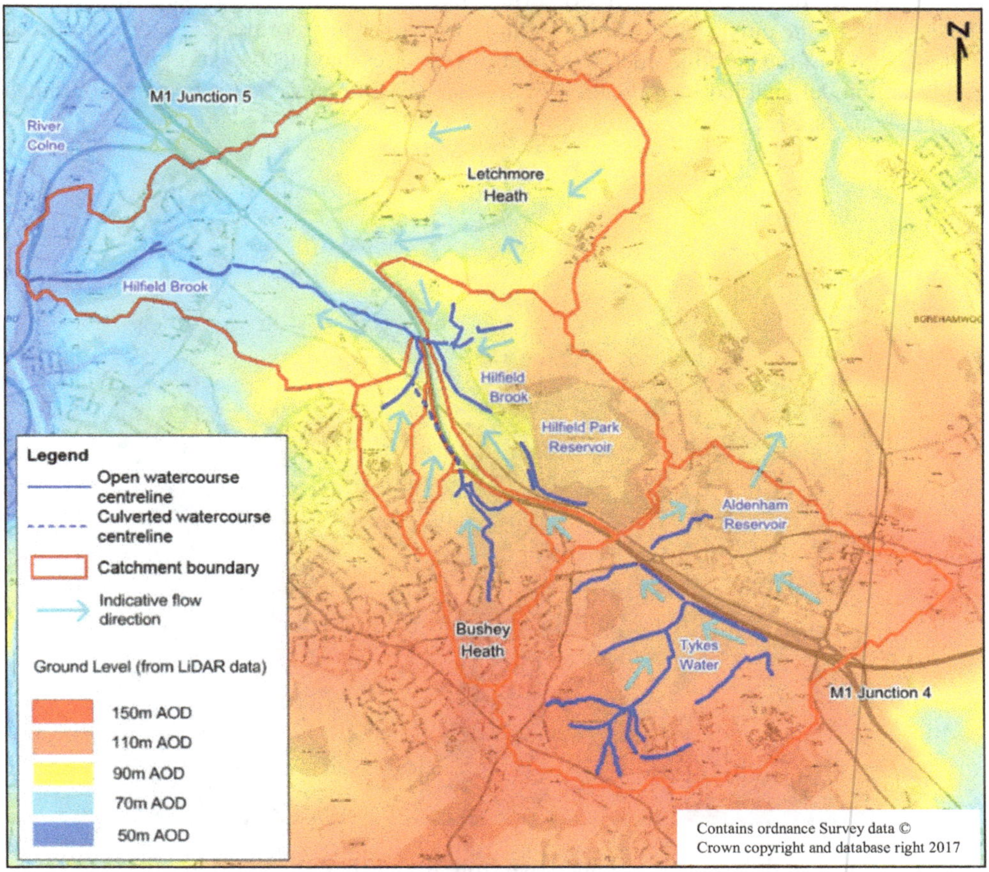

Figure 1: M1 study area catchment showing ground levels, water courses and flow paths.

the ground surface. The highways drainage infrastructure was imported from HE data, and the Bushey Heath urban catchment was represented using Thames Water drainage data. The EA's Hilfield Brook fluvial model was imported into the model including the structures, such as bridges and culverts.

During heavy rainfall events, the capacity of the highways drainage infrastructure on the M1 is exceeded. This, coupled with the surface water flow paths from the surrounding catchment and the carriageway's location in a cutting, means that flow runs down the motorway, causing significant disruption to drivers. There is a large flow path which originates from the Bushey Heath urban area, which is conveyed to the Bushey Heath drain by the surface water drainage infrastructure (Fig. 2). This flow is too great for the Bushey Heath culvert to convey safely downstream into the Hilfield Brook; therefore, it runs down the hill onto the M1. The baseline flood risk identified using ICM was verified using recorded flood information, including CCTV footage and photographs.

This updated model and recorded flood information differs from the EA's existing fluvial model, which solely models the fluvial inflow in the Bushey Heath drain and Hilfield Brook (Fig. 3). The existing EA model does not consider surface water input from the surrounding catchments. This EA fluvial model shows that the Bushey Heath culvert is able to convey the flow during a 0.1% AEP event. In contrast, the new ICM approach suggests that the capacity of the culvert is exceeded during lower intensity storm events. The improved understanding of the baseline flood risk locations, sources and mechanisms provided by the combined model enabled identification of conceptual options to better manage combined flood risk on the M1.

Figure 2: Photo of flood water running down from the Bushey Heath urban catchment onto the carriageway on 11–15th January 2008, facing north.

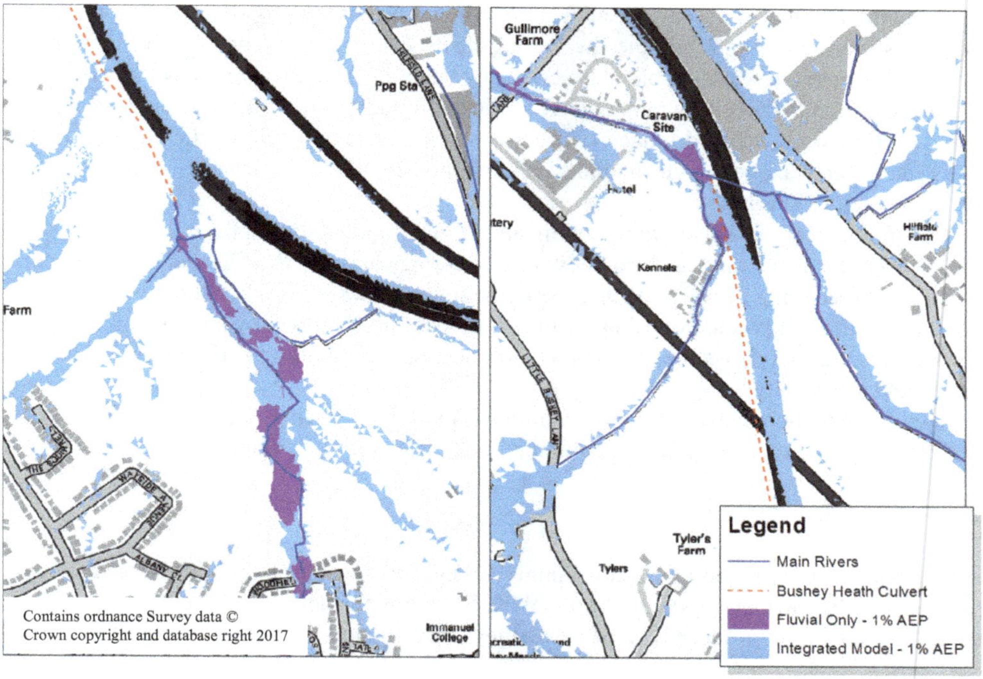

Figure 3: Comparison of modelled flood outlines between EA fluvial flood maps and the M1 combined model (annual exceedance probability (AEP) is the chance of a given flood event being exceeded in any given year).

2.2 Oxford flood alleviation scheme

According to the Flood Risk in Oxford policy paper from the EA [3], there are 4,500 properties in the city at a 1% (or higher) annual risk of flooding. Following the Oxford Flood Risk Management Strategy [4], the EA commissioned the construction of a river-only hydraulic model to quantify and develop plans to alleviate fluvial flood risk. This hydraulic model was built in ISIS-TuFlow; software was designed to simulate in-channel and surface routing hydraulics for the purpose of fluvial flood modelling. The design of the Oxford Flood Alleviation Scheme (OFAS) was undertaken using the results from the river-only model. The design comprises a two-stage alleviation channel and associated retention bunds costed at approximately £100 m [5]. The OFAS was designed to raise the current standard of protection to 1% annual risk.

Thames Water was interested to understand the effect that fluvial flooding had on their drainage system, and to what extent the proposed OFAS would benefit their network. They commissioned Atkins to use ICM to investigate the combined issue of fluvial, surface water and sewer flooding within the Oxford sewer catchment.

The OFAS integrated model revealed that a portion of the Thames Water network was affected by the inundation of flood water from fluvial sources. This was resulting in hydraulic overload of the sewer network. The integrated view of flood mechanisms, provided by the ICM approach, allowed Thames Water to update the source of flooding for certain assets in their risk database from 'sewer' to 'fluvial'.

The combined model was also used to assess the impact of the proposed OFAS on Thames Water's sewer network. While the fluvial-only ISIS-TuFlow model was suitable for fluvial FAS development, it was not able to describe any benefit to the urban drainage network. This was because the ISIS-TuFlow model only replicated flooding from the river channel with no interaction with the urban drainage network. As such, this approach precluded other potentially interested stakeholders from understanding the benefits of the scheme, thereby limiting PF opportunities (Fig. 4).

A comparison between existing flooding and OFAS flooding within the model showed that the OFAS reduces flooding and provides a considerable benefit to the Thames Water network, as a result of reducing fluvial water ingress to the sewer network. This benefit allowed Thames Water to understand what contribution they may be able to make towards part-funding the OFAS. This was only possible by using a combined view of flooding between fluvial, surface water, and sewer networks.

With the development of the OFAS combined model, all further optioneering can be carried out using the representation of combined risk provided by the model.

2.3 Rive catchment study

Atkins was commissioned by Surrey County Council to consider the flood risk at five locations within the town of Woking in Surrey, UK, following the delivery of the Rive Ditch Constraints Assessment project [5]. This assessment investigated the viability of multiple flood

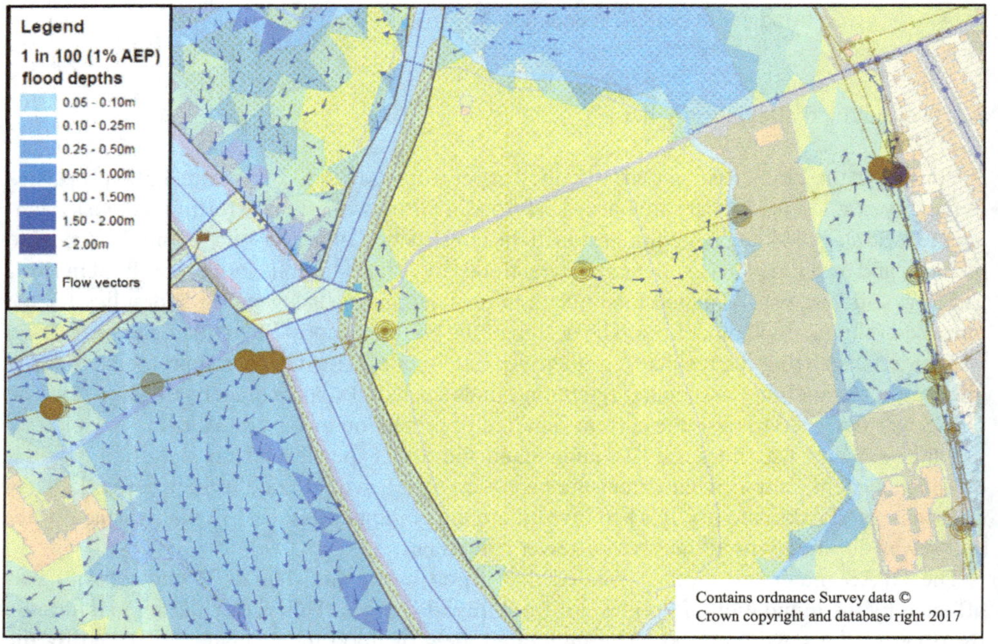

Figure 4: Model output showing integrated flooding between sewers and rivers in the Oxford integrated model.

attenuation schemes in the Rive catchment. Therefore, for the next stage, Atkins proposed to undertake the assessments collectively by developing an integrated catchment model. The use of ICM was chosen because it is more cost effective than using separate models for each location; it provides a more effective tool for making strategic flood management decisions; and it would be easier to develop for future studies.

To facilitate collaborative working on the Rive Catchment Assessment, a Project Board consisting of the following organizations was established:

- Surrey County Council
- Woking Borough Council
- Runnymede Borough Council
- Environment Agency
- Thames Water
- Network Rail.

The combined flood model represents the watercourses, canals, ground surfaces, buildings and foul and storm water sewers within the Rive Catchment. The model includes over 36 km of watercourse, 477 km of pipes (including culverts) and 8,663 manholes. This combined model was calibrated to recorded flood events and gauged flows to ensure a more robust representation of the combined flood mechanisms. A range of flood risk mitigation options were identified through stakeholder consultation and simulated in the model.

The use of ICM allowed for a better representation of combined flood risk, compared with the original fluvial-only flood models. This meant that various alleviation schemes could be taken forward for further option assessment, as the integrated model showed that there were more benefits than the fluvial-only models. Additionally, the presence of the Project Board with multiple stakeholders, as facilitated by an ICM approach, has also allowed schemes to be developed with full approval from the stakeholders.

By using ICM from the start of this project, a collaborative approach has been used from risk assessment through to option development. This enables all future assessments and developments to address multiple sources of flooding and all stakeholders' interests.

3 DISCUSSION AND CONCLUSION

The aim of this article is to outline the benefits of ICM and to determine whether these benefits are being realized in the UK.

The benefits of using ICM for the development of FASs have been explored using three case studies in this article. These examples show that an ICM approach can provide a better representation of combined flood risk in urban areas over traditional 'separate' models. This is due to the representation of the dynamic interaction between different hydraulic and hydrological systems allowed by using ICM. This has been demonstrated well by the M1 Junction 4–5 Flood Study, for which an ICM approach enabled the significant improvement of flood estimation over previous approaches, enabling the development of more effective flood alleviation options.

These examples also provide evidence that an ICM methodology can generate increased interest and awareness of a flooding issue from a wider range of stakeholders, thereby increasing PF opportunities. The OFAS and the Rive Catchment Study demonstrate how using ICM can provide a robust and fuller project stakeholder group which increases the likelihood of a scheme reaching the required PF target and, therefore, progressing to design and construction. The examples show that ICM simplifies the process of stakeholder engagement, given the wider view of flooding mechanisms, sources and receptors.

The ICM methodologies discussed in this article allowed the clients to better understand the risk to their (and others') assets meaning that flood alleviation options can be developed to protect against a broader range of combined risk. As such, ICM provides a methodology to build more robust benefit–cost ratios due to the efficiencies that can be made in the alleviation of multiple sources of risk. The increased BCR enables the alleviation scheme to be taken through to design and construction. Due to the combined ICM methodology, there is inherently a larger number of stakeholders engaged in a flooding issue, compared with separate modelling approaches. This allows increased PF opportunities, ensuring that the schemes are fully funded and taken through to construction. These benefits are significantly harder to achieve using separate or siloed modelling approaches.

While this article has shown evidence that an approach using ICM yields improved outcomes over separate modelling, there are still inhibitors to the wider adoption of ICM in the UK. An initial willingness is required from the client or regulatory body to invest in an ICM approach. Given the devolution of responsibilities for different flood source risks across different management bodies in the UK, it can be difficult to show the benefit of ICM beyond the remit of their responsibilities.

As set out in the Flood and Water Management Act 2010, Lead Local Flood Authorities are responsible for strategic flood risk management. As a result they are moving towards the use of ICM to open multiple funding streams for FAS development. In order to encourage more regulatory bodies to adopt the use of ICM, there needs to be a change in the industry mind-set regarding strategic flood risk management. If the EA is to reach the target of 15% PF contribution to its FASs by 2021, there needs to be increased cooperation between risk management authorities and stakeholders to provide a driver for stakeholder investment. This article has shown that ICM can provide a robust methodology to achieve this requirement. At present, the devolved responsibilities set out in the Flood and Water Management Act (2010) [2] do not encourage collaboration or the assessment of the risk as an integrated issue. The EA's 6-year investment programme states a target of reducing the risk of flooding to 300,000 properties by 2021 [1]. This number could be more easily achieved with the use of ICM as more properties could be protected from combined flooding with a single scheme, compared with protecting against single sources of flooding.

REFERENCES

[1] Defra. Reducing the risks of flooding and coastal erosion – an investment plan; 2014.

[2] UK Government. Flood and Water Management Act; 2010.

[3] Environment Agency. Flood Risk in Oxford policy paper, available at https://www.gov.uk/government/publications/flood-risk-in-oxford/flood-risk-in-oxford, 2016 (accessed 1 March 2017).

[4] Oxfordshire County Council. Local Flood Risk Management Strategy, available at https://www.oxfordshire.gov.uk/cms/content/oxfordshire-local-flood-risk-management-strategy, 2010 (accessed 1 March 2017).

[5] Oxfordshire County Council. Environment and Planning, available at https://www.oxfordshire.gov.uk/cms/content/oxford-and-abingdon-flood-alleviation-scheme, 2017 (accessed 1 March 2017).

[6] Atkins. Rive Ditch Constraints Assessment; 2013.

ECONOMIC, SOCIAL AND ENVIRONMENTAL VULNERABILITY TO DROUGHT IN THE NORTHWEST RIVER BASIN SYSTEM, MEXICO

DAVID ORTEGA-GAUCIN, HEIDY V. CASTELLANO & JESÚS DE LA CRUZ
Mexican Institute of Water Technology, Mexico

ABSTRACT

Vulnerability to drought is the degree to which a system is susceptible to damage by drought and incapable of coping with its adverse effects. This article presents a method to calculate drought vulnerability indices in the Northwest River Basin System, Mexico. The method is based on the concept of vulnerability developed by the Intergovernmental Panel on Climate Change (IPCC, 2007), which differentiates three components of vulnerability: degree of exposure, sensitivity and adaptive capacity. Each of these components is represented by a set of relative indicators at the municipal level that allow for the determination of three basic types of vulnerability: economic, social and environmental, which converge in overall vulnerability. The successful use of the method in the Northwest River Basin System shows that it can be applied to the rest of river basin systems in Mexico, and it can also be adapted to be used at the state or national level. The strength of the method lies in its approach as an objective analytic procedure that makes it possible to identify the most vulnerable municipalities from economic, social and environmental perspectives, which is useful in managing resources and efforts to reduce vulnerability to drought in the different regions of the country.

Keywords: adaptive capacity, degree of exposure, drought, sensitivity, vulnerability

1 INTRODUCTION

Drought is a natural phenomenon, an intrinsic part of climate to which every part of the planet is exposed, even its rainiest regions. Droughts are characterized by their inevitability, and they are still scarcely predictable [1]. The phenomenon lacks an epicentre or a definite trajectory, its progress is slow and gradual, and it covers vast geographical extents; as a consequence, identifying its spatiotemporal limits and providing help to affected populations are challenging endeavours.

Vulnerability to drought is inversely related to the degree of development in the affected areas: whereas developed countries can usually withstand drought without serious consequences, since they possess the structural and economic means to face it, drought can spell hunger, poverty, migration and reduced well-being and quality of life in marginalized areas. The technological level of a region or country, as well as its organizational, institutional and social performances, is a key element in vulnerability to drought. An example of these differences in vulnerability can be observed when comparing Mexican and American farmers living near the border between their countries, where they share similar physical and biological conditions in contrast with their different social, political, economic and historical backgrounds [2]. Clearly, the former are more vulnerable than the latter.

As a consequence, water scarcity due to drought can be a decisive factor for some settlements to be abandoned by their working-age inhabitants (especially in arid and semiarid areas in the North), who prefer to seek job opportunities in large cities or abroad. These 'ghost towns', where at best remain only women, children and old people who make their livings from the meager products of the land and the resources sent by their young who have migrated, are an irrefutable proof of the scope and impact caused by droughts [3]. Recent research demonstrates that when crop yields – especially corn – are reduced due to drought, a significant increase in migration from Mexico to the United States is observed [4].

DOI: 10.2495/EI-V1-N3-240-253

In this context, as an attempt to cope with the effects of drought by means of adequate risk management strategies, and in response to the severe drought that affected most of the country in 2011 and 2012, the Mexican federal government launched the National Program Against Drought (PRONACOSE, for its acronym in Spanish) with the goal of implementing a series of preventive actions to mitigate and reduce the vulnerability of populations to this natural phenomenon [5]. It is therefore important to assess vulnerability in each river basin system in the country bearing in mind that the risk of disaster by drought depends not only on the degree of rainfall scarcity and the duration and geographical reach of the phenomenon but also on vulnerability conditions that increase the chance of disaster when drought takes place.

This article describes our method for calculating drought vulnerability indices (VIs) in the Northwest River Basin System (Mexico) using relative indicators at the municipal scale. The method can be used by decision makers and officials responsible for developing public policies in identifying the most critical investments to reduce vulnerability and in delineating and proposing effective risk management actions from the economic, social and environmental perspectives of the problem.

2 METHOD

For a better understanding of the multiple dimensions of vulnerability, our method is based on the concept created by the Intergovernmental Panel on Climate Change [6], which explains vulnerability as a result of three different components: degree of exposure (DE), sensitivity (*S*) and adaptive capacity (AC). Degree of exposure is defined as the possibility of a system to change as a result of a potentially destabilizing situation, be it positive or negative; sensitivity refers to the extent to which a system responds to environmental variations and adaptive capacity has to do with the system's capability to fit or adjust itself to new conditions in its environment. Degree of exposure and sensitivity represent the potential impact (*I*) of the phenomenon (in other words, the expected amount of damage), and adaptive capacity is a measure of the extent to which potential impact can be avoided; therefore, vulnerability equals potential impact minus adaptive capacity, as expressed by the following equation:

$$V = f(I - \mathrm{CA}) = f(\mathrm{GE} + S - \mathrm{CA}). \tag{1}$$

Based on eqn (1), higher degrees of exposure and sensitivity result in higher vulnerability to drought, whereas higher adaptive capacity in comparison with the other two parameters represents lower vulnerability to drought.

Additionally, we must consider the different types of vulnerability: physical, economic, political, institutional and educational vulnerability, among others [7]. Accordingly, this study analyses economic, social, environmental and overall vulnerability to drought; what follows is a description of these concepts:

- *Economic vulnerability:* It is a consequence of an indirect relationship between income level and drought impact, which increases the risk of disaster due to insufficient funding for production, low wages, unemployment, underemployment and job instability, among others.
- *Social vulnerability:* This concept refers to the insecurity and defencelessness experienced by communities, families and individuals when their life conditions are threatened by drought. Poverty, marginalization and social backwardness, as well as difficult access to health care, education and leisure services inhibit the capacity to prevent, mitigate and promptly respond to a disaster situation due to hydrological deficit.

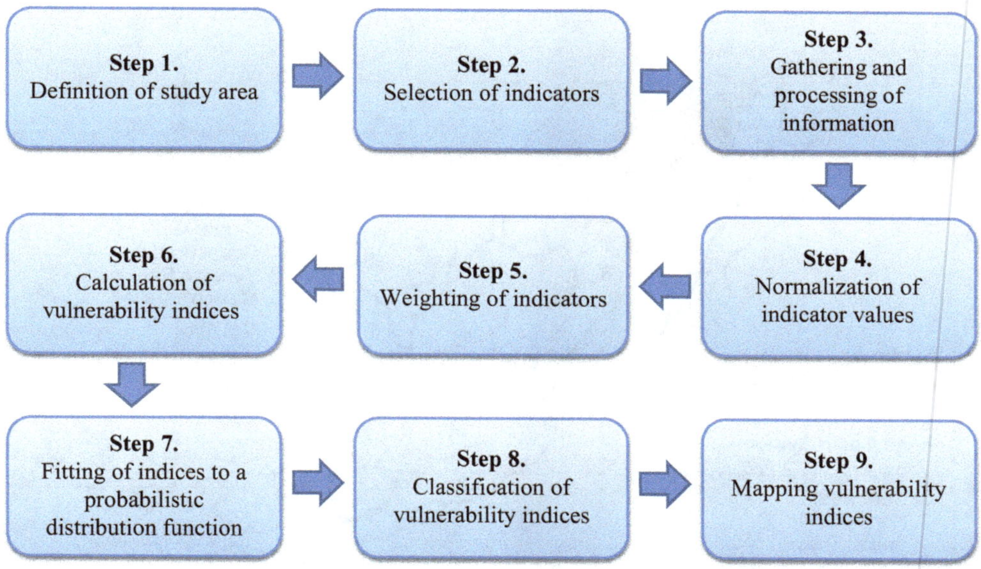

Figure 1: Steps of the method employed to calculate drought vulnerability indices in the Northwest River Basin System.

- *Environmental vulnerability:* This concept is related to an intrinsic susceptibility of the environment or its natural resources to be negatively affected by the lack of water; all living beings need certain environmental conditions for their development, but when nature can be deteriorated as a result of the destruction of environmental assets, ecosystems become highly vulnerable to hazards such as drought.
- *Overall vulnerability:* The three types of vulnerability previously described are integrated by this concept, in which economic, social and environmental factors converge.

Based on these concepts, we developed a nine-step method, which is shown in Fig. 1. The first step consists in selecting the study area; in steps 2–4, indicators are chosen, information is processed and indicators are normalized; in step 5, the specific weight of each indicator is determined; step 6 consists in calculating VIs by multiplying the normalized indicator by its specific weight; in step 7, indices are fitted to the beta distribution function; in step 8, indices are classified in five degrees of vulnerability and, finally, the vulnerability map for the river basin system in question is obtained in step 9.

The nine steps included in the proposed method are described in the following sections.

2.1 Definition of study area

The National Water Commission (CONAGUA, for its acronym in Spanish) has divided the Mexican territory into 13 river basin systems to better administrate the country's water (Fig. 2). This study selected the Northwest River Basin System, a notably arid region. This river basin system has a surface area of 197,586 km^2 (10% of the national territory), and it consists of 71 municipalities in the state of Sonora and 7 in the state of Chihuahua (78 total municipalities).

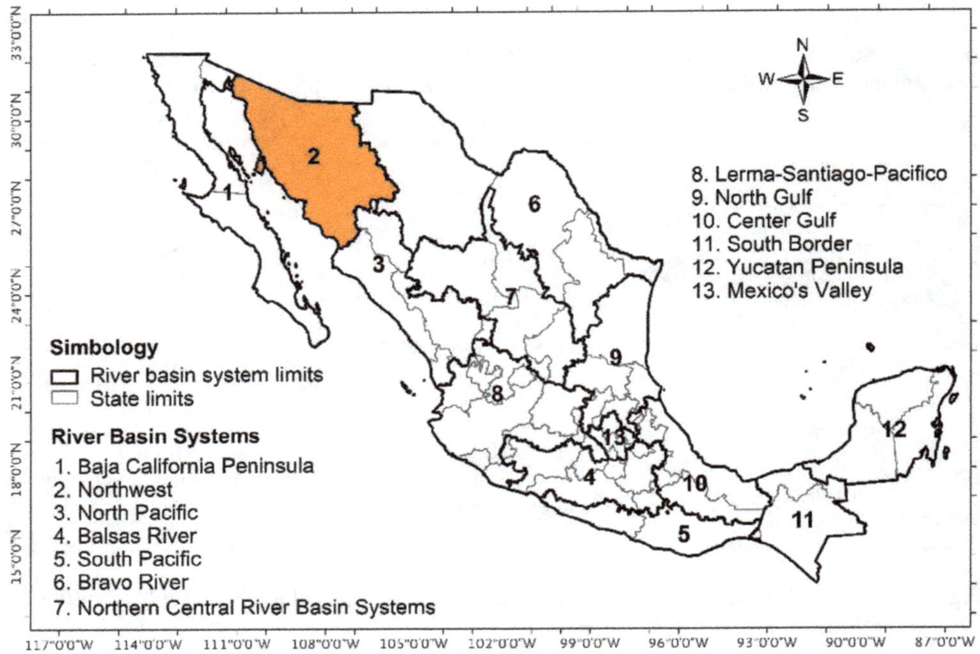

Figure 2: Boundaries of Mexican river basin systems and location of the Northwest River Basin System (grey).

2.2 Selection of indicators

Several criteria were taken into account to select the indicators, although the most determinant was direct relevance to the factor measured by the indicator and availability of municipal data (or how easily data could be calculated) in the form of figures, indices, rates or proportions obtainable from widely used national databases (INEGI, CONAPO, CONEVAL, etc.). In this way, 24 indicators were chosen and employed to build a matrix with the indicators grouped based on the three vulnerability components (exposure, sensitivity and adaptive capacity) and the four types of vulnerability (economic, social, environmental and overall), as shown in Table 1.

As shown in Table 1, overall vulnerability to drought is calculated using all the previously described indicators in the procedure explained below.

2.3 Gathering and processing of information

This process consists in gathering the necessary information to determine the value of each indicator considered for analysis based on the formulas and information sources presented in Table 2, which also lists the main information sources identified to calculate or obtain the indicators.

Table 1: Matrix showing the selected indicators, grouped by type and component of vulnerability to drought.

Type of vulnerability		Components of vulnerability		
		Degree of exposure (ED)	Sensitivity (S)	Adaptive capacity (AC)
		Indicators		
Overall	Economic	– Population density (persons/km^2) – Unemployed population in working age (%)	– Value of irrigation and rainfed agricultural production (thousands of pesos) – Value of livestock production (thousands of pesos)	– Surface rehabilitated for irrigation (ha) – Technified agriculture surface (ha) – Total length of rural roads (km)
	Social	– Population in poverty (%) – Population without health care insurance (%) – Illiterate population (%)	– Households without running water (%) – Households without drainage and flush toilet (%) – Households without electric energy (%) – Households with dirt floor (%)	– Annual per capita income (dollars) – Beneficiaries of Oportunidades programme (%) – Beneficiaries of Liconsa programme (%) – Average schooling years (dimensionless)
	Environmental	– Degree of watershed exploitation (dimensionless)* – Degree of aquifer exploitation (dimensionless)*	– Deforestation (% of critical forest area) – Surface affected by forest fires (ha)	– Re-forested surface (ha) – Protected natural areas (% of total surface)

*Due to their relevance in the analysis of drought, these two indicators are included in the three types of vulnerability: economic, social and environmental.

Table 2: Formulas and sources of information to calculate indicators.

Indicator	Formula and units of measurement	Information source
Population density (PD)	$PD = \frac{\text{No. of people}}{\text{Area (km}^2)}$; people/km^2	National Institute of Statistic and Geography (INEGI)
Unemployed population in working age (UPWA)	$UPWA = \frac{\text{Unemployed population}}{\text{Total population}}$; %	National Institute of Statistic and Geography (INEGI)
Population without health care insurance (PWHC)	$PWHC = \frac{\text{Pop. without health care insurance}}{\text{Total population}}$; %	National Evaluation Commission (CONEVAL)
Population in poverty (PIP)	$PIP = \frac{\text{Population in poverty}}{\text{Total population}}$; %	National Evaluation Commission (CONEVAL)
Illiterate population (IP)	$IP = \frac{\text{Illiterate population}}{\text{100 people of the same age}}$; %	National Population Commission (CONAPO)
Households without running water (HWRW)	$HWRH = \frac{\text{Households without running water}}{\text{Total households}}$; %	National Water Commission (CONAGUA)
Households without drainage and flush toilet (HWDT)	$HWDT = \frac{\text{Households without drainage}}{\text{Total households}}$; %	National Water Commission (CONAGUA)
Households without electric power (HWEP)	$HWEP = \frac{\text{Households without electric power}}{\text{Total households}}$; %	National Population Commission (CONAPO)
Households with dirt floor (HWDF)	$HWDF = \frac{\text{Households with dirt floor}}{\text{Total households}}$; %	National Population Commission (CONAPO)

(*Continued*)

Table 2: (*Continued*)

Annual per capita income (APCI)	$\text{APCI} = \frac{\text{Annual per capita income (dollars)}}{\text{Total population}}$; dollars / inhabitant	United Nations Development Programme (UNDP)
Mean years of schooling (MYS)	Years a person aged 24 years or older has spent in formal education	United Nations Development Programme (UNDP)
Beneficiaries of Oportunidades programme (BOP)	$\text{BOP} = \frac{\text{Beneficiaries of Oportunidades program}}{\text{Total population}}$; %	National Institute of Statistic and Geography (INEGI)
Beneficiaries of Liconsa programme (BLP)	$\text{BLP} = \frac{\text{Beneficiaries of Liconsa program}}{\text{Total population}}$; %	National Institute of Statistic and Geography (INEGI)
Total length of rural roads (TLRR)	TLRR; amount, km	National Institute of Statistic and Geography (INEGI)
Value of irrigation and rainfed agricultural production (VAP)	VAP, amount, thousands of pesos	Ministry of Agriculture and Livestock (SAGARPA)
Value of livestock production (VLP)	VLP, amount, thousands of pesos	Ministry of Agriculture and Livestock (SAGARPA)
Rehabilitated irrigation surface (RIS)	RIS, amount, ha	National Institute of Statistic and Geography (INEGI)
Technified agriculture surface (TAS)	TAS, amount, ha	Ministry of Agriculture and Livestock (SAGARPA)
Degree of watershed exploitation (DWE)	$\text{DWE} = \frac{\text{Annual volume of surface water extraction } (\text{hm}^3)}{\text{Average annual volume of runoff } (\text{hm}^3)}$; dimensionless	Official Journal of the Federation (DOF)

Indicator	Formula and units of measurement	Information source
Degree of aquifer exploitation (DAE)	$\text{GEA} = \frac{\text{Extraction volume (hm}^3\text{)}}{\text{Average annual recharge (hm}^3\text{)}}$; dimensionless	Official Journal of the Federation (DOF)
Deforestation (DF)	$\text{DF} = \frac{\text{Deforested area (km}^2\text{)}}{\text{Total area (km}^2\text{)}}$; %	National Forest Commission (CONAFOR)
Surface affected by forest fires (SAFF)	$\text{DF} = \frac{\text{Area affected by forest fires (km}^2\text{)}}{\text{Total area (km}^2\text{)}}$; %	National Institute of Statistic and Geography (INEGI)
Re-forested surface (RFS)	RFS, amount, ha	National Institute of Statistic and Geography (INEGI)
Protected natural areas (PNA)	$\text{ANP} = \frac{\text{Protected area (km}^2\text{)}}{\text{Total area (km}^2\text{)}}$; %	National Commission for Protected Natural Areas (CONANP)

2.4 Normalization of indicator values

Given that indicators were expressed in different units of measurement, their values were normalized so that all of them were expressed as dimensionless values between 0 and 1. To carry out this procedure, we first identified the indicator's functional relationship with vulnerability to drought. There are two possible types of functional relationships: according to the indicator value, vulnerability may increase or decrease. In the first case, the relationship is direct and is identified with symbol ↑; in the second case, the relationship is inverse and is indicated with symbol ↓. The indicators grouped under ED and *S* have a direct relationship (↑) with vulnerability, whereas the relationship is inverse for indicators grouped under AC (↓). Once we identified the positive or negative impact of the indicators on vulnerability, indicator values were normalized using the following equations:

The expression (↑) was used for indicators with a direct functional relationship:

$$X_i = \frac{x_i - x_{\min}}{x_{\max} - x_{\min}}. \tag{2}$$

In the case of indicators with an inverse functional relationship (↓), the expression was

$$X_i = \frac{x_{\max} - x_i}{x_{\max} - x_{\min}}. \tag{3}$$

where, in both equations, X_i is the normalized value of variable x_i, and $x_{\min}$ and $x_{\max}$ are the minimum and maximum values of dataset x_i, respectively.

2.5 Weighting of indicators

After obtaining the normalized indicator values, we calculated their weights by the method proposed by Iyengar and Sudarshan [8], which uses the following expression:

$$W_i = \frac{1}{(\sigma_i)\left(\sum_{i=1}^{n} 1/\sigma_i\right)}. \tag{4}$$

where W_i is the normalized indicator weight *i*; σ_i is the standard deviation of the set of values for indicator *i*; and *n* is the number of selected indicators.

This weighting method prevents abnormally large variations in one or many indicators to overshadow the contribution of the rest of the indicators.

2.6 Calculation of VIs

In order to obtain overall vulnerability indices (OVIs), the economic, social and environmental vulnerability indices (EVI, SVI and EnVI, respectively) were calculated using the following formula:

$$\text{EVI, SVI, EnVI} = \sum_{i=1}^{n} X_i W_i. \tag{5}$$

where X_i is the normalized value of indicator i; W_i is the weight of normalized indicator i; and n is the number of indicators associated with each type of vulnerability.

After that, OVIs were calculated assuming a weight W_i of 1/3 for each one of its components:

$$\mathrm{OVI} = \frac{\mathrm{EVI}}{3} + \frac{\mathrm{SVI}}{3} + \frac{\mathrm{EnVI}}{3}. \tag{6}$$

2.7 Fitting of indices to a probabilistic distribution function

The obtained VIs were classified into categories by fitting the indices to a beta probabilistic distribution following recommendations by Iyengar and Sudarshan [8], who assert the adequacy of this type of distribution for classification purposes, since it is generally skewed and it takes values within the interval (0, 1). The beta function probability density is given by

$$f(z) = \frac{z^{a-1}(1-z)^{b-1}}{\int_0^1 z^{a-1}(1-z)^{b-1}\,dz} .0 < z < 1;\ \text{and } a, b > 0 \tag{7}$$

where mean μ and variance σ^2 of variable z are determined by

$$\mu = \frac{a}{a+b}. \tag{8}$$

$$\sigma^2 = \frac{ab}{(a+b)^2(a+b+1)}. \tag{9}$$

The following equations are obtained when both expressions are algebraically simplified to find parameters a and b:

$$a = \alpha = \mu\left\{\frac{\mu(1-\mu)}{\sigma^2} - 1\right\}, \tag{10}$$

$$b = \beta = a\left\{\frac{1-\mu}{\mu}\right\}. \tag{11}$$

Using the previous two equations, parameters a (α) and b (β) were determined using an Excel® spreadsheet, which also allowed us to obtain the probability of occurrence of the VIs fitted to the beta probabilistic distribution.

2.8 Classification of VIs

VI_i were classified according to their probability of occurrence, first by multiplying the obtained probability values so as to express them as percentages, and after that by determining five equally sized percentile categories (20% each), which defined five degrees of vulnerability, as shown in Table 3.

Table 3: Vulnerability to drought classification categories.

Degree of vulnerability	Value of percentile
Very low	$0 < VI_i \leq 20$
Low	$20 < VI_i \leq 40$
Moderate	$40 < VI_i \leq 60$
High	$60 < VI_i \leq 80$
Very high	$80 < VI_i \leq 100$

2.9 Mapping VIs

VIs were mapped by ArcGis® software using National Geostatistic Framework [9] data as a baseline, which allows for the accurate reference of statistical information to its corresponding geographical locations, which in this case are municipalities.

3 RESULTS

In this section, results of the Northwest River Basin System are presented as an application example of the method described earlier. Figure 3 shows the economic, social, environmental and overall vulnerability maps. As can be appreciated in the maps, the most economically vulnerable municipalities (where population density is highest, as well as economic activity as a consequence) are located in the state of Sonora; among these is Hermosillo, the state capital, which is characterized by its intense industrial, agricultural and stockbreeding activities. The most socially vulnerable municipalities (those with highest poverty, marginalization and social backwardness indices) are located in the state of Chihuahua, along its border with Sonora, in the Sierra Madre Occidental: a region of steep topography that complicates access to basic sanitary services, drinking water, electric power and so on.

The maps also show that the most environmentally vulnerable municipalities, where ecosystems are most fragile in case of drought, are mainly located in the state of Sonora, along the cost of the Gulf of California, where aquifers have been overexploited as a result of excessive use of water for irrigation purposes, which has caused the intrusion of saltwater into phreatic zones. Finally, when the three types of vulnerability are taken together, the municipalities of Nogales and Hermosillo (Sonora) and Madera (Chihuahua) are found to be the most vulnerable; in overall terms, these are the municipalities where urgent actions are required to mitigate potential negative effects of drought in the future.

4 CONCLUSIONS

Despite the fact that drought cannot be prevented, there are ways to attenuate its impact and reduce the losses of affected populations. The best strategy to cope with drought should start by evaluating vulnerability, which will serve different purposes: to prioritize the implementation of preventive and mitigation measures, to understand where and how to improve adaptive capacities, to identify where and how exposure and sensitivity should be decreased, to optimize the distribution of resources and, more generally, to support better water management in the different river basin systems. The method described in this article, which was applied to the Northwest River Basin System, Mexico, can be applied to each of the 26 river basin systems in the country, and it could as well be used at the state or national level. The main advantage of the method is its approach as an objective analytic procedure that allows for the

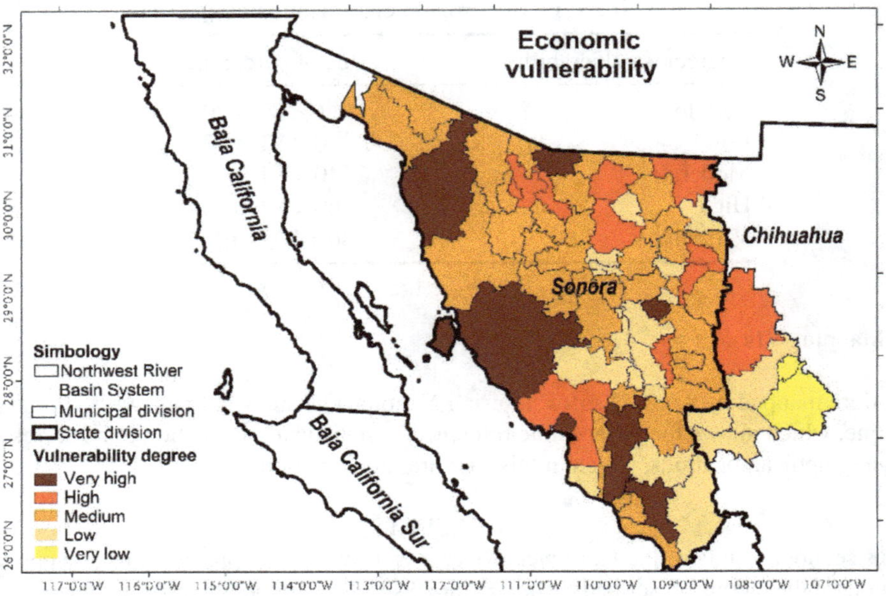

(a)

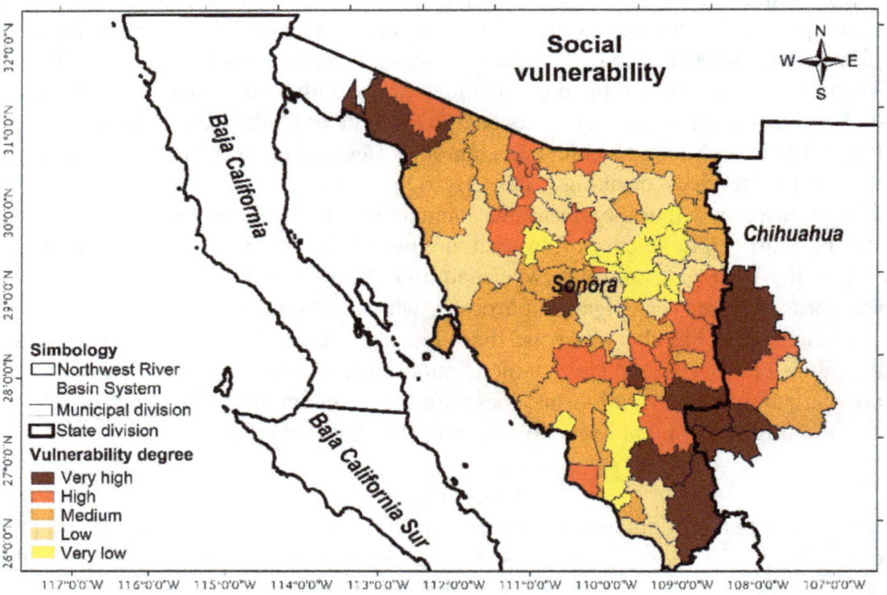

(b)

Figure 3: Economic (a), social (b), environmental (c) and overall (d) maps of vulnerability to drought in the Northwest River Basin System.

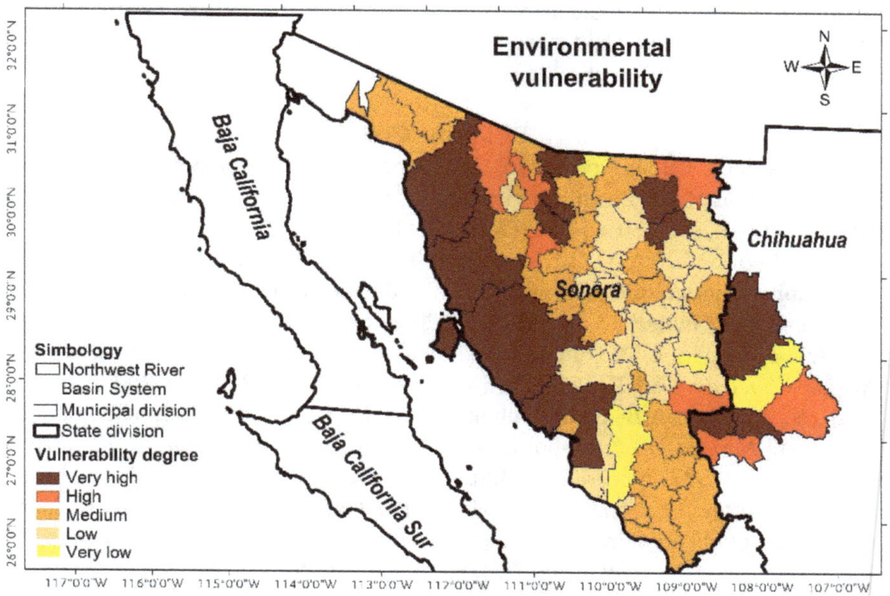

(c)

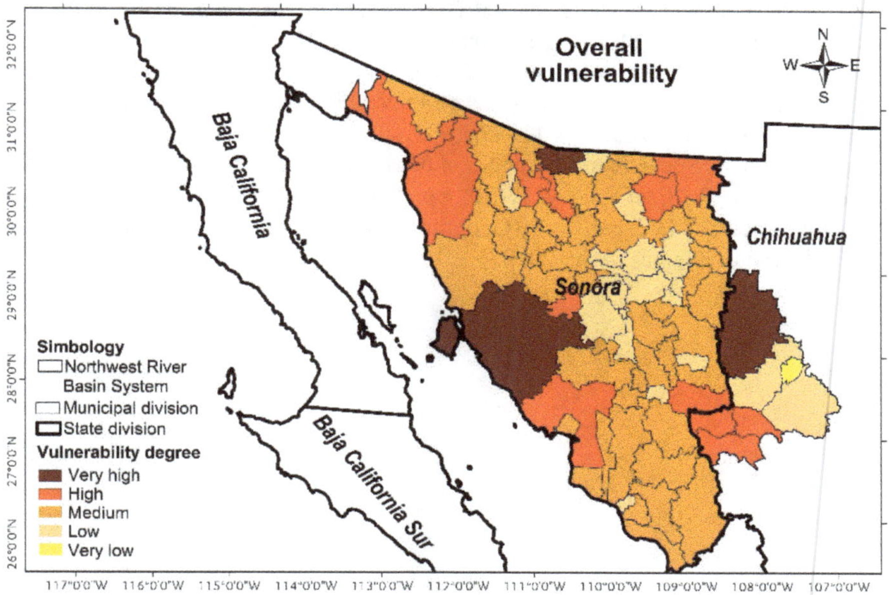

(d)

Figure 3: (*Continued*)

identification of highly vulnerable areas from economic, social and environmental points of view, which in turn enables a better management of resources and efforts to reduce vulnerability to drought in the different regions of the country.

REFERENCES

[1] Kim, T., Valdés, J.B. & Aparicio, J., Frequency and spatial characteristics of droughts in the Conchos River Basin, Mexico. *Water International*, **27**(3), pp. 420–430, 2002. DOI: 10.1061/(ASCE)1084-0699(2000)5:2(145).

[2] Vásquez-León, M., West, C.T. & Finan, T.J., A comparative assessment of climate vulnerability: agriculture and ranching on both sides of the US-Mexico border. *Global Environmental Change*, **13**, pp. 159–173, 2003. DOI: 10.1016/S0959-3780(03)00034-7.

[3] Velasco, I., Ochoa, L. & Gutiérrez, C., Sequía, un problema de perspectiva y gestión. *Región y Sociedad*, **XVII**(34), pp. 35–71, 2005. DOI: 10.22198/rys.2005.34.a615.

[4] Feng, S., Krueger, A.B. & Oppenheimer, M., Linkages among climate change, crop yields and Mexico-US cross-border migration. *Proceedings of the National Academy of Sciences*, **107**(32), pp. 14257–14262, 2010. DOI: 10.1126/science.1114722.

[5] Arreguín-Cortés, F., López-Pérez, M., Ortega-Gaucin, D. & Ibáñez-Hernández, O., La política pública contra la sequía en México: avances, necesidades y perspectivas. *Tecnología y Ciencias del Agua*, **7**(5), pp. 63–76, 2016.

[6] Panel Intergubernamental de Expertos sobre el Cambio Climático (IPCC), *Cambio Climático 2007, Informe de Síntesis*, Contribución de los Grupos de Trabajo I, II y III al Cuarto Informe de Evaluación del Panel Intergubernamental de Expertos sobre el Cambio Climático, WMO-UNEP: Ginebra, Suiza, 2007.

[7] Wilches-Chaux, G., La vulnerabilidad global. *Los desastres no son naturales*, comp. A. Maskrey, Red de Estudios Sociales en Prevención de Desastres en América Latina: Puerto Limón, Costa Rica, pp. 11–47, 1993.

[8] Iyengar, N.S. & Sudarshan, P., A method of classifying regions from multivariate data. *Economic and Political Weekly*, **17**(51), pp. 2048–2052, 1982.

[9] Instituto Nacional de Estadística y Geografía (INEGI), *Marco Geoestadístico Nacional*, INEGI: Aguascalientes, Ags., 2010.

LOCAL KNOWLEDGE-BASED WATER MANAGEMENT AND IRRIGATION IN THE WESTERN PAMIRS

ANDREI DÖRRE
Centre for Development Studies, Freie Universitaet Berlin, Germany

ABSTRACT
For many people living in Central Asia, the economic decline and societal disintegration of the post-Soviet space meant a dramatic socio-economic marginalization and pauperization. Against this background and accompanied by the liquidation of numerous state enterprises, many citizens lost their jobs and secure wages. Simultaneously, the state radically cut the provision of social services. The increased uncertainty led to a rising dependency on access to, as well as utilization and the decentralized management of, locally available resources and assets for one's own survival, especially in the countryside. Because of the immediate relationship between these challenges and the comparatively scarce availability of affordable food products, the collaborative management of scarce water sources and common operation of labour-intensive irrigation infrastructure for local agricultural food production are especially significant in arid high mountain regions such as the remote Western Pamirs in Tajikistan. By means of a micro-level study conducted in the Wakhan Valley, it will be shown how self-organized management practices and pragmatic technical solutions fed by local knowledge contribute to addressing the uneven and irregular spatiotemporal water supply for irrigation. The findings reveal that local-specific water management and irrigation arrangements prove to be not only an essential basis for local food production but also a central instrument for balancing interests within a community and a means of social organization. The study provides an example of where human and social assets such as local environmental, cultural and social knowledge; common decision-making; and coordinated action become key factors for equitable utilization of locally available natural resources on the one hand. On the other hand, the study shows how local knowledge-informed collaborative resource management contributes to community cohesion and individual survival in a society which is struggling with manifold societal challenges.
Keywords: Agriculture, Community-based natural resource management, Food supply, High Asia, Irrigation, Local knowledge, Pamirs, Tajikistan, Water.

1 MOBILIZING HUMAN AND SOCIAL ASSETS TO FACE SOCIAL CHALLENGES

In the course of the dissolution of the Soviet Union in 1991, societal disruptions accompanied by socio-economic upheavals, such as the liquidation of numerous state enterprises, have caused aggravated living conditions in the agrarian countries of Central Asia, which were formerly among the poorest of the Soviet Republics receiving substantial economic support from the political centre. Many people lost their jobs and secure incomes in the course of the restructuring of the command economy, including the agricultural farm sector consisting of state (*sovkhozy*) and collective (*kolkhozy*) enterprises. Simultaneously, the new states radically cut the provision of social services [1, 2]. In the countryside, manifold additional challenges, such as limited market integration, and monetary incomes lower than the national average exacerbated the already existing difficulties to make a living. The newly gained sovereignty was, therefore, perceived by many people as a 'burden of imposed independence' [3].

Against the background of the abolition of the Soviet economic system and the end of command structures strongly intervening in local affairs, on the one hand, and new state institutions often incapable and unaccountable in terms of providing effective administrative services, reliable social support and efficient resource management performances, on

DOI: 10.2495/EI-V1-N3-254-266

the other hand, the increased uncertainty raised the people's need to mobilize all resources and assets at hand to ensure their own survival. Informed by DFID's sustainable livelihoods approach [4] and Bourdieu [5], such assets can be categorized into human (e.g. knowledge, skills and physical capability), social (e.g. social networks, relationships of trust, memberships and affiliations) and material (e.g. natural resources such as land, water and forests, and infrastructure like irrigation canals and agricultural equipment), as well as financial (e.g. savings, income, and access to loans) forms of capital.

This article argues that both human and social assets purposefully mobilized and deployed at the local level are fundamental prerequisites for access to common natural resources, as well as their management and utilization. Popularized by the anthropologist C. Geertz [6], the concept of local knowledge (LK) being a specific kind of human capital can be seen as an inventory of spatial-historically bounded knowledge whose features are shaped by the socio-ecological conditions of the respective locality and the daily practices of its inhabitants. According to Mistry [7], LK is characterized, among other things, as follows: (i) LK cannot be transferred easily to another location because it is context specific; (ii) LK is generally transferred orally, as well as through demonstration, observation and imitation; (iii) Instead of being static, LK changes over time through adoption to novel situations, experimentation and learning; (iv) LK is widely shared, and this enables long-term application, and intergenerational transfer of LK; and finally, (v) LK emerges from people's daily performances and interactions within different life spheres that intersect one other, and, therefore, has to be seen as a holistic body of knowledge.

While environmental, cultural and social LK represents the most substantial part of customized approaches for the management and utilization of natural resources, as well as the operation of the related infrastructure, common decision-making is the process of the creation of the institutional framework for such assemblages of collaborative action. Accordingly, these local-specific arrangements go beyond immediate resource-related issues; they also have the potential of being effective means of social organization, and equitable instruments of balancing interests within the respective community. Against this background, LK became increasingly popular in the development discourse in recent years after frequent failures of blueprint top-down development measures. LK-informed development strategies that take into consideration local interests, experiences and practices were expected to produce more efficient and locality-specific responses to environmental and societal challenges on the ground and simultaneously empower the respective community [8, 9].

In the following paragraphs, the article centres its attention on the issue of LK-based collaborative water management and irrigation approaches in rural Central Asia using an example from the arid Western Pamirs in Tajikistan. LK-based management and operation of scarce water resources and irrigation infrastructure for agricultural food production are especially significant in this high mountain region due to the tense food supply situation caused by multiple societal and environmental factors. After outlining the food supply–related context of the research area, the paper introduces the case study of Shirgin Village, located in the Wakhan Valley at the upper Panj River right at the border with Afghanistan. It will be shown how, against the background of the absence of a superior body governing local water issues, sophisticated LK-informed practices and pragmatic organizational solutions contribute to addressing the problem of uneven and irregular spatiotemporal water supply for irrigation purposes. After presenting the natural and infrastructural components of the water supply and irrigation system of Shirgin, water management and irrigation-related LK, as well as decision-making bodies and processes, will be introduced.

The study has two goals. First, it tests the pragmatic subject-centred analytical perspective on LK for its explanatory power about local-specific resource management and utilization approaches. The second goal is to deliver an example where locally mobilized and deployed human and social assets become key factors for equitable utilization of locally available natural resources and related technical infrastructure; for community cohesion and individual survival in a society which is struggling with manifold societal challenges.

2 FOOD SUPPLY–RELATED CONTEXT OF THE WESTERN PAMIRS

The post-Soviet economic decline and societal disintegration described above were particularly prevalent in Tajikistan, where, in the course of a cruel civil war (1992–1997) accompanied by tremendous human and material losses, the socio-economic vulnerability of many people increased remarkably [10, 11]. In the course of the war, reliable food supply, as one of the essential preconditions for bare survival, came under threat in remote rural areas such as the arid high mountains of the Western Pamirs located in Tajikistan's Gorno-Badakhshan Autonomous Province (GBAP). The collapse of the national supply network, bad harvests and a blockade by government forces in 1992–1993 led to a severe famine that struck the whole population. For years, nearly all inhabitants relied on humanitarian assistance offered by the Aga Khan Foundation's 'Pamir Relief and Development Programme', which delivered staple food products like flour, oil and sugar from the organization's supply centre located in Osh, Kyrgyzstan [12–14, 19].

GBAP's food supply system structurally rests on two pillars. Local agricultural production, as the first pillar, is limited by the terrain and altitude of the mountainous environment. This economic activity is predominantly pursued at the household level by local farmers. Because of the environmental constraints of the cold and arid high mountain desert, the mainly Kyrgyz population of the eastern plateaux-like part of GBAP corresponding to the *nokhiya* (district) of Murghab is mainly practicing pasture-based animal husbandry. The Western Pamirs administratively including the *nokhiya* of Darvoz, Vanj, Rushon, Shugnon, Roshtqal'a and Ishkoshim are predominantly inhabited by Pamiri farmers belonging to the Isma'ili branch of Shi'a Islam (Fig. 1). They are pursuing an approach called mixed or combined mountain agriculture. This strategy combines the use of small land plots for food and fodder cultivation located mostly on plain debris cones and alluvial fans at the bottom of deep narrow valleys close to the farmers' residential centres, with animal husbandry seasonally utilising more or less distant mountain pastures. Due to annual precipitation lower than the agro-ecological threshold of 250 mm, this kind of land cultivation depends highly on irrigation [15–19]. Land plots suitable for cultivation in terms of flatness, soil properties, water availability and other characteristics are scarce due to the challenging conditions of the arid and steep mountainous environment. According to the Statistical Agency of Tajikistan [20], only 12,219 ha arable land existed in GBAP in 2013 for approximately 212,000 inhabitants. This corresponds to less than 0.06 ha per capita or just half of the national average. Local agricultural production covers GBAP's food demand only partially. Until today, the provision of externally produced staple foods and groceries is essential [12, 13, 21–23]. These imports represent the second pillar of GBAP's food supply system.

The challenging food supply situation in GBAP can be seen as the product of different factors. Natural hazards like rockfalls and other gravitational mass movements pose a continuous threat to the few existing roads connecting the region with the agricultural centres and economic growth poles of the country. Clouds and quickly shifting weather conditions impede the maintenance of regular flight operations between the country's capital Dushanbe and GBAP's main airstrip in Khorog. These conditions lead to a fragile and cost-intensive accessibility to the region and its comparatively weak integration into the national market.

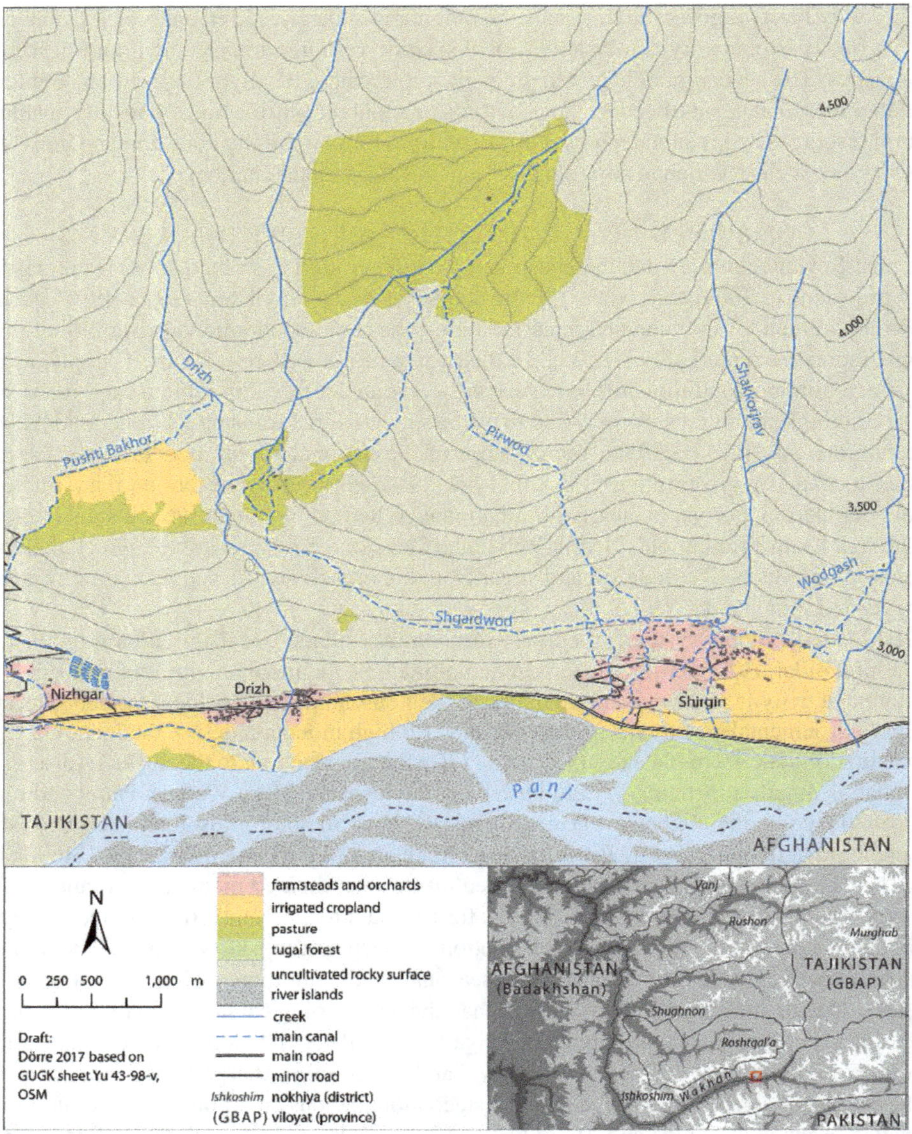

Figure 1: Spatial extension of the water supply system of Shirgin, and location of the research area in GBAP.

The high transportation costs contribute to prices for food, fertilizer, fuel and agricultural equipment, which are higher than in other regions of the country. For more than 51% of GBAP's population (which is the highest regional poverty rate in Tajikistan) monetary incomes lower than the national poverty line for monthly consumption (162 Tajik somoni (TS) per capita; ca. 20 US$) hamper access to these imported products [24]. Even basic products like flour are hard to afford when the price per kilogram is 3 TS or higher [25]. Finally, the de facto closed border with Afghanistan, international trade directed mainly to the capital and limited capacities of Tem reloading facility near Khorog for Chinese products imported from the Xinjiang Uyghur Autonomous Region of China represent serious obstacles for GBAP's exchange relations with, and its integration into, international food markets. Against this background, local agricultural food production gained importance after the dissolution of the Soviet Union and the following civil war.

3 WATER MANAGEMENT AND IRRIGATION IN SHIRGIN

The inhabitants of the arid valleys of the Western Pamirs have long-lasting experiences in cultivation based on proven water management skills and irrigation practices. Historically, thanks to local-specific resource-related knowledge, rules and practices, a remarkable share of the food products consumed was produced locally. However, the majority of the people lived on the breadline, and were highly vulnerable to crop shortfalls and other shocks [26]. The valley of the upper Panj River is a demonstrative example. While the land parcels were irrigated by water coming from the main river and its tributaries, the related infrastructure, such as channels and locks, was constructed and maintained by the communities [15, 17, 19, 27, 28]. After 1991, the knowledge and experiences accumulated over time proved to be highly valuable for adaptation to the new era with all its challenges.

The village of Shirgin (37°00'31"N 72°31'19"E) has an altitude of 2,800 m up to 3,000 m and is located in the Wakhan Valley on the right bank of the upper Panj River (Fig. 2). In administrative terms, Shirgin belongs to the *jamoat* (municipality) of Vrang, being part of Ishkoshim *Nokhiya*. With approximately 850 inhabitants living in 100-something households, Shirgin is considered to be of medium size compared to other settlements in the district [29].

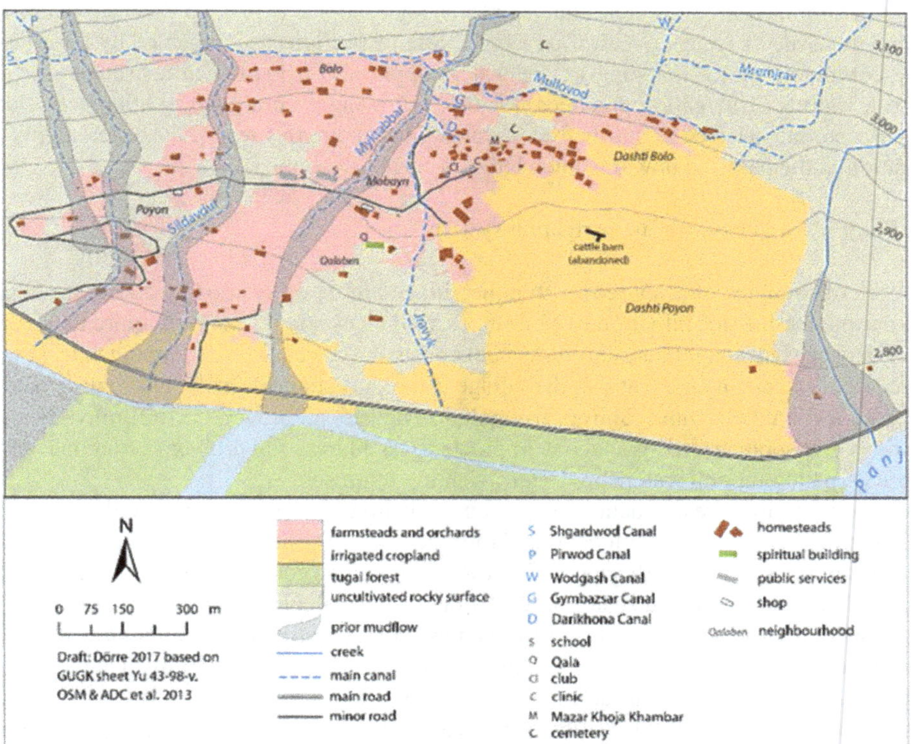

Figure 2: Canals and land use within the village limits of Shirgin.

Several reasons make Shirgin an appropriate case for the study of local-specific water management and irrigation arrangements.

The village is exposed to a couple of economic challenges concerning the food supply. The prices of staple foods offered in the few shops of the village are higher than in Khorog or Ishkashim. In August 2016 for instance, a 50 kg sack of flour was 170 TS, and pulses, an important protein source in the local cuisine, cost 8–10 TS/kg. At the same time, the few existing jobs, limited to the school, a kindergarten and a tiny clinic, are structurally underpaid. A teacher's monthly salary, for instance, is around only 450 TS. Additionally, the risk of mudflows poses a threat to the reliable provision of food products from Khorog, Ishkoshim and Vrang. Therefore, subsistence-oriented cultivation supplemented by animal husbandry is crucial for Shirgin's population.

Animal husbandry relies on seasonal pastures adjacent to the village, as well as beyond the Khargush Pass located 90 km to the East. Cultivation instead depends on irrigation due to the arid conditions, which are similar to the climate of the district centre Ishkoshim (approximate precipitation per year 100 mm with a peak in spring). At the same time, cultivation is restricted by the small area of suitable land plots (Fig. 2). With 111 ha of irrigated cropland and orchards covering 25 ha in total, only a bit more than one ha of arable land and a less than 0.25 ha of orchards are at each household's disposal on average [30].

The data for this analysis were gathered by the author during field stays in the spring of 2015 and the summer of 2016, as well as by a research assistant in the summer of 2015. Information on current environmental conditions, water management and irrigation arrangements was collected from knowledgeable persons involved in these issues and representatives of the local administration. Representatives of local households explained their daily activities around the issue of securing their food supply. Site inspections accompanied by local guides were combined with a mapping of the water supply and irrigation infrastructure, as well as observations of irrigation practices. Finally, following an oral history approach, elderly people were interviewed about their experiences and memories regarding water supply and irrigation in the past and how this knowledge is currently applied.

3.1 Shirgin's water supply and irrigation system

The lack of suitable water sources within the village limits, water scarcity during the vegetation period, the hillside situation of Shirgin, as well as its spatial extent, require an artificial water delivery and spatial distribution system consisting of canal heads tapping natural water sources, canals leading the water to the village, as well as channels and locks to distribute the water as needed. Over time, Shirgin's population developed proven LK-informed technical solutions and sophisticated management practices to address the problem of an uneven and irregular spatiotemporal water supply for irrigation purposes.

According to local respondents, the present irrigation infrastructure is based on historical constructions executed by the local population in the 19th century under the guidance of an Isma'ili religious leader, with support from Russian border guards in the early 20th century, and, finally, under Soviet rule. In socialist times, the canal system was used mainly for the irrigation of lands belonging to the collective farm 'Lenin', as well as for privately used plots near to the farmsteads [12, 31]. In that time, the infrastructure was maintained at the expense of the *kolkhoz*, which was shifted into a *sovkhoz* in 1977. The state farm was transferred in 1996 back into a *kolkhoz*, and, finally, to a so-called *khojagii dekhqoni* (farmers' association) named 'Mullo-i Bek', assembling most of the farmsteads of the three villages of Shirgin,

Nizhgar and Drizh (Fig. 1). Since then, the farmers have had to organize the supply and distribution of irrigation water by themselves, as well as the maintenance of the irrigation infrastructure.

3.1.1 Water tapping and supply-related infrastructure

Four main canals with different characteristics pipeline irrigation water from diverse natural sources to the agricultural lands used by Shirgin's inhabitants. These are from West to East: *Pushti Bakhor*, *Shgardwod*, *Pirwod*, and *Wodgash* (Figs. 1 and 2).

The *Pirwod* ('canal of the Pir') is regarded to be the oldest canal of Shirgin, stemming from the 19th century. According to local legend, the Isma'li religious leader *Pir Saidmamadsho* living in the *Qala* (castle) of Shirgin initiated and guided the construction of the approximately 6 km long structure. The canal head taps a snow- and glacier-fed creek at an elevation of approximately 4,050 m, and channels the water via a flagstone-lined, approximately 40 cm wide and deep construction using the natural gradient to a steep bluff located just north of the village. Here, the canal water creates an artificial waterfall. After the waterfall, another canal structure pipelines the water into Shirgin's distribution system (Fig. 3). Because of the altitude of the source, *Pirwod* works only temporarily from June to September. The rest of the year the canal head is blocked by ice. If the weather turns cold, windy and cloudy, the canal head can even freeze in summer. Such situations quickly lead to less irrigation water in the canal [32, 33].

Another canal initiated by a *Pir* is the shorter *Wodgash* ('canal's mouth'). This canal stems from a snow-fed natural pool located approximately 2 km to the Northeast of the village centre at an altitude of 3,350 m. From late April to June, the *Wodgash*, which was dug into the rocky surface, provides water to a couple of homesteads and land plots located in the Eastern

Figure 3: The *Pirwod* Waterfall (upper right) and *Shgardwod* canal (coming from the West, here from the left side) (Photograph: Dörre, August 2016).

Figure 4: Downslope vegetation lines mark the course of the *Wodgash* (*left*) and *Mremjrav* (*right*) canals (Photograph: Dörre, September 2016).

part of the village. In times of high discharge, the nearby subordinate *Mremjrav* takes over the water surplus and pipelines it to the consumers (Fig. 4). This solution is important for two reasons. First, the farmers can receive more water at the beginning of the cultivation period. Second, it reduces the risk of mudflows initiated by the steep *Wodgash*, by taking water out of this channel. When all of the snow feeding the pool is melted, the water level of the pool sinks and it cannot be tapped any more. In June, the *Wodgash* dries out. If the snow feeding the pool does not melt due to cold, cloudy or windy weather, the *Wodgash* can fell dry even in April or May [34–36].

At the time when the Russian Empire occupied the Pamirs at the end of the 19th century, the *Pirwod* and *Wodgash* canals could not provide enough water for Shirgin's growing population. In the course of the following search for new water sources, a tributary of the Drizh Creek located approximately 5 kilometres to the West was identified at an elevation of 3,400 m. The challenging task of digging the canal into the rocky surface was completed with support from the Russian border guards stationed at the nearby Langar Kikhn Post. In the early 20th century, the *Shgardwod* ('straight canal') project characterized by several masoned and long concreted sections was finished (Fig. 3). The canal works from April until October. The *Shgardwod* and *Pirwod* canals are regarded as Shirgin's main water lines [32, 33, 36, 37].

Finally, the fields of *Pushti Bakhor* ('delayed spring') located five kilometres outside of Shirgin to the West at 3,300 m were also developed in pre-Soviet times for the cultivation of fodder crops. In the course of this land acquisition project, a canal with the same name fed from the Drizh Creek was built, as well as a couple of sheep stables [32, 38, 39]. Today, local herders use the area as a summer pasture for sheep and goats, as well as for cattle [40].

3.1.2 Water distribution infrastructure

Within the limits of Shirgin, six main canals run through the neighbourhoods to distribute the irrigation water. They are locked, and can be opened only according to commonly defined schedules. These canals are, from West to East, *Sildavdur*, *Myktabbar*, *Jravyk*, *Darikhona*, *Gymbazsar* and *Mullovod* (Fig. 2). Hundreds of blocked side canals branch off towards clusters of cultivated fields, garden plots and orchards. The final stage of the irrigation system is tiny channels in the fields, gardens and orchards. These runnel- and furrow-shaped water lines are locked with small barrages made out of stones, iron sheets, branches, wool, textiles or animal hides, which can be removed quickly and easily on demand.

3.2 Water management and irrigation practices

To gain a better understanding, the study of local-specific water management and irrigation arrangements can be differentiated by looking specifically at the three subjects of decision making, the involved actors and the practices executed by them.

Against the background that a superior institution governing local water issues does not exist, Shirgin's central decision-making body is the village assembly that gathers every year during the celebration of *Shogun* (also known as the Persian New Year, *Nawruz*), the vernal equinox on March 21. At this occasion, the assembly elects the *Mirāb* ('water master'). He is responsible for securing the water supply through supervising the three main canals: *Pirwod*, *Shgardwod*, and *Wodgash*. In case he identifies damages or other problems with the infrastructure during his daily inspections, he announces an *Ashar* ('collective work'). As a century-long experience has shown, most damages occur in the winter, and have to be repaired before the start of the irrigation period. The maintenance regime for the *Shgardwod* canal is organized very pragmatically. The community agrees that each household is responsible for a specific section of the canal, and has to conduct small cleaning, maintenance and repair jobs to secure the functionality of the canal. Bigger damages require calling an *Ashar* [34, 36], 41]. Another duty of the *Mirāb* is the creation of irrigation schedules adapted to different stages of the cultivation period and the amount of available water. These schedules also have to be accepted by the leaders of the community and the village assembly. They will be briefly delineated below.

For several years, the same experienced man has occupied the position of the water master [33, 36]. This circumstance shows that the community is satisfied with his work and trusts his assessments. It also underlines the importance of accumulated knowledge, experience and mutual trust for the successful management and usage of common pool resources.

According to the flow of the water, so-called *Sardori-Deqkhonon* ('heads of farmers') are one management level below the *Mirāb*. They are responsible for the supervision and the proper handling of the locks blocking the main canals, as well as for the condition of the canals themselves. Each canal has four heads, which take turns and represent a different subgroup of up to ten farmers using the same canal. They are elected by their group and are responsible for irrigation practices according to the agreed schedules [33, 36]. Within these subgroups, the water division is organized by the members themselves. They continuously negotiate the time and the duration of their irrigation slot, and open and close the locks in accordance to the agreements. These arrangements are highly flexible, to quickly encounter shifting conditions. After the village assembly, the water master, the *Sardori-Deqkhonon* and the canal subgroups, individual farmers can be seen as the fifth and final tier of Shirgin's water management and irrigation arrangement.

Finally, the irrigation schedules adapted to different scenarios, in terms of two different stages of the cultivation period, and the amount of available water shall be outlined [33, 36].

One schedule is used during the beginning of the cultivation period from the middle of April until mid-May in years with normal water supply. All farmers are allocated to two main groups, each consisting of six subgroups of up to ten households. The six subgroups belonging to the first main group are entitled to irrigate their fields on even dates, the second main group on odd dates. The irrigation entitlement of a group comprises a day irrigation round lasting from 7 am to 7 pm, and a night irrigation round lasting from 7 pm until 7 am next morning. Accordingly, every subgroup has the right to irrigate their fields for two hours during the day, and two hours during the night. The water distribution within the subgroup is negotiated by its members.

In years without water scarcity, a second schedule is used from the middle of May until the harvest of legumes in August. Every neighbourhood receives irrigation rights valid for one specific day of the week. The neighbours organize the water distribution by themselves.

In years with water scarcity, a third schedule is used for the same cultivation period, as in schedule 2. Instead of whole neighbourhoods, canal-specific subgroups mentioned before receive 24-hour-long irrigation rights, and have to negotiate the water distribution within their groups. In case the water shortage is severe, the time slot for irrigation can be shortened. The responsible *Sardor* ('head') opens and closes the locks of the respective distribution channel his group is allocated to.

After the last field harvest has taken place in the middle or the end of August, a fourth irrigation approach is applied to irrigate remnant potato and vegetable beds and small legume sowings adjacent to farmsteads. At this time, the water demand is low and a regulation restricting the use of irrigation water is not necessary anymore. Every farmer can use as much water as needed and available.

4 CONCLUSION

The strength of Shirgin's community lies, on the one hand, in the outstanding water management and irrigation-related inventory of local knowledge comprising rich expertise about the environmental conditions of the surroundings of the village, its historical heritage, the detailed characteristics of the community and the place going beyond the immediately visible, as well as long-term personal experiences of the responsible managers, and the individual farmers. On the other hand, social assets such as mutual respect and support, trust and collaborative teamwork are characteristic of the relations between the different tiers of Shirgin's water management and irrigation approach. Therefore, it can be seen as a local-specific 'hydrosocial arrangement'.

This study provides an example of human and social assets becoming key factors for equitable utilization of locally available natural resources, as well as community cohesion and individual survival in a society which is struggling with manifold societal challenges. The findings support the argument that local knowledge-based approaches for the management of common pool resources have the potential to be not only a basis for resource utilization but also a central instrument of balancing interests within a community and a means of social organization.

ACKNOWLEDGEMENTS

Research for this study was kindly supported by the German Federal Ministry of Education and Research (BMBF). Research assistant Chorshanbe Goibnazarov gathered detailed information. Staying with his family during the field campaign provided an opportunity to gain invaluable insights into the daily life of the Wakhi people. Davlatali Sheronov introduced me to the peculiarities of Shirgin's water sector and enabled various meetings with numerous local respondents. The names of interview partners have been changed to ensure anonymity. Financial and practical support is highly acknowledged.

REFERENCES

[1] Anderson, K., Pomfret, R., *Consequences of Creating a Market Economy. Evidence from Household Surveys in Central Asia*, Edward Elgar: Cheltenham, Northampton, MA, 2003.

[2] Dudwick, N., Gomart, E., Marc, A. & Kuehnast, K., From Soviet expectations to Post-Soviet realities: Poverty during the transition. *When Things Fall Apart: Qualitative Studies of Poverty in the Former Soviet Union*, eds. N. Dudwick, E. Gomart, A. Marc & K. Kuehnast, The World Bank: Washington, DC, pp. 21–27, 2003.

[3] Mangott, G. (ed.), *Bürden auferlegter Unabhängigkeit: neue Staaten im post-sowjetischen Zentralasien*, Braumüller: Wien, 1996.

[4] DFID Department for International Development, UK, *Sustainable Development Guidance Sheet*, DFID: London, 1999.

[5] Bourdieu, P., The forms of capital. *Handbook of Theory and Research for the Sociology of Education*, ed. J. Richardson, Greenwood: New York, pp. 241–258, 1986.

[6] Geertz, C., *Local knowledge: Further essays in interpretive anthropology*, Basic Books: New York, 1983.

[7] Mistry, J., Indigenous knowledges. *International Encyclopedia of Human Geography*, eds. R. Kitchin & N. Thrift, Elsevier: Amsterdam, pp. 5:371–5:375, 2009.

[8] Barnes, T., Local knowledge. *The Dictionary of Human Geography*, eds. D. Gregory, R. Johnston, G. Pratt, M.J. Watts & S. Whatmore, Wiley-Blackwell: Chichester, pp. 422–423, 2009.

[9] Briggs, J. & Sharp, J., Indigenous knowledges and development: A postcolonial caution. *Third World Quarterly*, **25(4)**, pp. 661–676, 2004. DOI: 10.1080/01436590410001678915.

[10] Lynch, D., The Tajik civil war and peace process. *Civil Wars*, **4(4)**, pp. 49–72, 2001. DOI: 10.1080/13698240108402487.

[11] Gomart, E., Between civil war and land reform: Among the poorest of the poor in Tajikistan. *When Things Fall Apart: Qualitative Studies of Poverty in the Former Soviet Union*, eds. N. Dudwick, E. Gomart, A. Marc & K. Kuehnast, The World Bank: Washington, DC, pp. 57–93, 2003.

[12] Kreutzmann, H., *Ethnizität im Entwicklungsprozess. Die Wakhi in Hochasien*, Reimer: Berlin, 1996.

[13] Frank Bliss, F. & Mamadsaidov, M. *Socio-economic change in Gorno Badakhshan. A monitoring report of the MSDSP 1997-1998*, Mountain Society Development and Support Programme (MSDSP): Khorog, 1998.

[14] Sherbut, G., Kanji, N. & Hatcher, J., Linking past and future: Cross-border development and quality of life in the Badakhshans. *Central Asian Survey* **34(2)**, pp. 255–271, 2015. DOI: 10.1080/02634937.2014.976948.

[15] Bobrinskii, A.A., *Gortsy verkhov'ev Pyandzha (Vakhancy i Ishkashimcy): Ocherki Byta po Putevym Zametkam gr. A. A. Bobrinskogo*, A. A. Levenson: Moskva, 1908.

[16] Zarubin, I.I., *Kixn ailóq (Lüngar). In dafdar az baroji na visonidani naglnoji zaboni vaxi. I.XI.1935*, Institute of Oriental Manuscripts of the Russian Academy of Sciences: St. Petersburg, 121/1/362, pp. 1–4, 1935.

[17] Mukhiddinov, I., *Zemledelie Pamirskikh Tadzhikov Vakhana i Ishkashima v XIX - nachale XX v. (Istoriko-etnograficheskii ocherk)*, Izd. Nauka, Glavnaya redakciya vostochnoi literatury: Moskva, 1975.

[18] Kreutzmann, H., Pastoral practices in transition: Animal husbandry in high Asian contexts. *Pastoral Practices in High Asia. Agency of 'development' effected by modernisation, resettlement and transformation*, ed. H. Kreutzmann, Springer: Dordrecht/Heidelberg/New York/London, pp. 1–29, 2012. DOI: 10.1007/978-94-007-3846-1_1.

[19] Kreutzmann, H., *Pamirian Crossroads. Kirghiz and Wakhi of High Asia*, Harrassowitz: Wiesbaden, 2015.

[20] Agentii Omori nazdi Prezidenti Jumkhurii Tojikiston, *Mintakakhoi Jumkhurii Tojikiston 2014*, Agentii Omori nazdi Prezidenti Jumkhurii Tojikiston: Dushanbe, 2014.

[21] Baranov, P.A., Vnedrenie zemledeliya v vysokogornye zony Pamira i podnyatie produktivnosti ego na zapadnom Pamire. *Konferenciya po sel'skokhozyaistvennomu osvoeniyu Pamira, Leningrad, 23.–29. Yanvarya 1936 g.*, ed. Akademiya Nauk SSSR, Tipografiya AN SSSR: Leningrad, pp. 1–6, 1936.

[22] Breu, T., Hurni, H. & Wirth Stucki, A., Valley agriculture in the Western Pamirs. *The Tajik Pamirs. Challenges of Sustainable Development in an Isolated Mountain Region*, eds. T. Breu & H. Hurni, Centre for Development and Environment (CDE), University of Berne: Berne, pp. 20–21, 2003.

[23] Kreutzmann, H., Ethnic minorities and marginality in the Pamirian Knot. Survival of Wakhi and Kirghiz in a harsh environment and global contexts. *Geographical Journal*, **169**(**3**), pp. 215–235, 2003. DOI: 0016-7398/03/0003-0215/$00.20/0.

[24] Asian Development Bank, *Country Partnership Strategy: Tajikistan, 2016–2020, Poverty Analysis (Summary)*, ADB, available at https://www.adb.org/sites/default/files/linked-documents/cps-taj-2016-2020-pa.pdf (accessed 26 January 2017).

[25] Dusufbekov, Y. (Shopkeeper), *Personal communication,* 28 August 2016.

[26] Holzwarth, W., Segmentation und Staatsbildung in Afghanistan: Traditionelle soziopolitische Organisation in Badakhshan, Wakhan und Shegnan. *Revolution in Iran und Afghanistan*, ed. Berliner Institut für Vergleichende Sozialforschung, Syndikat: Frankfurt/Main, pp. 177–235, 1980.

[27] Andreev, M.S. & A.A. Polovcev, *Materialy po etnografii Iranskikh plemen Srednei Azii. Ishkashim and Vakhan*, Tipografiya Imperatorskoi Akademii Nauk: St. Petersburg, 1911.

[28] Olufsen, O., *Through the Unknown Pamirs. The Second Danish Pamir Expedition 1898-99*, William Heinemann: London, 1904.

[29] Administration of District Ishkoshim, *Shumorai akholi ba kholati 01.01.2016 dar nokhiyai Ishkoshim,* Administration of District Ishkoshim: Ishkoshim, 2016.

[30] Austrian Development Cooperation (ADC), Poverty Alleviation by Mitigation of Integrated High-mountain Risk (PAMIR), FOCUS. Humanitarian Assistance (FOCUS), Hilfswerk Austria International (HAI), *Nakshai idorakunii ofatkhoi tabii. Dekhai Shirgin – dzhamoati Vrang – nokhiyai Ishkoshim*, ADC, PAMIR, FOCUS, HAI: Khorog, 2013.

[31] Monogarova, L.F., *Preobrazovanie v bytu i kul'ture pripamirskikh narodnostei*, Nauka: Moskva, 1972.

[32] Mudoyorov, Kh. (former teacher; expert on local history), *Personal communication,* 16 June 2015 & 30 August 2016.

[33] Khirzoev, M. (former *kolkhoz* manager; water master), *Personal communication,* 16 June 2015 & 31 August 2016.

[34] Amonbekov, C. (local guide), *Personal communication*, 29 August–2 September 2016.
[35] Omadbekov, M. (elderly knowledgeable person), *Personal communication*, 15 June 2015.
[36] Deronov, Sh. (teacher), *Personal communication*, 28 August–2 September 2016.
[37] Invarov, A. (teacher), *Personal communication,* 9 May 2015.
[38] Mirzo, O., *Wakhan (A Scientific, Historic and Ethnographic Study)*, Irfon: Khorog, 2010.
[39] Astamov, R. (former *kolkhoz* shepherd), *Personal communication*, 31 August 2016.
[40] Milbekov, N. (shepherd), *Personal communication,* 29 August 2016.
[41] Gurmadov, N. (shepherd), *Personal communication*, 8 May 2015.

MONITORING OF SEEPAGES IN EARTHEN DAMS AND LEVEES

F. ČEJKA[1], V. BENEŠ[2], F. GLAC[3] & Z. BOUKALOVÁ[1]
[1]VODNÍ ZDROJE, a.s., Czech Republic
[2]G IMPULS Praha spol. s r.o., Czech Republic
[3]IL FAUT s r.o., Czech Republic

ABSTRACT

The aim of this article is to describe an innovation in the monitoring of performance of earth embankments, namely the use of sensing systems for full-time monitoring of dam and levee seepages. The monitoring system being developed by the companies VODNI ZDROJE (Czech Republic) and EGIS (Republic of Korea) within the EUREKA project Short Term Assessment and Mitigation of Flood Risks is based on the measurement of temperature changes in the embankment body by using the discrete and distributed temperature sensors. The article provides an overview of the activities performed at the reservoir earth dam in pilot site Vlčkovice v Podkrkonoší and describes the (i) preliminary activities necessary for the proper spatial distribution of sensors in the embankment body including the geophysical and geotechnical survey and (ii) collection and evaluation of the data.

Keywords: fibre optic distributed temperature sensors, flood risk assessment, geophysical monitoring system, levee management, seepage monitoring

1 INTRODUCTION

The risk for European population originating from possible failure of dams and structures designed for flood control (further referred to as embankments) is widespread throughout Europe. In many countries, the number of small dams (up to 15 m of height) significantly exceeds the number of the large ones. For instance, in the UK the humble earthen flood embankments are by far the most ubiquitous flood risk management assets along the coast and watercourses as stated by Normandale [1].

Diverse sensor types have been used for the monitoring of stability and seepage regime in large dams for decades. In recent years, however, the development in the field of information and communication technologies (ICT) and sensing technologies has facilitated their use for the monitoring of small earth accumulation and flood embankments. The use of fibre optic distributed temperature sensors (FO DTS) is regarded as having a great potential in embankment monitoring. In this area, research was conducted by Khan et al. [2] on the automatic detection of singularities (seepage flows) within a trial embankment using optical time-domain reflectometry. In a laboratory experiment, Huaizhi and Yeyuan [3] dealt with seepage detection including the evaluation of seepage intensity based on double-line and single-line heat source data analysis.

Experimental application of FO DTS on real dams is reported, among others, by Pyayt et al. [4] in the project FP7 Urban Flood including dams on the Rhine (Germany) and in Boston (UK). On the Rhine dam, real seepage flows were monitored using the FO DTS systems. Research teams from AGH University of Science and Technology in Cracow are dealing with further installations and data evaluations from FO DTS systems on a research polygon [5, 6].

The Short Term Assessment and Mitigation of Flood Risks (STAMFOR) project was focused on real site implementation of an innovative embankment safety monitoring

DOI: 10.2495/EI-V1-N3-267-278

system for full-time monitoring of embankment seepage flows by temperature sensing systems. The main objective of the project was not only to confirm the effectiveness of these sensing systems, which had been sufficiently accomplished in similar past projects (see above), but especially to integrate the installation of the sensing systems into the existing safety supervision procedures for earth embankments. The design of the spatial distribution of the sensors is obviously a vital precondition for collecting reliable data from the sensing systems. Such a design can only be made based on a comprehensive interpretation of historic and, if needed, additional surveys of the site. For these purposes, STAMFOR methodology was developed as part of the project to describe all the activities necessary for a successful design and implementation of the sensing systems including subsequent data evaluation.

2 STAMFOR METHODOLOGY

The STAMFOR methodology describes all the steps needed for a successful implementation of the temperature sensing technology. These include especially the operations associated with the site reconnaissance survey, the design of the sensing system distribution pattern, and the installation and setting of the system. The entire methodology can be divided into the following sequence:

I. **Research of available data sources** includes the collection and analysis of available information on the area of interest.
II. **Indirect survey** is used to locate the reference and/or anomalous areas that are suitable for long-term monitoring by the STAMFOR technology. It provides data for a preliminary description of the embankment body as well as for a preliminary location of the monitoring profiles. Within the STAMFOR methodology, the indirect survey methods include a geophysical investigation to locate the reference as well as anomalous embankment sections that are suitable or practical for long-term monitoring. A convenient investigation method is the application of the geophysical monitoring system (GMS) according to Boukalová and Beneš et al. [7, 8], as described in an article [9] as part of the STAMFOR project.
III. **Direct survey** using boreholes and pits provides detailed information particularly on the vertical structure of the embankment body in the reference and/or anomalous areas located in the previous step.
IV. **Final interpretation and design** provides information for precise positioning of the sensing technology in the embankment body based on expert multidisciplinary assessment and calculations of the embankment body model.
V. **Implementation** stage comprises all the operations necessary for an effective installation of the selected sensing technology, further for launching the STAMFOR system pilot and, after the calibration is done, the full operation stage.

It must be stressed that each step mentioned above forms an integral part of the STAMFOR system application methodology and needs to be carried out in a sufficiently qualified way to ensure the validity and relevance of the output information gathered from the sensing systems.

Based on the STAMFOR methodology, a general design for the positioning of the sensing systems in the embankment body was made (Fig. 1), which was further verified by pilot measurements on the site of Vlčkovice v Podkrkonoší.

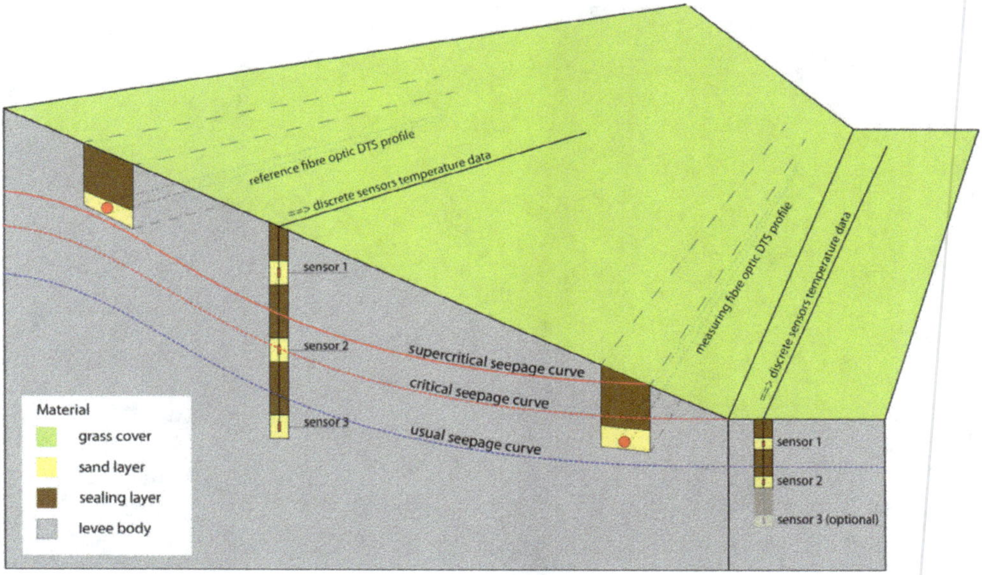

Figure 1: General spatial design of discrete and distributed sensors in the levee body.

3 PILOT INSTALLATION ON THE SITE OF VLČKOVICE V PODKRKONOŠÍ

In the project, the research was focused on two independent sensing systems – (1) borehole discrete temperature sensors (BDTS) and (2) FO DTS. For the pilot measurement, an earth dam of the Vlčkovice pond in Vlčkovice v Podkrkonoší having a height of 7 m (dam crest is located at approx. 299 m above sea level) and a length of 70 m was used. This pilot site was chosen mainly because there is a considerable seepage zone approximately at the midpoint of the height of the embankment body. Simple control over the water level of the pond enabled to activate and inactivate the seepage to monitor reactions of the sensors to the changes thus induced.

After the operations according to steps I and II of the STAMFOR methodology were carried out, monitoring profiles PF1 and PF2 (Fig. 2) were designed. Subsequently, step III was undertaken consisting in laying out and performing 4 core boreholes to determine the material composition within the embankment body. At each profile, the boreholes were drilled to a depth of approximately 11 m near the dam crest and to depths of 3.5 m and 6 m, respectively, at the dam toe. Based on core sampling and the determined geomechanical properties, a seepage model of the dam was then calculated in step IV and exact depths for the positioning of the sensing systems were suggested (Fig. 3). After the exploratory boreholes were cased and equipped with water-level probes, they were used for direct monitoring of water-level fluctuations within the embankment body. This control system served for the verification of data gathered from the two measurement systems. Figure 3 gives a layout of the sensor arrangement.

The identification and exact depths of the point temperature sensors at profiles PF1 and PF2 are as follows:

PF1

- MVT1_1, downstream face, temperature sensors at 1.4, 3.4 and 4.8 m bgl
- MVT1_2, dam toe, temperature sensors at 1.8 and 2.9 m bgl

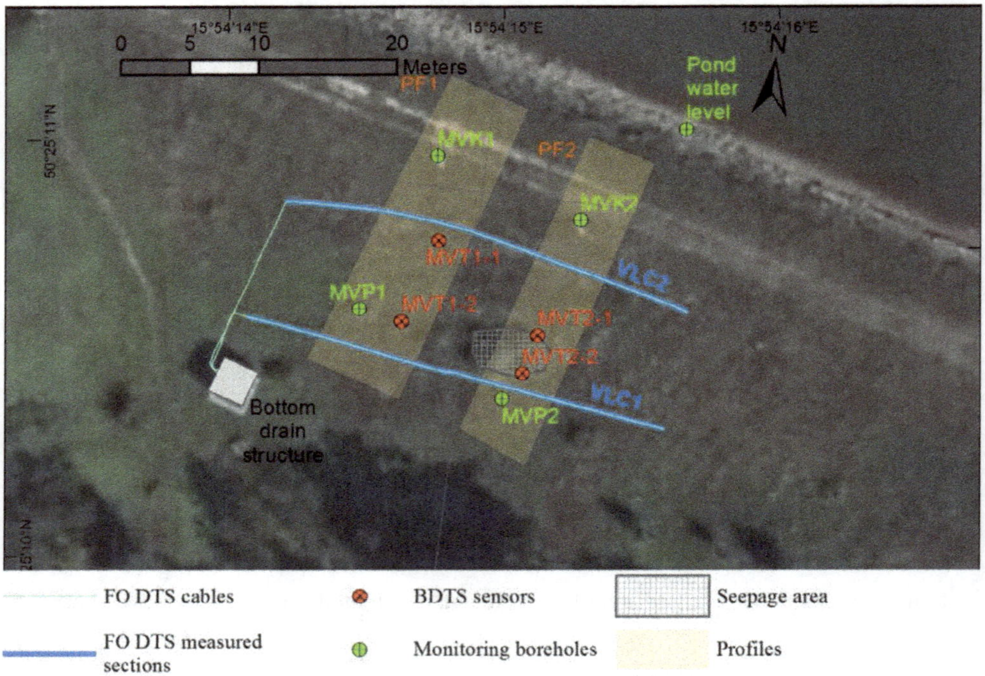

Figure 2: Positioning of the measurement equipment at the pilot site.

Figure 3: Seepage model of the dam body with depth indications for the positioning of point.

PF2

- MVT2_1, downstream face, temperature sensors at 1.0, 1.8 and 2.4 m bgl
- MVT2_2, dam toe, temperature sensors at 1.5 and 2.9 m bgl

The FO DTS cable lines were placed along the embankment body in two profiles (VLC1 and VLC2). VLC1 was directed through a seepage zone and used as a measurement profile, while VLC2 served as a reference profile. For protection, the cables were placed in a sand bedding of 20 cm in ditches 80-cm deep. Then they were covered with a 30-cm layer of compacted gravel of a 0/32 mm size and backfilled with the previously excavated earth, which was again compacted to minimize the impact of rainfall. On request of the dam manager, both profiles were sloped down in the western direction (further referred to as 'to the left') to drain off leaking water. This appeared to be an important aspect in the data evaluation stage (Sect. 4.3).

4 EVALUATION OF DATA FROM THE PILOT MEASUREMENT

After the installation of the BDTS system, the pilot trial was started on 06 October 2015 by emptying the Vlčkovice reservoir, which caused the water level to drop below the level critical for the activation of the seepage. From this moment, data on water-level and temperature fluctuations both in the reservoir and in the monitoring boreholes, further on air temperature, precipitation and temperature at each point temperature sensor were recorded in a 20-min step. In February 2016, the installation of the FO DTS cables was finished and data collection was started using a 5-min step.

The evaluation of the data from the BDTS system pilot measurement was done for the period of November 2015–March 2016, during which the water level in the reservoir rose beyond the critical level and thus the seepage was activated.

The evaluation of the data from the FO DTS system pilot measurement was done for the period of February 2016–October 2016, during which the water level again dropped below the critical level due to dry summer and thus the seepage was deactivated.

The following paragraphs give the measurement results of both the control system and the two measurement systems.

4.1 Measurement of the seepage conditions in the monitoring boreholes – control system

The water-level fluctuation curves for the monitoring boreholes reflect the water level changes in the reservoir. These reactions can be seen both at the dam crest and at the dam toe.

The analysis of the dependence of water levels in the monitoring boreholes on the pond water levels (Fig. 4) clearly shows that the dependence rate is not constant but significantly shifts at three height levels. Once the water level in the reservoir rises beyond approximately 296.64 m a.s.l., water level in profile PF2 starts to rise more rapidly. This rise becomes even more rapid, once the pond water level exceeds 296.88 m a.s.l., and continues until the pond water level reaches 297.10 m a.s.l.

From this point on, the reaction of profile PF2 decreases, while in profile PF1, a nearly steady rise continues from pond water level = 297.20 m a.s.l. These findings correspond to the occurrence of seepage zones observed on the downstream face of the dam, the results of the geotechnical investigation as well as the seepage model projections concerning the existence of permeable layers at 296.70 m a.s.l. The pond water level exceeded this critical level on 07 December 2015. On 21 December 2015 (pond water level at 297.34 m a.s.l.), an already fully developed seepage zone was observed on the downstream face of the dam during visual site inspection. On 10 July 2016, the pond water level again dropped below 296.70 m a.s.l.

4.2 BDTS system data evaluation

For BDTS system data analysis, the data from the period of 31 October 2015 to 31 March 2016 were collected.

We assume that if the dam is in an equilibrium state (without seepage or water-level fluctuations within the dam), the temperature changes of the temperature sensors in the dam

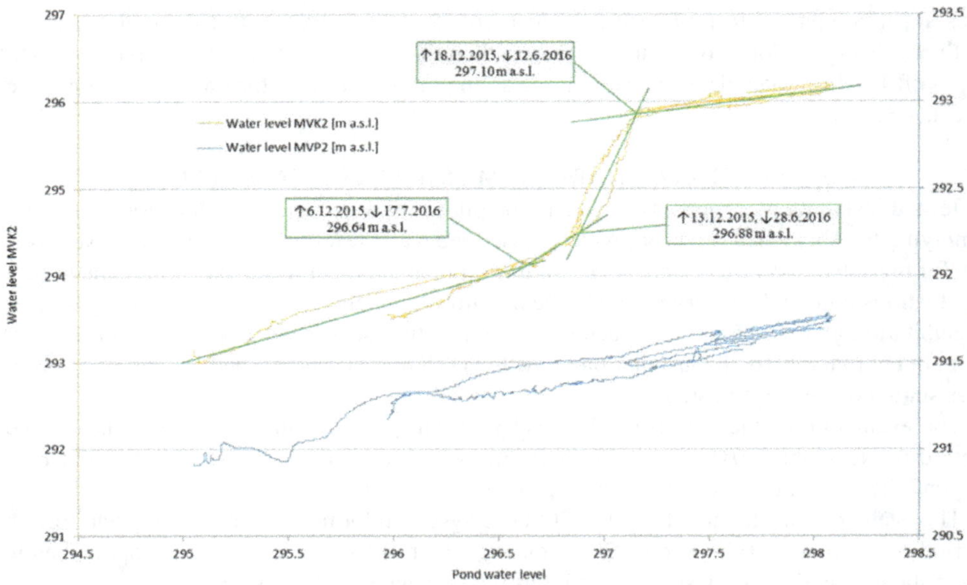

Figure 4: Dependence of water levels in the monitoring boreholes (profile PF2) on the pond water levels.

are influenced by air temperature (or water in the pond) and heat flux from Earth's interior (which we consider to be constant for our purposes). In case of an equilibrium state, the total rate of change (derivative) in the temperature behaviour as well as the amplitude of the local anomalies thus corresponds to atmospheric (climatic) influences and their intensity decreases with depth. Temperature changes spread from the surface to the depth with a time delay. Therefore, it is necessary to make a time shift for the temperature curves for analysis. The magnitude of the shift is to be determined based on the depth of each temperature sensor.

A deviation from the temperature behaviour in the dam body as described above may indicate a seepage flow (that is a water-level change within the dam), if such deviation shows a direction of change congruent with the temperature gradient between the existing sensor temperature and the temperature of water in the reservoir. If, on the other hand, such deviation has an opposite direction, it may be caused, e.g., by a change in groundwater level. In most cases, groundwater has a significantly different temperature from that of the pond water. It may also be a false anomaly, caused, e.g., by percolation of rainwater towards the sensor.

In this article, the sensors in boreholes MVT1_2 and MVT2_2 were used for the evaluation.

4.2.1 Evaluation of data from MVT1_2 sensors

In borehole MVT1_2, two temperature sensors at depths of 2.9 m bgl (measuring sensor) and 1.8 m bgl (reference sensor) were installed.

To adjust the temperature response of the measurement sensor to the reference sensor, to air temperature and seepage conditions in the dam (Fig. 5), an optimal correlation was found by shifting the time series forward by 3 days (reference sensor MVT1_2-1.8 m) and 20 days (air temperature sensor). The air temperature data were smoothed out by a moving average method using 255 hour-series.

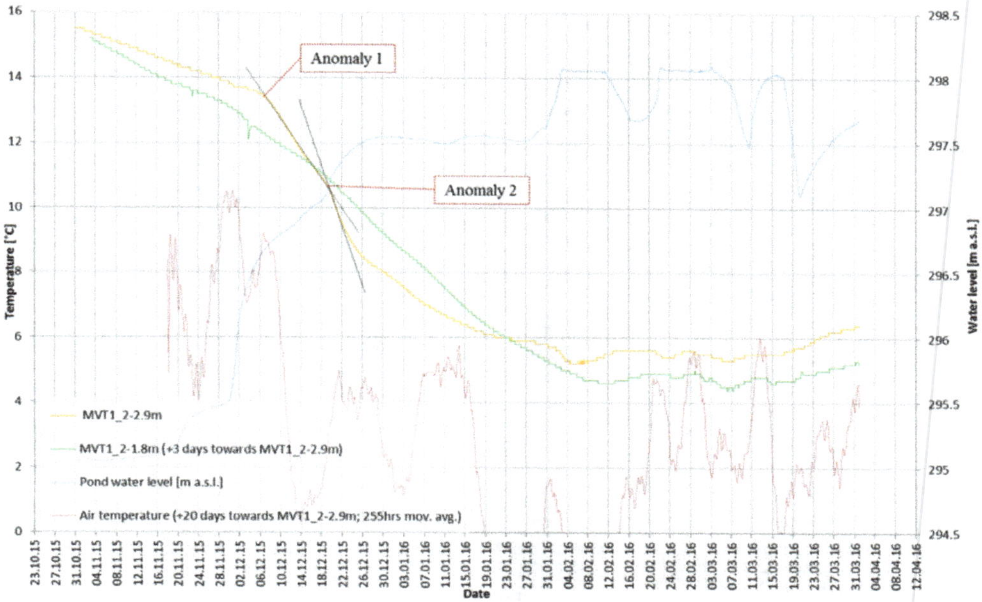

Figure 5: Comparison of the temperature time series in borehole MVT1_2 (sensors at 1.8 and 2.9 m bgl), air temperature and pond water level.

In the measurement sensor data series, two sharp deviations were found where the measurement sensor temperature data trend significantly diverged from that of the reference sensor (this sensor is assumed to be influenced by air temperature fluctuations). The first anomaly took place on 06 December 2015 at pond water level = 296.6 m a.s.l., the second on 19 December 2015 at pond water level = 297.15 m a.s.l.

It is assumed that the first anomaly was caused by a reactivation of the seepage regime in the dam after a long period of drought. At that time, the standard seepage curve reached the rock toe area of the dam.

The second anomaly may be attributed to an intensification of the seepage condition in the dam body, associated also with the observed seepage zone on the downstream face of the dam.

4.2.2 Evaluation of data from MVT2_2 sensors

In borehole MVT2_2, two temperature sensors at depths of 2.9 m bgl (measuring sensor) and 1.5 m bgl (reference sensor) were installed.

In this case, time shift was done only for air temperature (+20 days, smoothed out by a moving average using 255-h time series) to preserve the time–temperature dependences recorded by the BDT sensors. Additionally, pond water temperature fluctuations were included in the visualization of the measured data (Fig. 6).

In the measurement sensor data series, two sharp deviations were again found. In borehole MVT2_2 (located below the active seepage zone – see Fig. 2), the anomalies were detected at the measurement sensor on 01 December 2015 at pond water level = 295.7 m a.s.l., and on 03 December 2015 at pond water level = 296.4 m a.s.l. The first case may be accounted for

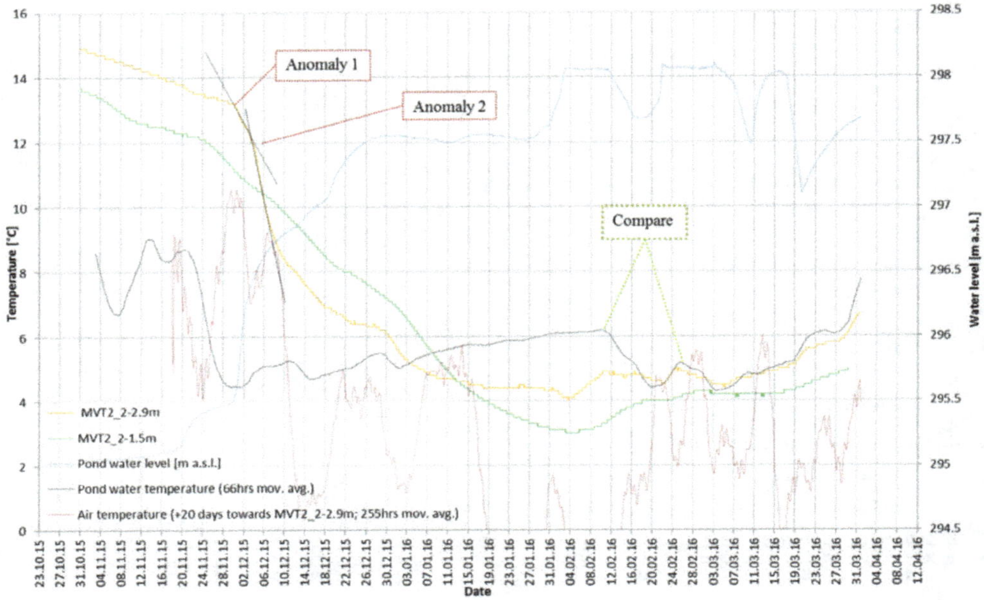

Figure 6: Comparison of the temperature time series in borehole MVT2_2 (sensors at 1.5 and 2.9 m bgl), air temperature, pond water temperature and pond water level.

by an activation of the usual seepage regime in the dam (note the quick response to the pond water-level rise). The second sharp deviation indicates a more intense flow of the cooler pond water, as it reached a permeable layer.

To comprehend the complexity of the investigated dependences (the explanation of which exceeds the scope of this article), compare the influences of pond water temperature and air temperature in the periods of 04–08 February 2016 and 24–28 February 2016, respectively (compare the temperatures of MVT2_2–2.9 m, air and pond water).

The comparison between the interpreted temperatures in boreholes MVT1_2 and MVT2_2 shows that in the MVT2_2 area, seepage appears considerably sooner and at lower pond water level. This fact corresponds to the existence of observed seepage zones on the downstream face of the dam in the proximity of this borehole.

4.3 Evaluation of data from the FO DTS system

Passive temperature measurements at VLC1 and VLC2 profiles were started on 12 February 2016 after a series of calibrations were carried out until the end of October 2016. The measurements using optical fibre cables did not coincide with the rise in water level in the Vlčkovický pond and the activation of seepage in December 2015. However, it did coincide with the inverse phenomenon, that is the water-level decline, which was caused by low precipitation in the spring and summer months of 2016, and thus with the deactivation of seepage in July 2016.

The data evaluation showed that for almost the entire period (with the exception of the 'temperature shift', see below) an anomalous behaviour of the left part of monitoring profile VLC1 took place that was relatively sharply delimited on the right. This behaviour was in contrast with that of the reference profile VLC2, which was consistent over the entire length and the whole measurement period. This anomalous behaviour was caused by a seepage flow occurring above VLC1 profile at local chainage 16–22. On the dam's manager request that the FO DTS profiles should also work as drains, the ditches were sloped in the direction to the lower local chainage (to the left). As a result, we can see a sharp delimitation of the seepage zone on the right side, while to the left the spatial delimitation of the seepage is blurred by water flowing along the preferential pathway of the drain containing the optical cable.

In the interpretation of data gathered from the FO DTS as well as BDTS systems, the interdependences between the pond water temperature, air temperature and the temperature of the dam body should always be kept in mind. Taking into consideration these facts, the four following indications of seepage could be identified on the site during the measurement period:

1. In winter (12–29 February 2016), water seeping through the dam body warmed up the relatively cooler dam body (Fig. 7)
2. In the period of the 'temperature shift' (at the beginning of April 2016 in our case – Fig. 8), the temperature of the saturated zone (influenced for the most part by pond water temperature) and that of the unsaturated zone (influenced for the most part by air temperature and the sun) come into balance.
3. In summer (Fig. 9), water seeping through the dam body cools down the relatively warmer body/surface of the dam.
4. The seepage was vanishing in the summer of 2016, when the pond water level was gradually declining due to low precipitation until it reached the critical level for the deactivation of the seepage. The measurement data show gradual disappearance of the described anomaly on the left side of profile VLC1 (Fig. 10), likely caused by the ceasing seepage flow. The turning point for the deactivation of the seepage appears to be the period between 25 and 30 July 2016, corresponding to pond water level = 296.5 m a.s.l.

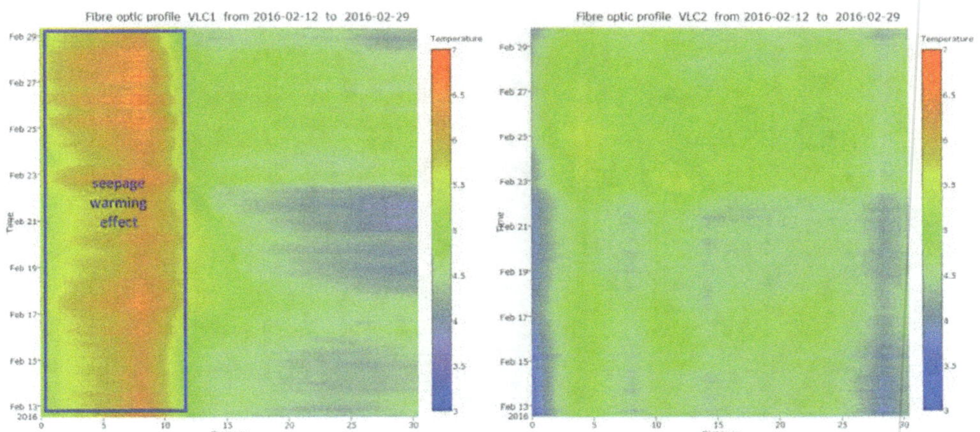

Figure 7: Comparison of FO DTS measurement and reference profiles in winter.

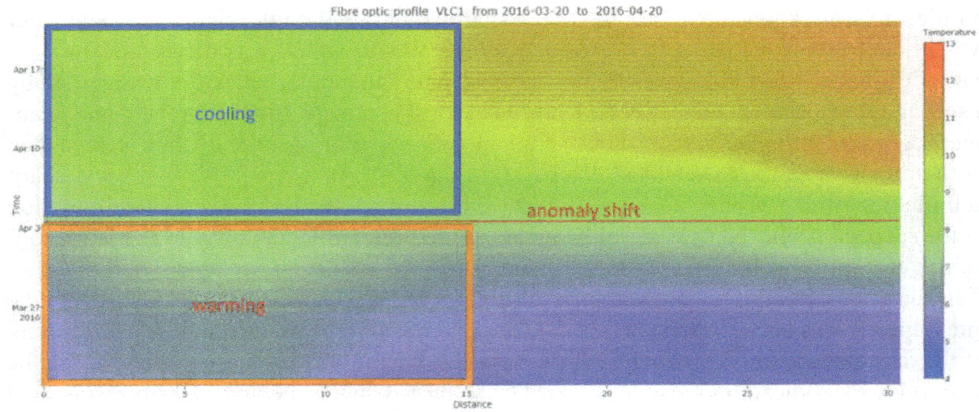

Figure 8: Measurement profile data visualization in the period of 'temperature shift'.

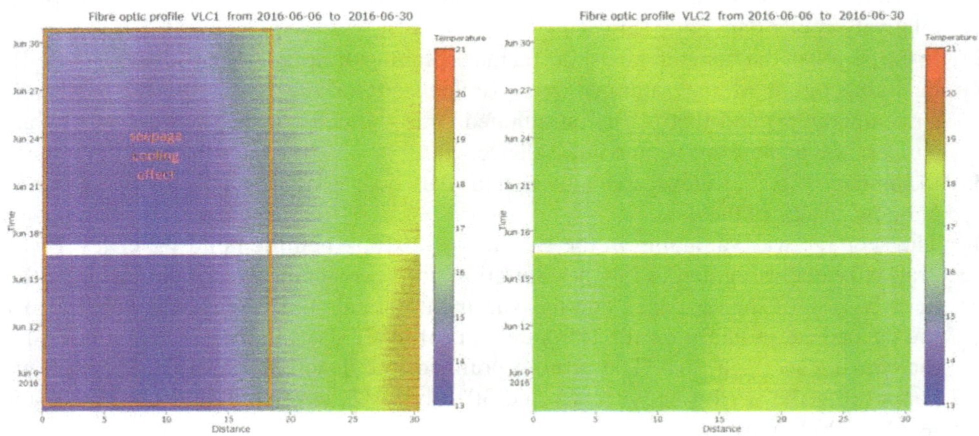

Figure 9: Comparison of FO DTS measurement and reference profiles in summer.

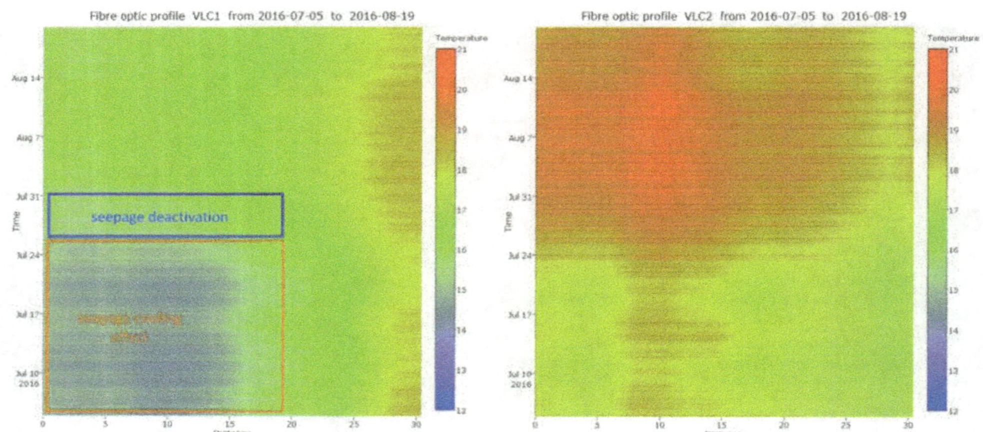

Figure 10: Comparison of FO DTS measurement and reference profiles in the period of seepage deactivation.

5 DISCUSSION OF THE RESULTS, LIMITATIONS AND CONCLUSION

Based on three independent measurement systems (control system of monitoring boreholes, BDTS and FO DTS) at the site of Vlčkovice v Podkrkonoší, critical water levels in the reservoir were identified, at which crucial changes in the seepage regime of the dam take place.

During the ascent as well as descent of the seepage curve, both the BDTS and FO DTS systems showed high correlation rate with the results of the evaluation of seepage regimes by the control system, which allows for the possibility of their application for various levels of monitoring of seepage regimes in dams. However, it must be again pointed out that the installation itself has to be preceded by thoroughly performed investigative operations, site-specific design of the spatial distribution of the sensors and subsequent pilot calibration of the system.

Based on the determined dependences, an algorithm was developed for the BDTS system as part of the project that automatically models temperature development trends at selected sensors. Based on the analysis of the degree and time of duration of the divergence between the measured and modelled development, this algorithm automatically generates three status values. The algorithm is calibrated using site-specific input parameters, which can further be specified during the operation of the system based on the gathered data.

The limitation of both systems appears at the time when the average air and water temperature trends intersect (period of the 'temperature shift' – see above). In temperate climate zones, this phenomenon occurs at least twice a year. During this time, seepage interpretation is made more difficult, if not impossible. However, for these periods there is the possibility of using heat-pulse FO DTS systems that work by analysing temperature fluctuations measured by periodically heated optical cables. These systems were also investigated in the STAMFOR project. However, the interpretation of the results exceeds the scope of this article.

Both the investigated systems feature specific advantages that determine the possibilities of their applications. Due to low investment costs, the BDTS systems can be specifically recommended for complementing and/or extending piezometer wells, while the FO DTS systems, due to their connectedness and lengths of the monitoring profiles (hundreds of meters up to several kilometres), represent an optimal solution for instances when monitoring systems are required for covering large distances or areas, such as for levees protecting areas of high-value assets or critical infrastructures.

ACKNOWLEDGEMENTS

The authors thank the E!8688 EUREKA project STAMFOR, financed by the Ministry of Education, Youth and Sports, Czech Republic and the company VODNÍ ZDROJE, a.s.

REFERENCES

[1] Normandale, D., Assessing flood defence integrity using geophysical monitoring techniques. *ADA Gazette.*, p. 19, Winter 2016.

[2] Khan, A.A., Vrabie, V., Mars, J.I., Girard, A. & D'Urso, G., Automatic monitoring system for singularity detection in dikes by DTS Data Measurement. *IEEE Transactions on Instrumentation and Measurement*, **59**(8), pp. 2167–2175, 2010. DOI: 10.1109/TIM.2009.2032880.

[3] Su, H., Kang, Y. & Huaizhi, S., Design of system for monitoring seepage of levee engineering based on distributed optical fiber sensing technology. *International Journal of Distributed Sensor Networks*, **9**(12), 2013. DOI: 10.1155/2013/358784.

[4] Payat, A.L. et al., Time-frequency methods for structural health monitoring. *Sensors*, **14**, pp. 5147–5173, 2014. DOI: 10.3390/s140305147.
[5] Balisa, B., Bartynski, T., Brzoza-Woch, R., Bubak, M., Harezlak, D. & Kasztelnik, M., Dedicated IT infrastructure for smart levee monitoring and flood decision support. *Proceedings of the FLOOD risk 2016 – 3rd European Conference on Flood Risk Management,* Lyon, E3S Web of Conferences 7, 2016.
[6] Dwornik, M., Krawiec, K., Franczyk, A. & Leśniak, A., Numerical modelling of levee stability based on coupled mechanical, thermal and hydrogeological processes. *Proceedings of the FLOOD risk 2016 – 3rd European Conference on Flood Risk Management,* Lyon, E3S Web of Conferences 7, 2016.
[7] Boukalová, Z. & Beneš, V., Dike breaks prevention as the process of flooding protection. *Proceedings of the International Association for Hydro-Environment Engineering and Research Congress*, 32nd IAHR Congress- Venice, 2007.
[8] Boukalova, Z., Benes, V., Cejka, F. & Hajek, T., Practical use of Geophysical monitoring system (GMS) focussed on WP3 activities. *Technical report of the FP7 project FloodProBE (WP3),* 2010.
[9] Čejka, F., Boukalová, Z. & Beneš, V., Short term assessment and mitigation of flood risks in the catchment scale. *WIT Transactions on Ecology and the Environment*, Vol. 197, ed. C.A. Brebbia, Wessex Institute of Technology, WIT Press: Southampton and Boston, 2015.

RACK LENGTH IN BOTTOM INTAKE SYSTEMS

LUIS G. CASTILLO[1], JUAN T. GARCÍA[1], PATRICIA HARO[2] & JOSÉ M. CARRILLO[1]
[1]Departamento de Ingeniería Civil, Universidad Politécnica de Cartagena, Spain
[2]Departamento de Ingeniería Civil y Ambiental, Escuela Politécnica Nacional, Ecuador

ABSTRACT

Bottom intake systems, made by racks longitudinally disposed in the flow direction, are classically used to derive mean flows in continuous mountain rivers with intensive sediment transport. They are known as Caucasian or Tyrolean Intakes. Nowadays, works are also being addressed to study these intakes in order to derive flash floods from ephemeral rivers in semiarid zones to collect runoff and improve the availability of resources. In the design of bottom intake systems, the length of rack is an important variable. This article includes several formulations that come from experimental works in laboratory and in prototype. The objective is to complement and update those works, including some empirical expressions proposed and experimental works developed since 2010 at the Universidad Politécnica de Cartagena. Rack length values for different flow rates are presented in dimensionless form (L/h_c – required rack length over critical depth) for the spacing area ratio over total area (void fraction m) of values 0.12, 0.28 and 0.60. These lengths present important differences depending on the conditions for which they were obtained, for example type of bars (circular or flat), spacing between the bars or the slope adopted by the rack. Values of the dimensionless rack length considering obstruction factors are also analysed. The Froude similitude criterion has been considered to obtain prototype values from the laboratory experimental conditions.

Keywords: bottom racks, design conditions, rack length, sediment transport, void ratio.

1 INTRODUCTION

Krochin [1] describes mountain streams as that which has strong longitudinal slopes, intense short-term precipitation, large flow variation and high sediment content with scarce fines (clean water in dry season).

On the other hand, semiarid zones are characterized by having low precipitation levels with huge temporal and spatial variability, low soil moisture, lack of vegetation cover, steep morphology and great sediment transport capacity by runoff [2].

Tyrolean or Caucasian Intakes have been used in mountain river basins due to their flow derivation capacity with moderate-high sediment transport, which facilitates their operation and maintenance when they are located in places of difficult access [1].

In semiarid basins, bottom rack intakes could be implemented in the upper zones in order to derivate flash floods with high sediment transport capacity during the wet season. Detailed studies of design parameters are required in order to optimize the structure sizing and guarantee design flow derivation [2–4].

2 DESIGN CONSIDERATIONS

Bottom intake systems are usually composed of a rack with parallel bars arranged longitudinally to the streamflow and located with a certain inclination (tan θ) over an intake gallery (built into a weir), or in the bottom channel, that facilitates flow derivation.

Principal factors of intakes design are: the ratio of clear opening area and total area of the rack (void ratio, m), the bars geometry, the rack slope – that reduces rack occlusion and facilitates self-cleaning, the approximation conditions and flexion resistance capacity of steel profiles (considering blows or abrasion caused by materials transported by avenues) among others.

DOI: 10.2495/EI-V1-N3-279-287

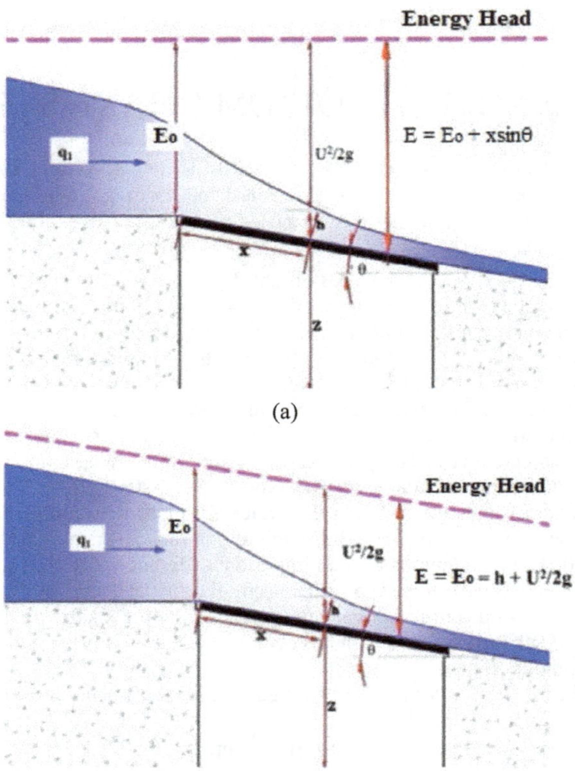

(a)

(b)

Figure 1: Bottom rack scheme considering: (a) horizontal energy level; (b) constant specific energy – energy level parallel to the bottom rack (García [2]).

Bottom racks behaviour have been analysed by different authors like it is mentioned in García [2]. Intake racks systems present three-dimensional and spatially varied flow with high turbulence level. Some hypotheses are adopted in order to simplify setting parameters:

- Hydrostatic pressure distribution along the rack.
- One-dimensional flow with decreasing flow.
- Energy conservation equation.
- Discharge coefficient experimentally set.

Regarding the loss of energy, some studies assume that it is negligible (horizontal energy level) as shown in Fig. 1a, while in other cases, it is based on the constant specific energy hypothesis along the rack (Fig. 1b), assuming critical depth at the beginning of the rack.

3 BOTTOM RACK LENGTH

This work exposes several formulations from different researches in order to calculate the wetted rack length necessary to derive the design flow in bottom intake systems. These formulae come from experimental work in laboratory and in prototype that are collected in the bibliography. The results obtained from its application are shown in Figs 2–4 for the cases of void ratio $m = 0.60$ ($b_1 = 0.15$ m and $b_w = 0.10$ m), $m = 0.28$ ($b_1 = 0.039$ m and $b_w = 0.10$ m) and $m = 0.12$ ($b_1 = 0.014$ m and $b_w = 0.10$ m), respectively, being b_1 the spacing between bars and b_w the bar wide. Flat T-shaped, flat prismatic, circular and prismatic rounded bars are considered; and 20% of slope. In the case of 0.60 void ratio, values can be compared with those measured in prototype [5] for circular bars and 20% of slope.

Figures 2–4 distinguish different kinds of rack lengths: L is the total rack length for deriving a design flow; L_1 the maximum length reached by water in the slit between two bars; and L_2 the envelope of lengths measured in prototype for different flow values (envelope of L) proposed by Drobir [5]. Different types of bars used in each formulation setting are shown in these figures, while the laboratory setup and detail of equations are collected in the Appendix.

In case of $m = 0.60$ (Fig. 2) and circular bars, dimensionless rack length as a function of critical flow (L/h_c) oscillates from 1.70 to 2.60. Envelope of Drobir is located over this group of curves with the value of 3.70, for specific flow of $q = 1.00$ m³/s/m. In the case of T-shaped bars, L/h_c varies from 2.70 to 4.10 when sediment occlusion is not considered. At the top of the Fig. 2 are the values of L/h_c equal to 4.40 and 5.40, which correspond to 15% and 30% of occlusion factor proposed in the equation of Krochin [1], respectively.

For $m = 0.28$ and circular bars, considering the specific flow of 1.00 m³/s/m, dimensionless rack length as a function of critical flow (L/h_c) oscillates from 2.90 to 4.30. In the case of T-shaped bars, L/h_c varies from 5.30 to 8.10 when sediment occlusion is not considered. At the top of the Fig. 3 are the values of 9.50 and 11.30, which correspond to L/h_c with 15% and 30% of occlusion factor, respectively. Krochin [1] proposes f values of up to 30% for channels with high sediment transport and this has been experimentally proven in Castillo et al. [3].

Finally, for $m = 0.12$ and circular bars, dimensionless rack length as a function of critical flow (L/h_c) oscillates from 4.40 to 8.80, for specific flow of $q = 1.00$ m³/s/m. For T-shaped

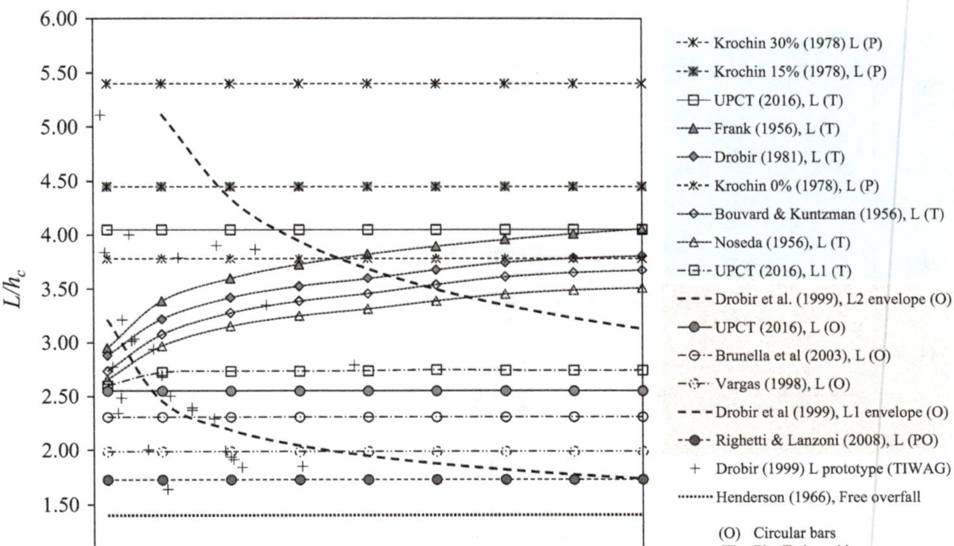

Figure 2: Dimensionless wetted rack length as a function of critical flow depth, h_c, approximation proposed in literature for racks with void ratio $m = 0.60$.

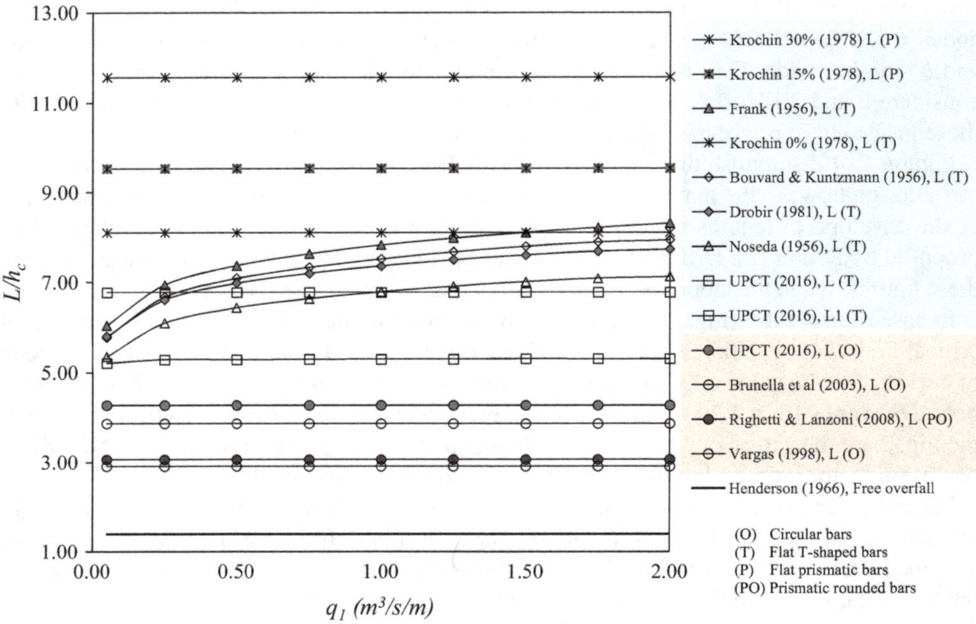

Figure 3: Dimensionless wetted rack length as a function of critical flow depth, h_c, approximation proposed in literature for racks with void ratio $m = 0.28$.

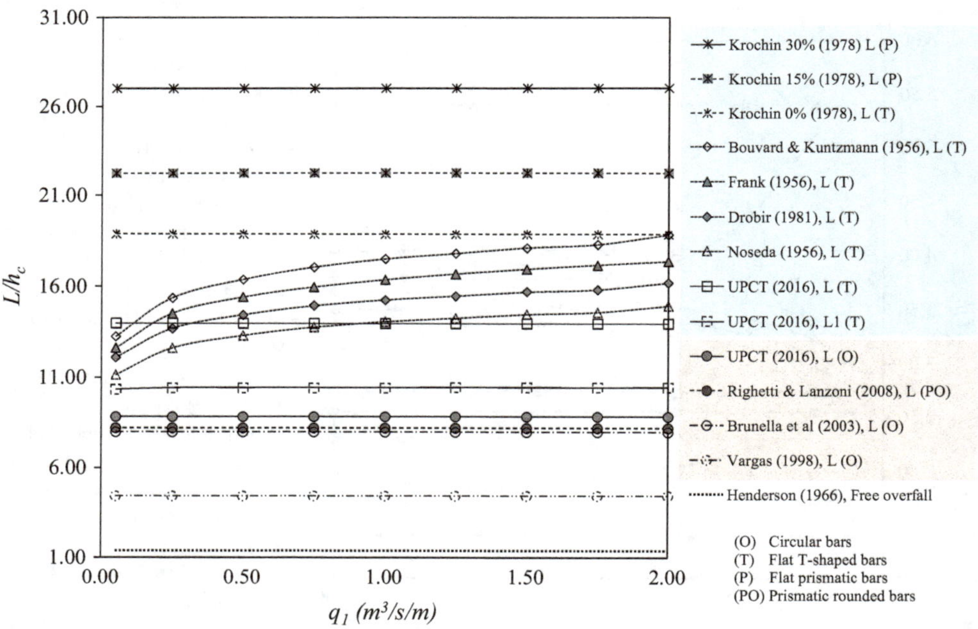

Figure 4: Dimensionless wetted rack length as a function of critical flow depth, h_c, approximation proposed in literature for racks with void ratio $m = 0.12$.

bars, dimensionless rack length as a function of critical flow varies from 10.40 to 18.90 when sediment occlusion is not considered. At the top of the Fig. 4 are the values of 22.20 and 27.00, which correspond to dimensionless rack length as a function of critical flow with 15% and 30% of occlusion factor, respectively.

It is worth mentioning that bottom racks with $m = 0.12$ could be used for small hydropower plants that need to capture small amounts of sediment to avoid damaging the turbine blades.

As shown in the previous figures, there are important differences in the dimensionless length of rack proposed by the researchers considered. This is due to variation of the parameters established for each experimental work. These parameters are: void ratio (normally, this value is set to $3/4(d_{90})$ of riverbed sediment granulometry), bar shapes, rack slope or approximate conditions of the flow at the beginning of the rack.

Due to these differences, a summarizing figure is proposed where average results presented in Figs 2–Fig. 4 are shown. In order to generate Fig. 5, the representative curves were grouped, considering the following factors:

- Occlusion factor of the bottom rack. Formulae that considered the transport of sediments effect (Krochin 15% and Krochin 30% [1]) were separated from those made with clear water.
- Type of bar. The average value of dimensionless wetted length, L_{prom} (T), was calculated using the formulae proposed by the different authors. In this way, Krochin 0% [1], UPCT-T [2], Frank [6], Drobir [5], Bouvard & Kuntzmann [7] and Noseda [8] formulae were considered for T-shaped bars. The expressions from UPCT-O [2], Brunella et al. [9] and Jimenez & Vargas [10] were used for circular bars, L_{prom} (O).
- Void ratio value was multiplied by each average dimensionless wetted length in function of the critical depth.

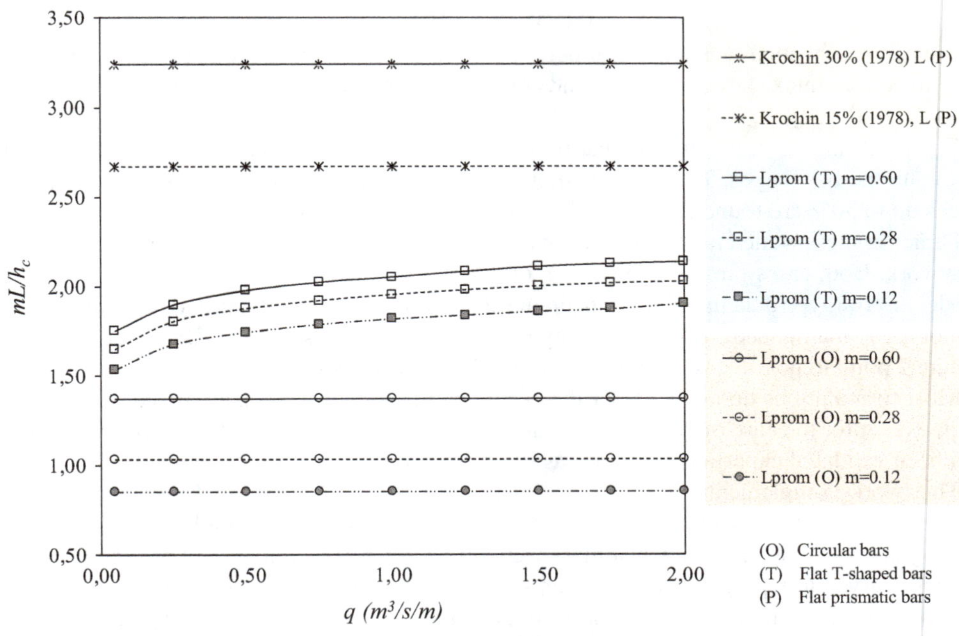

Figure 5: Dimensionless wetted rack length as a function of critical flow depth, h_c, multiplied by void ratio, m; considering bar shapes and occlusion coefficient.

Table 1: Details of laboratory models used by Orth [11], Drobir [5], and Jiménez & Vargas [10] in their research.

Scale type	Orth [11] 1:5	Jiménez & Vargas [10], Drobir [5] 1:10
Length scale	$L'/L = 1/5$	$L'/L = 1/10$
Velocity scale	$V'/V = 1/2.34$	$V'/V = 1/3.16$
Specific flow scale	$q'/q = 1.18$	$q'/q = 31.62$

Analysing the results presented in Fig. 5, Krochin's expressions for 15% and 30% occlusion factors in the bottom rack show no variation with the void ratio, so the figure includes only two lines without distinction of m. For the average dimensionless wetted lengths for T-shaped bars, the difference between the minimum and maximum void ratio is around 11% for a specific flow rate of 1 $m^3/s/m$; while for circular bars, the difference between void ratios reaches 38% between $m = 0.12$ and 0.60 for the dimensionless factor mL/h_c.

Some researches were performed in laboratory models, which have specific scale characteristics (Table 1), where Froude similitude hypothesis is assumed.

From this hypothesis, no differences should be found between the length calculated from model dimensions and that obtained using the same equations, for prototype dimensions, calculated according to Froude-similitude scale factors (Table 1). Prototype measurements [5] point to possible scale effects that should be further studied. With information available so far, it is concluded that equations proposed may be used at model or prototype scale without the appearance of any differences. To compare the model and prototype data, we used always the same void ratio for both cases.

4 CONCLUSIONS

Dimensionless wetted rack length for deriving a design flow is presented as a tool which can help in bottom intake systems design and complements previous works (Castillo & Lima [4], Castillo et al. [12]).

The most important parameters that influence the value of wetted rack length are the void ratio and the bar shapes. Thus, in function of the kind of bar shapes (circular or flat) differences up to 50% are found for $m = 0.60$; up to 89% for $m = 0.28$ and 110% for $m = 0.12$.

Dimensionless wetted rack length was calculated using prototype and model parameters for this work. Both cases present similar L/h_c ratios, so the equations collected may be used for model or prototype data in order to obtain the same results while void ratio in both model and prototype is maintained. Anyway, prototype measurements presented by Drobir et al. [5] in Figure 2 point to possible scale effects that should be further studied.

Most formulations do not consider the occlusion of the rack as a result of sediment transport, except for the case of Krochin [1] that proposes an obstruction factor up to 30%, which has been validated experimentally by Castillo et al. [3].

This work complements and updates what was made by Castillo & Lima [4] in 2010, where a figure was presented that collected the existing wetted rack lengths proposed in the bibliography. In this case, dimensionless wetted rack lengths as function of critical flow are shown. Curves groupings by type of bar shapes are also presented; and the formulae and experimental conditions for which were obtained are summarized in the Appendix. In the same way, formulae are updated with new found expressions, such as Jimenez & Vargas [9] and García [2].

APPENDIX

Table A1: Summary of wetted rack lengths to derive an approximation flow proposed in the literature (Castillo et al. [12]).

Reference	Experimental data	Wetted rack length L (m)
Noseda [8]	$q_{1\max} = 100$ l/s; $B = 0.50$ m; $0.16 < m < 0.28$; T-shaped bars; Slope: 0–20%	$L = 1.1848\frac{E_o}{C_{qh}m}$; $C_{qh} = \overline{C_{qh}(h)}$; $C_{qh}(h) = 0.66m^{-0.16}\left(\frac{h}{l}\right)^{-0.13}$
Bouvard & Kuntzmann [7]	Information taken from Noseda [8]	$L = \frac{1}{2m} \; j + \frac{1}{2j^2} \; \text{arcsen}\sqrt{\frac{j}{j+\frac{1}{2j^2}}} + 3\sqrt{\frac{1}{2j}} + \frac{0{,}303}{m^2} + \frac{2j^3 - 3j^2 + 1}{4j^2} \; tg\theta \quad h_1\cos\theta$; $j = \frac{h_1}{h_c} = 1$; $m = C_{qo}m$; $C_{qo} = 0{,}82$
Frank [6]	Information taken from Noseda [8]	$L = 2.561\frac{q_1}{\lambda\sqrt{h_o}}$; $\lambda = mC_{qh}\sqrt{2gcos\theta}$; $C_{qh} = \overline{C_{qh}(h)_{\text{Noseda}}}$
Krochin [1]	Information from Melik Nubarov [13] Prismatic flat bars	$L = \left(\frac{0.313q_1}{\left(C_{qE}k_1\right)^{3/2}}\right)^{2/3}$; $k_1 = (1-f)m$; $C_{qE} = C_0 - 0.325\tan\theta$; $C_0 = 0.50$
Drobir [14]	Information taken from Frank [6] and Bouvard & Kuntzmann [7]	$L = \left(\frac{0.313q_1}{\left(C_{qH}k_1\right)^{3/2}}\right)^{2/3}$; $2cos\theta x^3 - 3x^2 + 1 = 0$; $C_{qh} = \overline{C_{qh}(h)_{\text{Noseda}}}$
Drobir et al. [5]	$q_{1\max} = 20$ l/s; $B = 0.50$ m; $m = 0.60$ circular bars; Slope: 0–20%	$L_1 = 0.9088q_1^{0.4993}$; $L_2 = 1.7205q_1^{0.4296}$; L_{TIWAG}

(*continued*)

Table A1: (*Continued*)

Reference	Experimental data	Wetted rack length L (m)
Brunella et al (2003)	$q_{1\max} = 100$ l/d; $B = 50$ cm Circular bars; Two setups: $m = 0.352$ $m = 0.664$ Slope: 0–51°	$L = \frac{0.83E_0}{C_{q0}m}$; $C_{q0} = 0.87$
Righetti & Lanzoni [15]	$q_{1\max} = 37.5$ l/s; $B = 25$ cm Prismatic rounded bars; $m = 0.20$; $b_1 = 0.50$ cm; $b_w = 2.0$ cm	$Q = C_{q0}mBL\sqrt{2gE_0}\ \frac{a}{2}\frac{L}{E_0}F_{E0}\ +$ $1\left\{\tanh\ b_o\left(\sqrt{2} - F_{E0}\right)\right\}^{b_a}$ $F_{E0} = \frac{U_0}{\sqrt{gE_0}}$; $a = -0.1056$; $b_0 = 1.50$ $b_a = 0.478$
Jimenez & Vargas [10]	$q_{1\max} = 40$ l/s; $B = 55.2$ cm; circular bars Slope: 0–20º; Two setups: $m = 0.33$ and $m = 0.50$	$L = k_2\sqrt{\frac{2\cos\theta q_1^2}{mgh_1}}$; $k_2 = 1.10$
García [2]	$q_{1\max} = 77.7$ l/s; $0.16 < m < 0.28$; $0.0057 < b_1 < 0.0117$; T-shaped bars; Slope: 0–33%	L $L \approx 1.3\frac{h_c}{C_{qo}m}(1 + 0.3tg\theta)$; $C_{qo} = a \cdot e^{-0.77m}$ $a = 1.43$ $a = 0.90$ Circular bars T-shaped bars $\text{L1}: \frac{\text{dh}}{\text{dx}} = \frac{2mC_{qH}\sqrt{(E_o + x\text{sen}\theta)(E_o + x\text{sen}\theta - h\cos\theta)} + h\text{sen}\theta}{3h\cos\theta - 2(E_o + x\text{sen}\theta)}$ $C_{qE} \approx \frac{0.58e^{-0.75\frac{x}{h_c}m}}{(1 + 0.90tg\theta)}$
Henderson [16]	–	Free overfall: $L = 1.40h_c$

Note: For this table, it is considering that h_1 is the flow depth at the beginning of the rack; h_c the critical depth; $h_2 = 0$; $E_o = E_{min}$, the specific energy head at the beginning of the rack; B the channel width; q_1 the specific approximation flow; θ the angle of rack with horizontal; b_1 the space between two bars; b_w the bar width; l the resultant of $(b_1 + b_w)$; F_{Eo} the Froude Number at the beginning of the rack; C_{qh}, C_{qE} and C_{qo} the discharge coefficients in function of flow depth, energy head and static, respectively; a, b_o, b_a and k_2 proper coefficients for the equations where they belong; and f the obstruction factor (%).

ACKNOWLEDGEMENTS

The authors are grateful for the financial support received from the Seneca Foundation of Murcia Region (Spain) through the project 'Optimización de los sistemas de captación de fondo para zonas semiáridas y caudales con alto contenido de sedimentos. Definición de los parámetros de diseño'. Reference: 19490/PI/14.

REFERENCES

[1] Krochin, S., *Diseño Hidráulico*. Segunda. Escuela Politécnica Nacional: Quito, 1978.

[2] García, J.T., *Estudio experimental y numérico de los sistemas de captación de fondo*. Ph.D. Thesis, Universidad Politécnica de Cartagena, 2016 (in Spanish).

[3] Castillo, L.G., García, J.T. & Carrillo, J.M., Experimental and numerical study of bottom rack occlusion by flow with gravel-sized sediment. Application to ephemeral streams in semi-arid regions. *Water*, **8**, p. 166, 2016. DOI: 10.3390/w8040166.

[4] Castillo, L.G. & Lima, P., Análisis del Dimensionamiento de la Longitud de Reja en una Captación de Fondo. *Proceedings of the XXIV Congreso Latinoamericano de Hidráulica*, Punta del Este, 2010.

[5] Drobir, H., Kienberger, V. & Krouzecky, N., The wetted rack length of the Tyrolean weir. *Proceedings of the IAHR-28th Congress*, Graz, 1999.

[6] Frank, J., Fortschritte in der hydraulic des Sohlenrechens. *Bauingenieur*, **34**, pp. 12–18, 1959 (in German).

[7] Bouvard, M. & Kuntzmann, J., Étude théorique des grilles de prises d'eau du type (En Dessous). *Houille Blanche*, **5**, pp. 569–574, 1956 (in French). DOI: 10.1051/lhb/1954049.

[8] Noseda, G., Correnti permanenti con portata progressivamente decrescente, defluenti su griglie di fondo. *L'Energia Elettrica*, **33**, pp. 41–51, 1956 (in Italian).

[9] Brunella, S., Hager, W.H. & Minor, H.-E., Hydraulics of bottom rack intake. *Journal of Hydraulic Engineering*, **129**, pp. 2–10, 2003. DOI: 10.1061/(ASCE)0733-9429(2003)129:1(2).

[10] Jiménez, Ó & Vargas, O.A., Some experiences in the performance of bottom intakes. *Proceedings of the International Symposium on Hydraulic Structures*, IAHR: Guayana, 2006.

[11] Orth, J., Chardonnet, E. & Meynardi, G., Étude de Grilles Pour Prises d'Eau du Type « En Dessous ». *La Houille Blanche*, **3**, pp. 343–351, 1954 (in French). DOI: 10.1051/lhb/1954034.

[12] Castillo, L.G., García, J.T. & Carrillo, J.M., Influence of rack slope and approaching conditions in bottom intake systems. *Water*, **9**, p. 65, 2017. DOI: 10.3390/W9010065.

[13] Melik-Nubarow, S.G., Type perfectionnè de prise d'eau à grille horizontale. *Revue*, N10–N11, 1939 (in French).

[14] Drobir, H., Entwurf vonWasserfassungen im Hochgebirge. *Österreichische Wasserwirtsch*, **11**, pp. 243–253, 1981 (in German).

[15] Righetti, M. & Lanzoni, S., Experimental study of the flow field over bottom intake racks. *Journal of Hydraulic Engineering*, **134**, pp. 15–22, 2008. DOI: 10.1061/(ASCE)0733-9429(2008)134:1(15).

[16] Henderson, F.M.N., *Open Channel Flow*, MacMillan: New York, NY, 1966.

MICROPOLLUTANTS IN STORMWATER DISCHARGE IN THE SWIST RIVER BASIN

FRANZ MICHAEL MERTENS, EKKEHARD CHRISTOFFELS, ANDREA FRANZISKA BRUNSCH & JENS WUNDERLICH-PFEIFFER
Erftverband, Bergheim, Germany

ABSTRACT
A study was conducted from September 2010 to September 2012 at the outlet of a stormwater basin in the city of Meckenheim, near Bonn (Germany). For this purpose, a strategy was developed to collect samples from the stormwater basin located directly upstream from the point of discharge into the Swist river. Numerous locally applied pesticides, frequently used pharmaceuticals, industrial chemicals and flame retardants were detected. Median discharge load for selected micropollutants ranged from 4.0 to 19.9 µg/s. The results showed that treatment may be necessary in the case of rainwater discharge from separate sewer systems to reduce the substance load in the receiving river.
Keywords: micropollutants, monitoring, retention soil filter, separate sewer system, stormwater sewer

1 INTRODUCTION

In contrast to combined sewer systems, separate sewer systems are used to collect and transport wastewater and rainwater in separate channels. Wastewater is carried to a wastewater treatment plant, while rainwater is infiltrated locally or discharged into the nearest river. To reduce hydraulic peaks in the receiving river, stormwater basins can be interposed in the stormwater sewer system. Due to the obligation in the German state of North Rhine-Westphalia to build separate sewer systems in areas under current development [1], the proportion of separate sewer systems within urban drainage systems is here increasing. For this reason, comprehensive analysis of pollutants in discharged rainwater is necessary to estimate the load in the receiving watercourse and to plan adequate measures when necessary.

Micropollutant loads in discharged rainwater from separate sewer systems have already been the subject of various studies. Birch et al. [2] and Zgheib et al. [3] detected pesticides and phthalates in rainwater. Burkhardt et al. [4] found high concentrations of biocides leached from building materials. To broaden knowledge on micropollutant loads in the collected rainwater in separate sewer systems, a project was funded by the Ministry for Climate Protection, Environment, Agriculture, Conservation and Consumer Protection of the State of North Rhine-Westphalia to study rainwater discharge of a representative separate sewer system in the Swist river basin, a sub-basin of the Erft river basin. The rainwater was analysed for micropollutants along with other substance groups significant for water management from September 2010 to September 2012.

2 URBAN DRAINAGE SYSTEM, SURROUNDING LANDSCAPE AND RECEIVING RIVER

The stormwater sewer under study drains the northeastern part of the city of Meckenheim, 12 km southwest of the city of Bonn (Table 1). The mixed residential and commercial drained area served by the stormwater sewer covers a surface of 0.74 km^2. Meckenheim is located in the Swist sub-catchment, which is dominated by farmland interspersed with villages. There is no noteworthy contribution to the wastewater coming from industry. The study area, as a town section adjacent to farmland, is in its spatial structure typical for the Erft basin, so

DOI: 10.2495/EI-V1-N3-288-297

Table 1: Characteristics of the northeast Meckenheim separate sewer and the receiving Swist river at the point of discharge (SSO = separate sewer outflow).

Catchment Area	
City	Meckenheim; North Rhine-Westphalia
Runoff area	0.74 km^2
Development	Largely residential
Population served	Approx. 7,900
Stormwater basin	
Operation	Mechanical
Max. throttle outflow	1,100 L/s
Storage volume	3,650 m^3
River	
Receiving water course	Swist (lowland river)
Mean runoff at the SSO	158 L/s

results should be transferrable. A stormwater basin is installed in the sewer just before the outlet into the Swist river to provide hydraulic relief.

3 SAMPLING PROCEDURE AND STUDY SIZE

A specifically adapted automatic sampling system was installed (Fig. 1) to permit sampling during stormwater events. In these cases, manual sampling was not feasible due to the low predictability of such events. At the outlet of the stormwater tank just before the discharge point into the Swist, an automatic sampling device with twelve 2.7-L bottles as well as a 200-L barrel was installed. The 200-L volume was necessary to examine for parasites, which pose a high risk of infection when present, even at low concentrations. The microbiological results will be published in a separate manuscript. An ultrasound-altimeter was installed in the stormwater tank to register the water level, which was used to estimate the discharge volume. The management of the sampling system and data recording was done using a programmable logic controller.

The beginning of a discharge event was registered by the ultrasound-altimeter, and the sampling procedure was activated. The sampling device took a number of 12-min composite samples from every event. With 12 bottles a maximum duration of sampling of 144 min per discharge event was possible. At the beginning of an event the 200 L barrel was filled automatically in a time span of 15 min for microbiological examinations.

A total of 62 stormwater discharge events of widely varying duration (see Fig. 2) occurred during the study period. Though some lasted longer than the 144-min maximum sampling time, it was not considered important to measure for a longer time interval because then the contents would have been relatively diluted. The maximum possible number of samples (755) can be calculated by multiplying the total number of events (62) by the length of sampling intervals (12 min). However, numerous discharge events were shorter than 144 min (see Fig. 2). The actual number of samples taken and analysed came to 379. In some cases, flow volume and concentrations of substances could not be measured.

The samples were stored under cool, dark conditions until they could be picked up for examination by the research team. Notification of time and location of a discharge event was

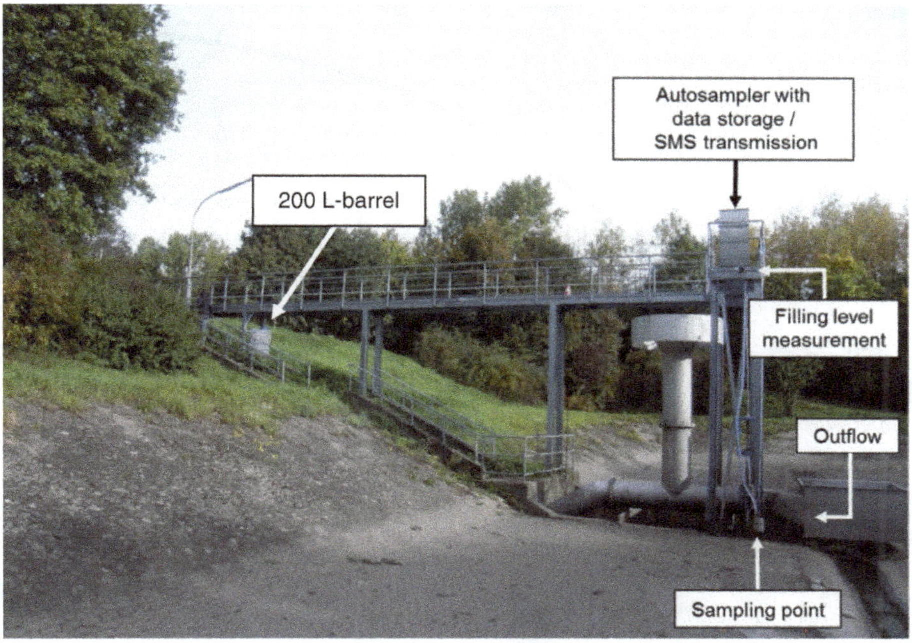

Figure 1: Sampling system in the stormwater basin of Meckenheim [5].

sent directly to the staff by SMS. Prompt pickup and processing of the samples was of particular importance as a consequence of unstable chemical compounds in the samples. From the obtained samples, 45 micropollutants were analysed throughout most of the time length of the project. Due to a laboratory expansion during the project period, the sampling strategy and the parameter set were changed. For this reason, a portion of samples was analysed as one composite sample representing a whole discharge event, but with an extended scope of parameters. Here, analysis was performed to test for the presence of 137 different micropollutants (pesticides, pharmaceuticals, flame retardants, industrial chemicals and the stimulant caffeine).

4 VOLUME AND DURATION OF DISCHARGE EVENTS

The data required to calculate flow volume and duration of discharge could be collected for 46 of the 62 events. Cluster analysis (SPSS 14©) was conducted to group the discharge events into five specific classes based on similarities in volume and discharge duration [6]. With this method 45 events were assigned to one of the five clusters. One event could not be assigned to any class and was thus declared as an outlier. Nonlinear regression analysis yielded representative hydrographs for each cluster.

The cumulative discharge volume corresponding to the duration of discharge for 45 analysed events as well as the regression curves representing each of the five clusters are depicted in Fig. 2. The results show that despite high variability of rainfall events, certain regularities in the behaviour of discharge were found. The similarities of the results within each cluster were based on comparable duration of discharges and similar flow volumes.

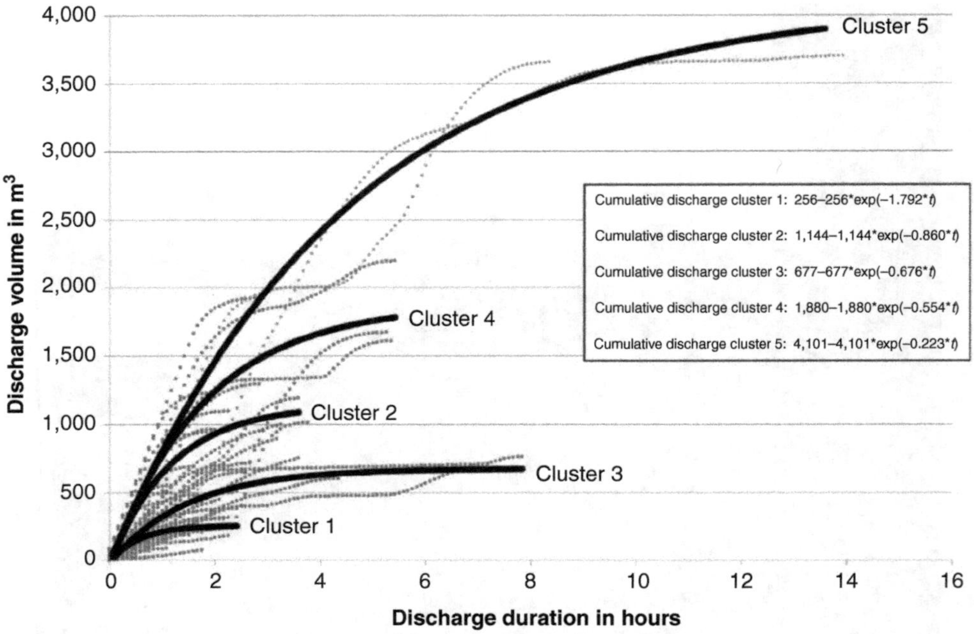

Figure 2: Five typical classes of discharge events depicted as cumulative discharge volume depending on the flow volume and duration as a result of cluster analysis ($n = 45$, dotted line), and the associated regression curves (solid line), coefficients of determination: cluster 1 = 0.28; cluster 2 = 0.74; cluster 3 = 0.66; cluster 4 = 0.68; cluster 5 = 0.96.

Table 2: Characteristics of discharge types as a result of cluster analyses.

	Duration (average hh:mm)	Discharge volume (average value m^3)	Number
Cluster 1	Short (01:30)	Low (245)	22
Cluster 2	Short (02:30)	Moderate (1,080)	8
Cluster 3	Medium (04:00)	Low (730)	9
Cluster 4	Medium (04:45)	Moderate (1,850)	4
Cluster 5	Long (11:00)	High (3,690)	2

The median discharge duration and discharge volume were calculated for all events within each cluster. Using median values, three groups of duration and discharge volume could be differentiated: short duration (<3 h), medium duration (<3 to 6 h) and long duration (>6 h); low volume (<1,000 m^3), moderate volume (>1,000 to 2,000 m^3) and high volume (>2,000 m^3). The median values are listed in Table 2. The table shows that most of the discharge events were short and the discharge volume in 31 of 45 events was low. However, there were two events in which duration was long and volume was high (cluster 5).

5 MICROPOLLUTANTS

The collected samples were analysed for numerous parameters including micropollutants. In the following the frequencies of detection and concentrations of micropollutants, the calculated load and the pathways of entry will be presented and the results will be discussed against the background of the regulatory status.

5.1 Frequencies of detection and concentrations

Micropollutants were detected in 56% of the discharge samples from the storm sewer under study. Figure 3 shows the percentage of events in which a targeted substance was detected in at least one sample. Among pesticides, Mecoprop (37% of all events) and MCPA (2-methyl-4-chlorophenoxyacetic acid, 18% of all events) were found the most frequently. In addition, concentrations of 13 other pesticides were detected in a range from 2% to 7% of all events. The maximum concentrations of pesticides were between 0.03 µg/L (terbuthylazine) and 5.23 µg/L (MCPA).

Besides pesticides, pharmaceuticals were found in the discharge of the storm sewer (Fig. 3). In particular, the non-steroidal anti-inflammatory drugs Ibuprofen and Diclofenac

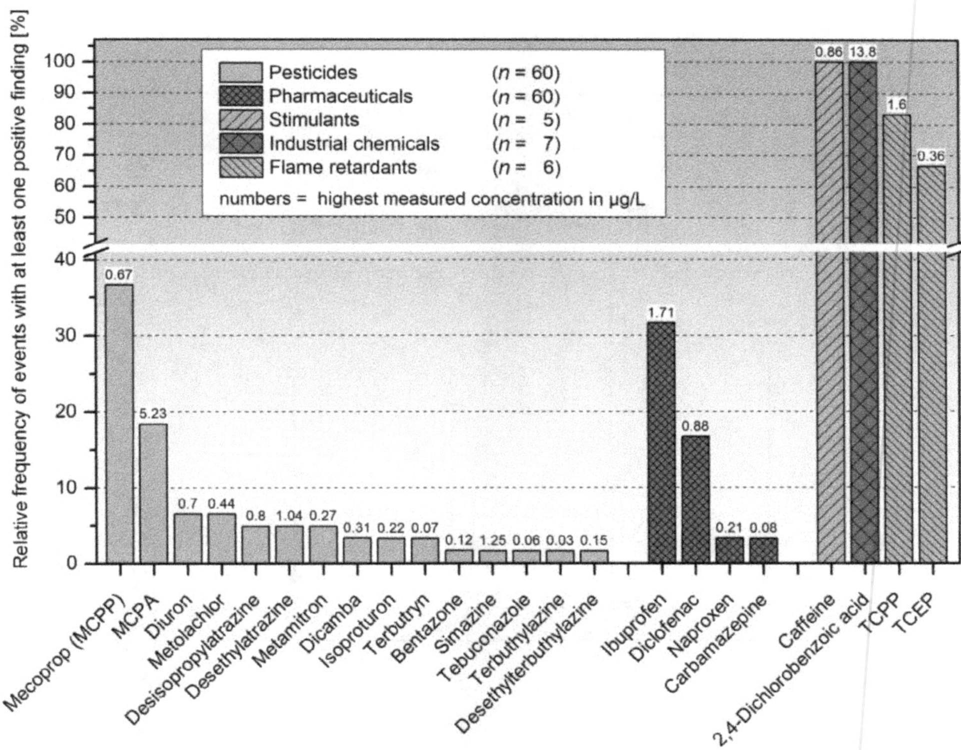

Figure 3: Micropollutants in the discharge of the Meckenheim storm sewer: relative frequency of events with at least one positive finding for each substance, listed by substance groups (*n* = number of events in which each substance group was analysed).

were detected in 32% and 17% of events, respectively. The maximum concentrations were between 0.08 µg/L (Carbamazepine) and 1.71 µg/L (Ibuprofen).

In the samples examined under the extended parameter set, micropollutants were found which belonged to additional substance groups including the stimulant caffeine, industrial chemicals and flame retardants (Fig. 3, far right side). The industrial chemical 2,4-dichlorobenzoic acid and the stimulant caffeine were detected in all events under study. Maximum concentrations were 13.80 µg/L for 2,4-dichlorobenzoic acid and 0.86 µg/L for caffeine. The flame retardants Tris(2-chloroisopropyl)-phosphate (TCPP) and Tris(2-chloroethyl)-phosphate (TCEP) were found in 83% and 67% of all investigated events, respectively. The maximum concentrations were 1.60 µg/L (TCPP) and 0.36 µg/L (TCEP).

5.2 Load

Besides the frequency of occurrence and concentrations of micropollutants in the storm sewer discharge, the discharged load is of particular interest for the evaluation of water contamination in the receiving river. Figure 4 shows the discharged load for selected micropollutants. The calculation was made using the average concentration of all samples of the individual events and the average volume of discharge during the sampling period. Only events that show substance concentrations above the detection limit were considered. The calculated

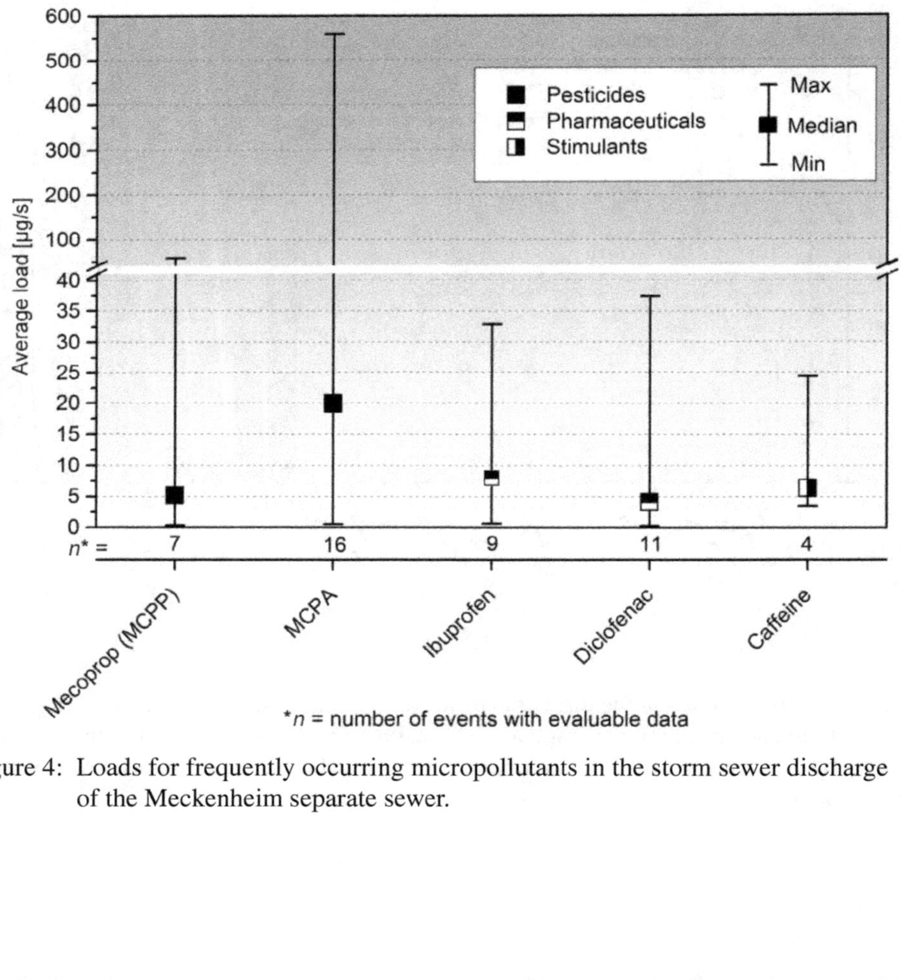

Figure 4: Loads for frequently occurring micropollutants in the storm sewer discharge of the Meckenheim separate sewer.

Table 3: Occurrence of positive findings depended on the winter/summer half-year.

	Pharmaceuticals	Pesticides		
Positive findings	Diclofenac	Ibuprofen	MCPA	MCPP
Positive findings[en]	42[en]11%	68[en]18%	73[en]19%	83[en]22%
Portion of all samples (n = 379)				
Winter half-year[en]27 of 62 events (44%)	35 (83%)	53 (78%)	50 (68%)	47 (57%)
Summer half-year[en]35 of 62 events (56%)	7 (17%)	15 (22%)	23 (32%)	36 (43%)
Related to half-year				
Winter half-year	9%	14%	13%	12%
Summer half-year	2%	4%	6%	10%

median loads for some of the most frequently occurring micropollutants (see Fig. 3) were between 4.0 and 19.9 µg/s, with maximum loads above 500 µg/s for MCPA.

The calculated micropollutant input loads were used to estimate the load of the receiving watercourse stemming from the Meckenheim storm sewer. It should be noted that the occurrence (positive findings) of the detected pharmaceuticals and pesticides depended on the respective times of examination. Pharmaceuticals and pesticides were found more frequently during the two winter half-years (1 November–30 April) covered by the study. By contrast, industrial chemicals and the stimulant caffeine were found in the storm sewer discharge nearly continuously. Table 3 shows the pharmaceuticals and pesticides under study which were detected with more than 40 positive findings of the 379 analyses conducted during the monitoring period from September 2010 to September 2012. About 83% of Diclofenac findings stemmed from the winter half-year, but only 17% from the summer half-year, although more separate sewer outflow events occurred during the summer half-year (56%) than during winter half-year (44%). The same is true for Ibuprofen with 78% and 22%, respectively. The pesticides MCPA and MCPP also show more positive findings during the winter half-year than in the summer half-year, but not as strongly as for pharmaceuticals.

The winter half-year period was selected to estimate the expected range of the discharge input load for Ibuprofen. Based on the collected data from the monitoring period, it was determined that in sum 4 h stormwater from the Meckenheim storm sewer was discharged into the Swist during a winter period. To generate a best-case scenario for the winter half-year, it was assumed that all discharge events of this half-year period belong to cluster 1 (short duration, low discharge volume, see Table 2). For the worst-case scenario, it was estimated that in this period all discharge events belong to cluster 4 (medium duration, moderate discharge volume). With median concentrations representing winter half-year for Ibuprofen of 0.31 µg/L, Ibuprofen input load for the winter half-year was calculated to be in the range from 196 mg (best case) to 519 mg (worst case).

5.3 Pathways of entry

Pathway of entry will be discussed in reference to the most frequently occurring micropollutants listed in Fig. 3. One of the most important causes of the detected pollution in stormwater discharge appeared to be faulty connections of wastewater piping to the stormwater sewer. The visible residues (sanitary articles) of domestic wastewater in the stormwater tank provided a clear sign of contamination by this means. Such an assumption was reinforced by the high number of positive findings of the stimulant caffeine, which is a typical indicator for domestic wastewater. Faulty connections also explain the pollution of the collected stormwater with pharmaceuticals. The two most often found pharmaceuticals were Ibuprofen and Diclofenac, which are widely available 'over the counter'. The active ingredients of the two analgesics are partly excreted after application or washed off when applied externally.

Pesticide pollution of stormwater from the separate sewer system under study likely originated either from farmyard runoff or from incorrect private use of pesticides on paved areas. The pesticides that occurred most frequently (MCPA, Mecoprop) are used in agriculture and in private homes as weed killers. The two agents are commercially available together with other substances such as Dicamba, 2,4-D, Clopyralid or Fluroxypyr as a combined preparation – marketed as 'lawn weed free' [7]. Another source of pollution with Mecoprop is bitumen roof sheeting, in which this substance is used as an agent to prevent against rooting [8]. The herbicide Diuron, which has not been authorized for agricultural use since 2007, is added to building materials to prevent growth of moss on exterior wall paints. Biocides used in building materials are washed out during rain events and thus enter the storm sewers. Burkhardt et al. [4] proved the occurrence of leaching from building materials in both laboratory and field studies. They concluded that storm sewer water from catchments with predominantly new or newly refurbished buildings is especially likely to contain biocides such as Mecoprop, Diuron and Terbutryn.

The numerous positive findings for flame retardants such as TCEP and TCPP are due to their ubiquitous distribution in the environment. These substances are also in use as additives to many everyday items as well as in building materials. Klemmlein et al. [9] have demonstrated the steady emissions of flame retardants by outgassing from everyday items. For example, a TCPP emission of up to 140 $g/m^2/h$ was measured from assembly foams. Rainwater pollution is thus driven in part by wash-off from paved areas, in part by atmospheric deposition.

5.4 Regulatory status

Some of the frequently detected micropollutants mentioned in Section 5.3 are of regulatory concern. MCPA and Mecoprop are listed in the German Federal Surface Water Ordinance [10] as river-basin-specific pollutants with an EQS (Environmental Quality Standard) of 0.1 μg/L as an annual average. The biocide Diuron has also been classified in this listing as a priority substance. A maximum concentration of 1.8 μg/L and an annual average EQS of 0.2 μg/L should be respected for this substance. Although pharmaceuticals have not yet been listed in European or national regulations, the painkiller Diclofenac is placed on the watch list of substances for which monitoring data should be collected to facilitate identification of appropriate measures against risk from the substances in question [11].

6 CONCLUSION

This study shows that water from the rainwater channel in separate sewer systems can be significantly polluted with micropollutants. In particular, surface waters with low mean runoff could be influenced considerably by pollutant loads discharged from stormwater sewers, possibly resulting in non-compliance with legal requirements (e.g. German Federal Surface Water Ordinance [10]). For this reason, a need for action is at hand wherever a further build-out of urban drainage systems is undertaken. Two options exist to reduce pollution arising from stormwater discharge. The first is to eliminate faulty connections from wastewater to the storm sewer. The other is to perform advanced treatment of the collected rainwater by means of retention soil filters [12]. Retention soil filters can also be a practical solution with which the hydraulic stress in rivers brought about by stormwater can be significantly attenuated.

ACKNOWLEDGEMENTS

This study was conducted in the framework of the research and development project 'Examination and Evaluation of Measures to Reduce Physico-chemical and Hygienic-microbiological Loads in Rivers with the Example of the Swist River' funded by the Ministry for Climate Protection, Environment, Agriculture, Conservation and Consumer Protection of the federal state of North Rhine-Westphalia. The authors are grateful for the financial support of the project. Thanks also to B. Smallberg and colleagues from Erftverband and the Institute for Hygiene and Public Health.

REFERENCES

[1] MURL (Ministerium für Umwelt Raumordnung und Landwirtschaft), *Niederschlagswasserbeseitigung gemäß § 51a des Landeswassergesetzes RdErl. d. Ministeriums für Umwelt, Raumordnung und Landwirtschaft IV B 5 – 673/2-29010/IV B 6 – 031 002 0901 v. 18.5.1998 (Collecting and Draining Rain Water according to § 51a Water Act of North Rhine-Westphalia, Circular of the Ministry of Environment, Planning and Agriculture IV B 5 – 673/2-29010/IV B 6 – 031 002 0901 v. 18.5.1998),* Düsseldorf, 1998.

[2] Birch, H., Mikkelsen, P.S., Jensen, J. & Lutzhoft, H.-C.H., Micropollutants in stormwater runoff and combined sewer overflow in the Copenhagen area, Denmark, *Water Science & Technology*, **64**(2), pp. 485–493, 2011. DOI: 10.2166/wst.2011.687.

[3] Zgheib, S., Moilleron, R. & Chebbo, G., Priority pollutants in urban stormwater: Part 1 – Case of separate storm sewers, *Water Research*, **46**(20), pp. 6683–6692, 2012. DOI: 10.1016/j.watres.2011.12.012.

[4] Burkhardt, M., Kupper, T., Hean, S., Haag, R., Schmid, P., Kohler, M. & Boller, M., Biocides used in building materials and their leaching behavior to sewer systems, *Water Science & Technology*, **56**(12), pp. 63–67, 2007. DOI: 10.2166/wst.2007.807.

[5] Swist IV, Abschlussbericht, *Überprüfung und Bewertung von Maßnahmen zur Reduzierung der chemisch-physikalischen und hygienisch-mikrobiologischen Belastungen von Fließgewässern am Beispiel der Swist (Review and assessment of measures to reduce the chemical-physical and hygienic-microbiological load of the river waters, using the example of Swist)*, Erftverband and Institut für Hygiene und Öffentliche Gesundheit der Universität Bonn im Auftrag des MKULNV NRW: Bergheim/Bonn, p. 113, 2012.

[6] Christoffels, E., *Monitoring und Modellanwendung – Entwicklung eines Immissionsinventars am Beispiel der Erft,(Monitoring and Model Application – Development of an Immission Inventory with the Example of the Erft River),* ISBN 978-3-00-024445-2, Erftverband: Bergheim, 2008.

[7] BVL (Bundesamt für Verbraucherschutz und Lebensmittelsicherheit), Pflanzenschutzmittelverzeichnis (Register of plant protection), available at http://www.bvl.bund.de/DE/04_Pflanzenschutzmittel/01_Aufgaben/02_ZulassungPSM/01_ZugelPSM/psm_ZugelPSM_node.html (accessed 16 April 2015).

[8] Wittner, I., Dynamik von Biozid- und Pestizideinträgen (Dynamics of biocide and pesticide input), *Eawag News*, **67d**(Juni), pp. 8–11, 2009.

[9] Kemmlein, S., Hahn, O. & Jann, O., Emissionen von Flammschutzmitteln aus Bauprodukten und Konsumgütern (Emissions of flame retardants from building material and consumer products), *Umweltforschungsplan des Bundesumweltministeriums für Umwelt, Naturschutz und Reaktorsicherheit*, Forschungsbericht 29965, 2003.

[10] OGewV, *Oberflächengewässerverordnung vom 18. März 2016 (BGBl. I S. 1429), (German Federal Surface Water Ordinance),* 2016.

[11] European Commission, Proposal for a Directive of the European Parliament and of the Council amending Directives 2000/60/EC and 2088/105/EC as regards priority substances in the field of water policy, COM (2011) 876 final 2011/0429 (COD), available at http://eurlex.europa.eu/LexUriServ/site/de/com/2011/com2011_0876de01.pdf (acessed 12 April 2014).

[12] Mertens, F. M., Christoffels, E., Schreiber, C. & Kistemann, T., Rückhalt von Arzneimitteln und Mikroorganismen am Beispiel des Retentionsbodenfilters Altendorf (Retention of pharmaceuticals and microorganisms using the example of the retention soil filter Altendorf), *Korrespondenz Abwasser Abfall*, **59**(12), pp. 1137–1143, 2012. DOI: 10.3243/kwe2017.03.002.

WATER QUALITY MONITORING IN URBAN BASINS AS SUPPORT FOR WATER RESOURCES MANAGEMENT: A CASE STUDY FROM SOUTHERN BRAZIL

NÍCOLAS R. FINKLER[1], JARDEL COCCONI[2], TAISON A. BORTOLIN[3], LUDMÍLSON A. MENDES[4], VANIA E. SCHNEIDER[3]
[1]University of São Paulo, Brazil
[2]Federal University of Santa Catarina, Brazil
[3]University of Caxias do Sul, Brazil
[4]Federal University of Sergipe, Brazil

ABSTRACT

One of the main challenges of water management in developing countries is to control the impact of the urban environment on the natural environment. Identifying sources of pollution in an urban watershed is a critical first step towards providing more integrated environmental planning, proper wastewater disposal and public water supplying. Thus, in this study we assessed 5-year water quality data from six urban river basins in Southern Brazil. In addition to the principal component analysis (PCA), three indexes were evaluated individually: Water Quality Index (WQI), Toxicity Index (TI) and CCME WQI framework (CCME WQI). In order to evaluate the effect of land use, the monitoring sites were assessed according to the urbanization criteria. The application of PCA revealed the existence of six components, explaining 73.78% of data variation. The component that explains most of the variation in water quality (30.80%) is associated with domestic wastewater. The second component showed a strong dependence (29.44%) on industrial activities such as electroplating and metalworking in determining the water quality, while the other components are related to certain industrial and agricultural activities. Likewise, the application of WQIs demonstrated similar results to the PCA. WQI and TI showed scenarios of concern regarding public supply. CCME WQI presented a significant disparity between the assessed watersheds and the Brazilian legal framework goals. Studies in this field significantly contribute to the establishment of environmental licensing criteria, by demonstrating patterns and environmental features. In addition to it, one can identify which watersheds demand greater attention with respect to control and recovery of proper environmental conditions. Furthermore, it can provide support for revisions in urban and watershed planning, especially in qualitative aspects eluding conflicts over water use in future scenarios.
Keywords: Caxias do Sul, multivariate analysis, Toxicity Index, urban waters, Water Quality Index, water resources management

1 INTRODUCTION

Water quality pollution often works as a result of multiple land use purposes in a water body, and such pollution sources can be classified as point or diffuse sources. Whereas point source pollution is easily identified (e.g. an industrial effluent contribution), non-point sources (e.g. agrochemical drainage or urban runoff) are of mapping difficulty, when one does attribute contamination to a single root [1–4].

The spatiotemporal evaluation of water quality in an urban watershed provides more integrated planning with regard to environmental issues and public water supply. It also allows better understanding of the relationships between land use and water embedment, and to further identification of point source pollution. Technical data for non-point source recognition and mapping also become reliable, allowing an incisive discussion over the multiple water demands and uses [5–8].

In this context, at the beginning of 2009 the project 'Quali-quantitative Monitoring of Urban Watersheds in Caxias do Sul, RS' was created. The main goal was to collect and process water quality data from urban waters in the municipality as well as to, simultaneously,

DOI: 10.2495/EI-V1-N3-298-311

detail land use and cover of the region, seeking to link these topics in annual reports. There were essentially three campaigns for monitoring and sampling, the last of which was conducted in July 2014.

The present study aims at showing the current water resource diagnosis in the city of Caxias do Sul by calculating and reviewing Water Quality Index (WQI), Toxicity Index (TI) and CCME WQI framework (CCME WQI), as well as some statistical results through principal component analysis (PCA) of the monitoring data.

2 MATERIALS AND METHODS

2.1 Study area

The city of Caxias do Sul is located in the southern Brazilian state of Rio Grande do Sul, metropolitan region of Serra Gaúcha, with a total area of 1,644.3 km^2. The estimated population of the city in 2014 was 470,223 inhabitants, 96% of which reside in urban areas and 4% in the countryside. Demographic density is estimated to be 265 inhabitants/km^2. It is possible to observe a wide range of economic activities in the Serra Gaúcha region: mechanical engineering industries; textile sector; plastic manufacturing industries; horticultural farming; pig farming; and poultry and livestock [9].

Water parting and topographical limits split the area of the city into two water basins: the Taquari-Antas Watershed and the Caí Watershed. The total drainage areas for the main urban catchments are listed in Table 1.

Table 2 presents the framework of waters from each watershed within the city limits considering the already established classes and water uses according to Ref. [10]. Classes range

Table 1: Drainage areas for the main urban catchments (*monitored).

Sub-basins	Microbasin area (km^2)	Urban area (km^2)	% of area in urban perimeter	% of drainage area in urban perimeter	% of urban perimeter in basin area
Faxinal Stream*	118.9	26.53	22	11	Antas River 65
São Marcos River	403.54	26.53	7	11	
Herval Stream	77.59	70.872	91	30	
Maestra Stream*	53.68	48.49	90.33	20	
Samuara Stream	22.95	17.35	76	7	
Curuçu Stream	86.511	3.869	0	2	
Tega River*	291.47	123.231	42	52	
Biazu Stream	48,74	3.88	8	2	
Loreto Stream	30.49	2.35	8	1	
Ouro Stream	153.53	2.36	2	1	

(*Continued*)

Table 1: (*Continued*)

Sub-basins	Microbasin area (km²)	Urban area (km²)	% of area in urban perimeter	% of drainage area in urban perimeter	% of urban perimeter in basin area
Pinhal Stream*	87.08	28.52	33	12	Caí River 35
Belo Stream*	162.44	50.88	31	21	
Espelho Stream	46.35	15.2	33	6	
Pena Branca Stream	22.01	14.76	67	6	
Piaí River*	328.7	29.96	9	13	

Table 2: Framework targets for watersheds.

Watershed		Framework targets		Legislation (resolution)
		Intermediate	Final	
Belo Stream		Class 3	Class 2	CRH 50
Pinhal Stream		Class 3	Class 3	CRH 50
Faxinal Stream		Class 3	Class 2	CRH 121
Maestra Stream		Class 2	Class 2	CRH 121
Piaí River	Stretch E	Class 1	Class 1	CRH 50
	Stretch W	Class 3	Class 3	
Tega River		Class 2	Class 2	CRH 121

from 1 to 4 according to the pollution levels and restrictions of the water. Class 1 allows less restrictive water uses and presents significantly better water quality than class 4. Also, local legislation must consider improvement of the water bodies as targets to be achieved within the given years. The embodiment targets for the sub-basins inserted in both the Taquari-Antas and Caí Watershed are shown in Table 2 [11, 12].

In order to properly meet the goals of the monitoring project and to improve its importance over the years, the sampling point grid has been changed constantly. Figure 1 shows the watersheds assayed and the totality of sampling points from all monitoring campaigns, as well as spatial classifications with regard to the urbanization criteria (Table 3).

2.1 Tools for results analysis

This study analysed 28 physical, chemical and microbiological variables. During the three phases of the project, a volume of about 18,000 results were generated in the analysed watersheds. To enable the understanding of the results, multivariate statistical techniques and WQIs were used, whose calculation methodologies are presented later.

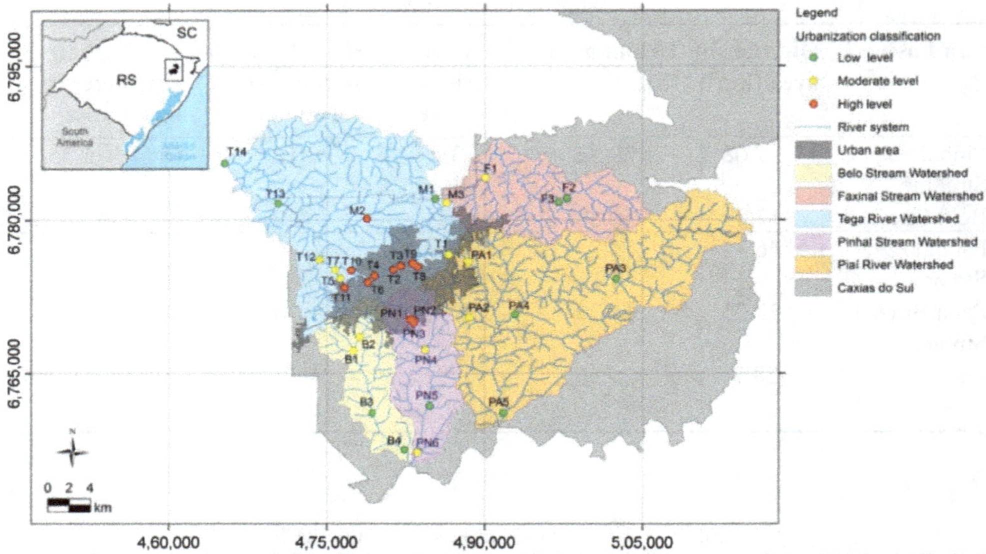

Figure 1: Location and characterization of sampling points in each watershed (T, Tega; PN, Pinhal; B, Belo; M, Maestra; F, Faxinal; PA, Piaí).

Table 3: Criteria for spatial classification of the monitoring points.

Classification	Features
High level of urbanization	Heavily urban surroundings Significant changes of the natural landscape Industrial sector (>100 m)
Moderate level of urbanization	Moderate urban surroundings Moderate changes of the natural landscape Industrial sector (>1 km)
Low level of urbanization	Small urban centres Little or no change of the natural landscape Absence of the industrial sector (1 km)

PCA has been widely used in the interpretation of complex water quality data and other environmental problems. PCA may be defined as a multivariate analysis technique that transforms the set of data into a new system of variables. Within this new arrangement, the principal components (PC) are linear functions of the original set of variables, which are not correlated. The projection of the greatest variation from the sample generates the first PC, the projection of the second greatest variation, the second PC and so on [13].

The application of PCA has been an effective tool in water resources management and pollution control, in addition to being useful in identifying possible factors of influence in river basins [14]. Some researchers [6, 14–16] faced a large set of data and found PCA as a suitable tool in the evaluation and identification of water quality.

WQI was initially created by the National Sanitation Foundation of the United States and modified by the State Sanitation Company of São Paulo (CETESB). It is calculated by the product operator equation, using the curves of importance of nine water quality parameters developed by CETESB [17], which represent the variation in the water quality produced by the possible measures of the parameter. The results are compared in a water quality classification table, with value intervals (Table 4). Studies [2, 4, 5, 8] used the index to assess water quality.

An index often used to represent the potential toxicity of a particular water sample is the TI. According to Ref. [18], a form of TI classification used by the Institute of Water Management of Minas Gerais characterized toxicity as low, medium or high. The evaluation is performed by comparing the concentrations obtained with the standards set by the CONAMA Resolution 357 [10]. The calculation of TI in this study considers the following parameters: ammoniacal nitrogen, nitrates, phenol, cyanides, lead, chromium, nickel and zinc. The worst situation identified at the point of sampling, considering all the parameters used in the TI calculation, defines the contamination category of the period in consideration [18]. The classification range is presented in Table 5.

The CCME WQI was developed by the Canadian Council of Ministers of the Environment [19], with the aim of providing a tool for evaluating water quality data, incorporating the most convenient parameters from each institution and ease of understanding. Studies such as [20–23] used this methodology in the assessment of water quality for various uses. The index was proposed by Amaro [24] in the assessment of water quality in the PCJ (Piracicaba, Capivari and Jundiaí rivers, SP – Brazil) basins and is currently used by the National Water Agency (ANA – Brazil) to evaluate compliance with the established quality objectives from Brazilian rivers [25]. Three ranges of values are established for the index, which characterize compliance with water quality objectives defined by the framework, as shown in Table 6.

Table 4: Water quality classification.

Category	Values
Excellent	$90 < WQI \leq 100$
Good	$70 < WQI \leq 90$
Regular	$50 < WQI \leq 70$
Poor	$25 < WQI \leq 50$
Very poor	$WQI \leq 25$

Table 5: Interpretation of the conditions for toxicity contamination.

Contamination	Representative colour	Relationship between the concentration and the class threshold value (quality standard)
Low	Green	$C \leq 1,2.P$
Mid	Yellow	$1,2.P < C \leq 2.P$
High	Red	$C > 2.P$

P = standard value (class limit), according to Ref. [10]; C = concentration of the parameter.

Table 6: CCME WQI classification categories.

Categories	Description
80 ≤ CCME WQI ≤ 100 Compliant	Most or all measurements are within the quality standards defined by the framework
45 ≤ CCME WQI < 80 Outlying	Measurements are often at odds with the quality standards defined by the framework
CCME WQI < 45 Non-compliant	Most or all of the measurements are not in compliance with the limits of the framework class

Among the quality parameters that were sampled, 14 are defined in the CONAMA Resolution 357 [10] within the maximum or minimum limits of the framework, depending on the type of use of the bodies of water. In order to calculate the CCME WQI in this study, the intermediate quality goal established by the aforementioned resolutions was considered for each river basin, as outlined in Table 2.

The WQI and TI results are presented based on the spatial division considering urbanization criteria (Table 3), as well as the breakdown by monitoring phase. The results of the CCME WQI are presented by river basin, representing the frequency that each category obtained at the end of the monitoring period.

3 RESULTS AND DISCUSSION

The PCA revealed the existence of six PCs, explaining 73.78% of total variation (Table 7). We used 18 water quality parameters, which are present in most of the PCA analysis used as baseline studies. Reference [26] classified the factorial loads corresponding to the absolute composition of the PC in values ranging from significant (>0.75), medium (0.50–0.75) and irrelevant (0.30–0.50). In water quality parameters study, which present great data variability, values greater than 0.50 are commonly used to define PC loads [5, 7, 27].

The analysis of components generated with the PCA suggests that the variation in the water quality stems mostly from domestic sewage discharges without proper treatment. It is possible to identify within the composition of PC1 the parameters of conductivity, total Kjeldahl nitrogen, ammoniacal nitrogen, total phosphorus, biochemical oxygen demand (BOD), thermotolerant coliforms and chemical oxygen demand (COD), which correspond to 30.80% of the variation of the data and are strictly related to domestic pollution.

PC2, PC3 and PC4, corresponding to 29.44% of the total variation, had as heavier weights the metals used in the most representative industrial activities in the region, electroplating and metal mechanics. Furthermore, the presence of COD composing PC2 and pH in PC3 suggests that there is a correlation between COD, pH and the presence of these metals in the determination of the water quality in the city rivers.

PC5 and PC6, composed of cyanide and phenol parameters, respectively, indicate that industrial activities such as metallurgy, plastics manufacturing and electrical components, as well as the food industry, influence the determination of water quality. There is still a parcel related to the use of pesticides, which has such elements in its composition. It is worth noting the toxicity these components present to aquatic ecosystems, being the implementation of control systems important, bearing in mind the potential contamination to the aforementioned.

Table 7: Factorial loads with rotation of the variables for the identified PC.

Parameter	Principal component					
	1	**2**	**3**	**4**	**5**	6
Conductivity	**0.959**	0.056	0.067	−0.004	0.009	−0.076
Total Kjeldahl nitrogen	**0.954**	0.065	0.108	−0.037	0.025	−0.113
Ammoniacal nitrogen	**0.918**	0.007	0.121	−0.073	−0.013	−0.162
Total phosphorus	**0.866**	0.239	−0.08	0.019	−0.036	0.087
BOD	**0.776**	0.232	−0.097	0.142	0.088	0.206
Thermotolerant coliforms	**0.648**	0.079	0.089	0.283	0.331	0.226
COD	**0.621**	**0.646**	−0.013	0.231	0.033	0.102
Aluminium	0.013	**0.894**	0.059	0.258	−0.072	0.016
Lead	0.176	**0.734**	−0.202	−0.297	0.131	0.066
Nickel	0.002	−0.036	**0.844**	−0.038	0.137	−0.126
pH	0.459	0.003	**0.536**	0.118	−0.386	0.154
Chrome	0.080	0.003	0.048	**0.836**	0.070	−0.074
Zinc	0.193	0.324	−0.134	**0.506**	0.036	−0.423
Cyanide	0.037	−0.023	0.076	0.087	**0.821**	0.148
Phenol	0.188	0.183	−0.050	−0.171	0.123	**0.674**
Dissolved oxygen	−0.453	−0.206	0.480	0.186	−0.438	0.113
Total solids	0.198	0.300	0.210	−0.089	−0.143	−0.335
Water temperature	0.383	0.032	0.442	−0.428	0.278	−0.395
Total	**5.545**	**2.154**	**1.597**	**1.549**	**1.295**	**1.142**
% Variance	**30.804**	**11.965**	**8.870**	**8.604**	**7.195**	**6.343**
% Accumulated	**30.804**	**42.770**	**51.640**	**60.244**	**67.439**	**73.782**

*Variable rotation method: VARIMAX with Kaiser normalization.

The absence of certain parameters within the PC loads, such as dissolved oxygen, total solids and water temperature, can be indicative that the variance of these is more strongly interlinked to natural rather than anthropic processes. However, the importance of these in regard to maintaining environmental balance is unquestionable, and its monitoring and quantification should be considered in any project where water quality is being monitored.

Figure 2 illustrates the evolution of WQI of the rivers, during the three monitoring phases. Note that the points analysed presented a predominance of waters classified as regular and/or poor at all stages of monitoring. In moderate and highly urbanized areas, the result of poor water quality is a reflection of domestic sewage discharge *in natura* from point sources. Other point sources such as industry, which do not have adequate wastewater treatment, or do not meet the minimum discharge requirements, contribute significantly to this definition.

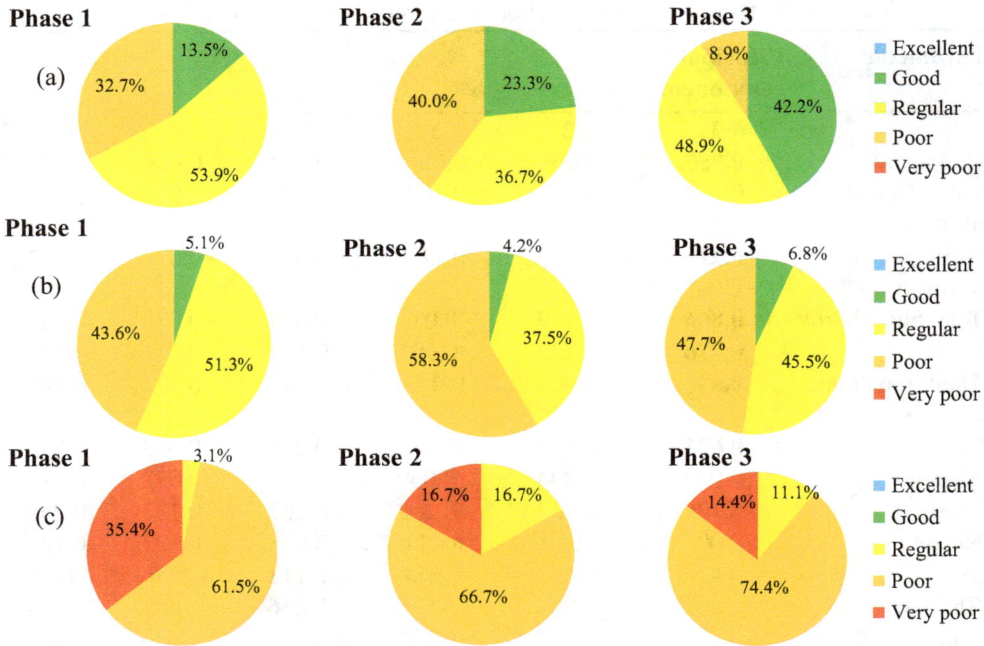

Figure 2: WQI evolution during monitoring. (a) Low degree of urbanization; (b) moderate degree of urbanization; and (c) high degree of urbanization.

At the same time, despite having lower domestic and industrial wastewater pressure, waters from areas with a low degree of urbanization were classified as regular and poor for most part of the monitoring period. The results showed that despite being located in peri-urban areas, water quality at these points suffer the influence from upstream urban areas. Here, it can be noted the presence of the self-purification process, even in small degree, given that the BOD concentration is mitigated at all points in these areas compared to those with a larger anthropic presence. Activities such as agriculture and livestock, which are characterized by their significant contribution of nutrients, also contribute to the variation of water quality at these sites.

An analysis considering only the results from the indexes in the monitoring phases demonstrates that there was an improvement in water quality over time (the WQI improves at the points of high and low degrees of urbanization). This may be influenced by the change of monitoring points between the three phases. However, it should be considered that the sewage service in the municipality (SAMAE) increased the percentage of sewage treated during the three phases. Figure 3 illustrates the evolution of TI in urban watersheds of Caxias do Sul, across the three monitoring phases.

Figure 3 shows the waters of Caxias do Sul are in risk with regard to the level of toxicity displayed. As expected, the points under a strong urban presence displayed high toxicity. It is interesting to note the sites with a lesser pressure from urbanization, likewise, demonstrated high levels of toxicity, particularly associated with nitrate, ammoniacal nitrogen, cyanide and phenol. The first two indicate the process of organic nitrogen degradation, originating from domestic sewage without treatment at points upstream [18]. Cyanide and phenol

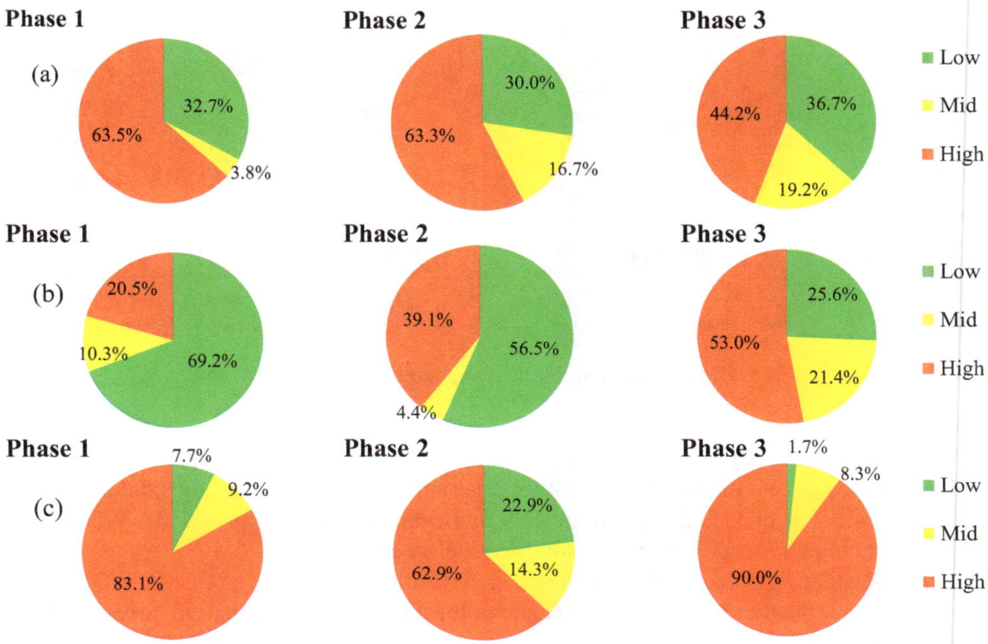

Figure 3: TI evolution during monitoring. (a) Low degree of urbanization; (b) moderate degree of urbanization; and (c) high degree of urbanization.

contamination may be associated with the transport of industrial areas located upstream, or even resulting from point application of agrochemicals.

The results exhibit some concerns in regard to the quality of municipal water. As there was no improvement of TI during the monitoring period, it can be inferred that actions such as the installation of a sewage treatment plant in the basin of Tega River (2012), among other non-structural improvements (licensing and environmental monitoring), did not generate the desired effect. Still, it is worth pointing out that basins such as those found in the Faxinal River and the Maestra Stream are located in catchment areas for public supply, and thus the parameters considered in the calculation of TI may be present in water which is provided to the population.

The overall CCME WQI results are presented in Fig. 4 and Table 8. A relevant point concerning the results of CCME WQI is that it affects a quality index that has as reference the class targets of where the river is situated. In other words, it is not a simple depiction of the quality such as with the WQI or the TI. Its interpretation should not be made considering only qualitative aspects, but also taking into account the goals proposed for the watershed commission.

The results presented in Fig. 3 show that the basin of the Tega River is the furthest away from its quality goals. The motives are diverse, being BOD, ammoniacal nitrogen, nitrate, phosphorus, thermotolerant coliforms, nickel, chromium, zinc, phenol and cyanide the parameters situated at furthest from the limits established for class 3, reflecting both domestic and industrial contamination. At some points of monitoring during phases 1 and 2, it was not possible to observe an improvement, indicating that quality problems are long-standing.

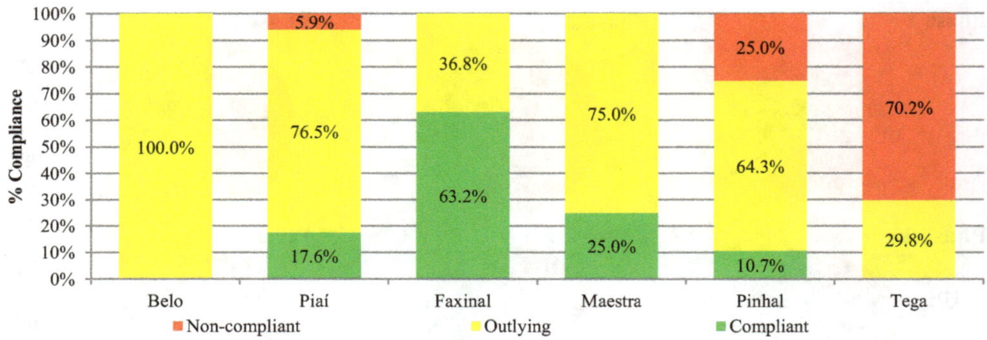

Figure 4: CCME WQI results in the urban basins during monitoring.

Moreover, even if it is possible to observe a slight improvement of the index values in the upstream–downstream direction, especially for parameters reflecting organic contamination, parameters such as nickel, chromium, phenol and cyanide decrease the CCME WQI in the basin, this being a reflection of the self-purification process.

A similar situation occurs with the watershed of the Pinhal stream. The CCME WQI results are likewise a reflection of both industrial and domestic water contamination, given the high level of urbanization in the upstream portion of the basin, in addition to greater restriction of the intermediate framework (Class 2) in relation to Tega River (Class 3). At points further away from the urban area, the CCME WQI presented compliance to the framework. However, there was a worsening in the quality of the stream, resulting from the presence of phenol, chrome and also an excess of BOD, resulting in a reduction of the index. This change seems to be related to the activities carried out in the district of Villa Cristina, the area drained by the water resource.

The Piaí River presents a peculiar situation, given that there are two quality targets in place for the river basin: Class 3 for the rivers of the western section and Class 1 for the rivers of the eastern section, this being evaluated only in phase 3. Therefore, both the analysed points are outlying from the framework, with point 1 being in worse conditions (CCME WQI between 68.7 and 75.3) than point 2 (CCME WQI between 77.0 and 86.8). Point 1 has been presenting

Table 8: CCME WQI results of each watershed by the monitoring phase.

Monitoring	Classification	Belo (%)	Piaí (%)	Faxinal (%)	Maestra (%)	Pinhal (%)	Tega (%)
Phase 1	Compliant	0.0	0.0	*	0.0	0.0	0.0
	Outlying	100.0	0.0	*	100.0	0.0	40.0
	Non-compliant	0.0	100.0	*	0.0	100.0	60.0
Phase 2	Compliant	0.0	0.0	0.0	0.0	0.0	0.0
	Outlying	100.0	100.0	100.0	100.0	100.0	57.1
	Non-compliant	0.0	0.0	0.0	0.0	0.0	42.9
Phase 3	Compliant	0.0	20.0	66.7	33.3	12.5	0.0
	Outlying	100.0	80.0	33.3	66.7	66.7	26.4
	Non-compliant	0.0	0.0	0.0	0.0	20.8	73.6

a worsening quality since the analysis began. The other points of the watershed, located in streams with greater human impact, showed slight improvement during the monitoring. This evolution seems to be the result of improved treatment, or ending of activities of industries in the region. This becomes evident after observing that for both points, nickel was one of the elements which was largely responsible for the reduced CCME WQI values in phase 1, which was not found in point 3 and found substantially reduced in point 4.

The analysis of target achievements for Maestra and Faxinal Streams is relevant because two water supply springs are located within these watersheds. The Maestra River basin depicted low values of CCME WQI. It is worth noting that the monitoring point is located in a spring, which should present CCME WQI values above 90. The results therefore show evidence of faecal contamination (thermotolerant coliforms, BOD, phosphorus, nitrate and ammoniacal nitrogen). In Faxinal Stream was observed better water quality, especially in points 1 and 2, monitored during phase 3 and classified in all campaigns as in compliance. Though point 3, analysed as of phase 2, presents problems of faecal contamination, with high levels of thermotolerant coliforms, BOD, phosphorus, ammoniacal nitrogen and nitrate, it also presents problems of industrial contamination, as is evidenced by the presence of cyanide, phenol, and in one of the campaigns chrome. Throughout the entire period, the point is classified as an outlier.

The Belo watershed is classified throughout the entire monitoring period as outlying. The region of the springs presents high levels of urbanization, exposing their quality in the downstream direction. The quality problems revealed in the four monitored points are due to faecal contamination, with high values of phosphorus, coliforms and nitrogen (ammoniacal form in point 1 and nitrate in the others), but the presence of cyanide, phenol and nickel were also evident.

4 CONCLUSIONS

This study presented an analysis of data pertaining to the water quality in the city of Caxias do Sul, obtained from a monitoring programme whose goal is to provide decision makers subsidies for managing water resources. The analyses performed on the water quality data revealed watersheds containing higher levels of contamination, primarily due to the urban occupation. Also, it demonstrated the principal parameters and processes that impact rivers.

The application of PCA has shown that the component which explains most of the variation in water quality (30.80%) is associated with domestic sewage lacking proper treatment. The main component includes parameters of conductivity, total Kjeldahl nitrogen, ammoniacal nitrogen, total phosphorus, BOD, thermotolerant coliforms and COD. The second component explained 29.44% of the water quality variation and suggests activities such as electroplating and metalworking impact the water quality. The other components show a strong dependence on industrial activities, in addition to certain agricultural activities when determining the quality of water.

The application of WQIs demonstrated similar results to the PCA. Factors such as the disposal of domestic and industrial effluents without proper levels of treatment significantly impact rivers in Caxias do Sul. The WQI and the TI exhibited such a situation for all watersheds, even including situations of great concern in relation to the public supply. The CCME WQI, in turn, showed how far the analysed basins are from the predicted framework goal, suggesting demand for proposed joint actions between the municipality and basin committees.

The assessment of the compliance with the framework goals shows the effectiveness of this management tool is not achieved solely with the definition of usage classes. Once the monitoring results are compared with the quality standards set for each class, essential results were obtained to support the structuring of management actions on behalf of public officials in a municipality, whose sanitation characteristics resemble the vast majority of Brazilian municipalities, that is, having a low percentage of domestic sewage treatment, and ineffective discharge monitoring of industrial effluents.

Currently, one of the main challenges of water management is to control the impact the urban environment has on the natural environment. Studies such as this one significantly contribute, for example, to the establishment of environmental licensing criterion, by demonstrating patterns and characteristics of the environment, in addition to identifying which watersheds deserve greater attention with respect to control and recovery of environmental conditions. Further, studies such as this can provide support for revisions in Municipal Master Plans and Watershed Plans, especially with regard to qualitative aspects, thus avoiding conflicts over water use in future scenarios.

REFERENCES

[1] Vega, M., Pardo, R., Barrado, E. & Deban, L., Assessment of seasonal and polluting effects on the quality of river water by exploratory data analysis. *Water Research*, **32**(12), pp. 3581–3592, 1998. DOI: 10.1016/S0043-1354(98)00138-9.

[2] Almeida, M.A.B. & Schwarzbold, A., Avaliação Sazonal da Qualidade das Águas do Arroio da Cria Montenegro, RS com Aplicação de um Índice de Qualidade de Água (IQA). *Revista Brasileira de Recursos Hídricos*, **8**(1), pp. 81–97, 2003. DOI: 10.21168/rbrh.v8n1.p81-97.

[3] Singh, K.P., Basant, A., Malik, A. & Jain, G., Artificial neural network modeling of the river water quality – A case study. *Ecological Modelling*, **220**(6), pp. 888–895, 2009. DOI: 10.1016/j.ecolmodel.2009.01.004.

[4] Bertossi, A.P.A., Cecílio, R.A., Neves, M.A. & Garcia, G.O., Qualidade da água em microbacias hidrográficas com diferentes coberturas do solo no sul do Espirito Santo. *Revista Árvore*, **37**(1), pp. 107–117, 2013. DOI: 10.1590/S1415-43662010000100008.

[5] Toledo, L.G. & Nicolella, G., Índice de qualidade de água em microbacia sob uso agrícola e urbano. *Scientia Agricola*, **59**(1), pp. 181–186, 2002. DOI: 10.2134/jeq1998.00472425002700020006x.

[6] Fan, X., Cui, B., Zhao, H., Zhang, Z. & Zhang, H., Assessment of river water quality in Pearl River Delta using multivariate statistical techniques. *Procedia Environmental Sciences*, **2**, pp. 1220–1234, 2010. DOI: 10.1016/j.proenv.2010.10.133.

[7] Guedes, H.A.S., Silva, D.D., Elesbon, A.A.A., Ribeiro, C.B. M., Matos, A.T. & Soares, J.H.P., Aplicação da análise estatística multivariada no estudo da qualidade da agua do Rio Pomba, MG. *Revista Brasileira de Engenharia Agrícola e Ambiental*, **15**(5), pp. 558–563, 2012. DOI: 10.2134/jeq2004.0337.

[8] Damasceno, M. da C.S., Ribeiro, H.M.C., Takiyama, L.R. & Paula, M.T. Avaliação sazonal da qualidade das águas superficiais do Rio Amazonas na orla da cidade de Macapá, Amapá, Brasil. *Revista Ambiente e Água*, **10**(3), pp. 598–613, 2015. DOI: 10.4136/ambi-agua.1606.

[9] Instituto Brasileiro de Geografia e Estatística (IBGE), Gross domestic product – Brazil 2014, available at http://www.ibge.gov.br/english/estatistica/economia/pibmunicipios/2014/default.shtm, 2014 (accessed 25 August 2017).

[10] Brasil. CONAMA Resolution 357, available at http://www.mma.gov.br/port/conama/res/res05/res35705.pdf, 2005 (accessed 25 August 2017).

[11] Rio Grande do Sul. CRH Resolution 121, available at http://www.sema.rs.gov.br/upload/arquivos/201708/22162317-resolucao-crh-121-2012-aprova-enquadramento-aguas-superficiais-bacia-taquari-antas-07-01-1.pdf, 2012(accessed 25 August 2017).

[12] Rio Grande do Sul. CRH Resolution 50, available at http://www.sema.rs.gov.br/upload/arquivos/201708/22151357-resolucao-crh-50-2008-aprova-enquadramento-das-bacias-dos-rios-cai-pardo-tramandai-e-lago-guaiba.pdf, 2008 (accessed 25 August 2017).

[13] Jolliffe, I.T. *Principal Component Analysis*, Springer-Verlag: New York, NY, 2002.

[14] Wang, X., Cai, Q., Ye, L. & Qu, X., Evaluation of spatial and temporal variation in stream water quality by multivariate statistical techniques: A case study of the Xiangxi River basin, China. *Quaternary International*, **282**, pp. 137–144, 2012. DOI: 10.1016/j.quaint.2012.05.015.

[15] Coletti, C., Testezlafl, R., Ribeiro, T.A.P., de Souza, R.T.G. & Pereira, D. de A., Water quality index using multivariate factorial analysis. *Revista Brasileira de Engenharia Agrícola e Ambiental*, **14**(5), pp. 517–522, 2009. DOI: 10.1590/S1415-43662010000500009.

[16] Pinto, U. & Maheshwari, B.L., River health assessment in peri-urban landscapes: An application of multivariate analysis to identify the key variables. *Water Research*, **45**(13), pp. 3915–3924, 2011. DOI: 10.1016/j.watres.2011.04.044.

[17] Companhia de Tecnologia de Saneamento Ambiental (CETESB), *Relatório Anual de Qualidade das Águas Interiores no Estado de São Paulo*, available at http://aguasinteriores.cetesb.sp.gov.br/wp-content/uploads/sites/32/2013/11/02.pdf, 2009 (accessed 25 August 2017).

[18] Von Sperling, M., *Princípios do tratamento biológico de águas residuárias*: *Introdução à qualidade das águas e ao tratamento de esgotos*, DESA-UFMG: Belo Horizonte, Brazil, 1996.

[19] Canadian Council of Ministers of the Environment (CCME), Water Quality Index: Technical Report. In: *Canadian Water Quality Guidelines for the Protection of Aquatic Life*, available at http://www.ccme.ca/files/Resources/calculators/WQI%20User's%20Manual%20(en).pdf, 2001 (accessed 25 August 2017).

[20] Cude, C.G., Oregon Water Quality Index a tool for evaluating water quality management effectiveness. *Journal of the American Water Resources Association*, **37**(1), pp. 125–137, 2001. DOI: 10.1111/j.1752-1688.1978.tb02261.x.

[21] Terrado, M., Barceló, D., Tauler, R., Borrell, E., Campos, S. & Barceló, D. Surface-water-quality indices for the analysis of data generated by automated sampling networks. *Trends in Analytical Chemistry*, **29**(1), pp. 40–52, 2010. DOI: 10.1016/j.trac.2009.10.001.

[22] Hurley, T., Sadiq, R. & Mazumder, A., Adaptation and evaluation of the Canadian Council of Ministers of the Environment Water Quality Index (CCME WQI) for use as an effective tool to characterize drinking source water quality. *Water Research*, **46**(11), pp. 3544–3552, 2012. DOI: 10.1016/j.watres.2012.03.061.

[23] Abukila, A.F., Assessing the drain estuaries water quality in response to pollution abatement. *Water Science*, **29**(1), pp. 1–18, 2015. DOI: 10.1016/j.wsj.2015.02.002.

[24] Amaro, C.A., *Proposta de um índice para avaliação de conformidade da qualidade dos corpos hídricos ao enquadramento*, available at www.teses.usp.br/teses/disponiveis/3/3147/tde-11082009-121147/pt-br.php, 2009 (accessed 25 August 2017).

[25] Agência Nacional das Águas (ANA), *Panorama da Qualidade das Águas Superficiais do Brasil*, ANA: Brasília, 2012.

[26] Liu, S., Manson, J.E., Stampfer, M.J., Hu, F.B., Giovannucci, E., Colditz, G.A., Hennekens, C,H. & Willett, W.C., A prospective study of whole-grain intake and risk of type 2 diabetes mellitus in US women. *American Journal of Public Health*, **90**(9) (1 September, 2000): pp. 1409–1415, 2003. DOI: 10.2105/AJPH.90.9.1409.

[27] Helena, B., Pardo, R., Vega, M., Barrado, E., Fernandez, J.M. & Fernandez, L., Temporal evolution of groundwater composition in an alluvial aquifer (Pisuerga River, Spain) by principal component analysis. *Water Research*, **34**, pp. 807–816, 2000. DOI: 10.1016/S0043-1354(99)00225-0.

CONTROLLING ENVIRONMENTAL POLLUTION IN THE URBAN WATER CYCLE

EMMA FÄLTSTRÖM
Sweden Water Research, Sweden

ABSTRACT

With the responsibility for stormwater, drinking water and wastewater, the water and wastewater sector handles a considerable amount of environmental pollution. In Sweden, the sector has developed strategies for reducing and controlling environmental pollution through a practice called 'upstream work'. 'Upstream work' consists of different strategies (e.g. mapping out industries, water sampling in the pipe system and information campaigns to the public) to hinder chemicals and other pollution from reaching the wastewater treatment plant. In this contribution, 'upstream work' is compared to the wider concept of source control to evaluate if there are any aspects of source control that can help develop 'upstream work'. Further, examples of tools, methods and approaches that facilitate source control and 'upstream work' are presented. A tool, method or approach can be used in different parts of the process of controlling environmental pollution and provide assistance with different challenges. Therefore, the efforts/measures are categorized based on which process step they address (identification, quantification, prioritization and control). Further, tools, methods and approaches that can be performed by the water and wastewater sector are highlighted. The categorization provides practical ways to work on controlling environmental pollution for organizations that currently work with source control and/or upstream work and for the organizations that have not started the process yet. Source control can provide insights to develop upstream work, especially for diffusing pollution. By developing 'upstream work' to include the entire urban water cycle, important steps towards a more sustainable urban water management are taken.

Keywords: drinking water, pollution abatement, pollution prevention, source control, stormwater, wastewater, water and wastewater sector, water management, water pollution, upstream work

1 INTRODUCTION

Environmental pollution from a wide range of sources is concentrated in urban areas [1]. With the responsibility for stormwater, drinking water production and wastewater treatment, the water and wastewater sector has to handle a significant amount of this pollution. In Sweden, the water and wastewater sector has developed strategies for reducing and controlling environmental toxins through a practice called 'upstream work'. According to the Swedish Water and Wastewater Association, 'upstream work' means to "hinder chemicals and other pollution at source in order to avoid introduction to the water cycle or cycle of nutrients" ([2], author's translation). The first 'upstream work' initiatives arose in the beginning of the 1960s and consisted of identifying and addressing releases of harmful substances from industries that damaged the pipe system and the treatment process at wastewater treatment plants (WWTP) [3, 4]. The term 'upstream work' was introduced in the beginning of the 1990s and referred to measures to increase sewage sludge quality [5] and it has had positive results [6]. In recent years, the positive effects of 'upstream work' on water quality in recipient water have obtained more attention and the first systematic attempts of 'upstream work' aimed at raw water have been introduced [7]. 'Upstream work' as a concept and practice is widely known in the water and wastewater sector in Sweden, but outside the sector and in the international community it seems fairly unknown. Source control is the most similar concept, although it is wider and can include strategies for different types of water as well as air, soil and sediments [8].

DOI: 10.2495/EI-V1-N3-312-322

Two ongoing EU projects are focusing on source control. NonHazCity (2016–2019) is a partnership between 18 cities around the Baltic Sea. The project will identify and prioritize sources of environmental toxins, specifically targeting households and other small emitters. The main aim is to reduce and prevent environmental toxins at source to avoid introduction to the sewer system and ultimately the Baltic Sea, with a particular focus on reducing consumption [9]. The SOLUTIONS project (2013–2018) focus on emerging pollutants and the project aim at developing a conceptual framework to support water quality policies with new approaches and tools for monitoring. The development of tools for identification as well as prioritization of water pollution is stressed in the project [10]. However, there are already established tools and approaches that can be useful for source control and 'upstream work' but it can be difficult for professionals to know when to use them [11].

In this contribution, the similarities and differences between the concepts *source control* and *upstream work* are examined, with the aim of understanding the aspects of source control that can help develop 'upstream work'. 'Upstream work' has traditionally mainly targeted wastewater (especially from industries) but this is changing. Amneklev [12] argues that 'upstream work' should be widened and applied to more areas, include more actors and tackle more substances. As source control is a broader concept, it might give some insights into how the practice of 'upstream work' can be expanded to include the whole urban water cycle. Further, I present an overview of some available approaches, methods and tools that can facilitate prevention and abatement of environmental pollution/environmental toxins in water, particularly suitable for use in the water and wastewater sector. I intend to answer the following research questions:

1. What similarities and differences do the concepts 'source control' and 'upstream work' have and what aspects of 'source control' can be used to develop 'upstream work'?
2. What efforts are available, internationally and in Sweden particularly, to facilitate control of environmental pollution that can be utilized by the water and wastewater sector?

International literature on source control and Swedish literature on 'upstream work' were examined to find definitions and examples of practical measures to be able to compare the concepts. As documentation on 'upstream work', especially the history, is rather limited, information is largely based on oral resources from professionals and grey literature. The overview of efforts to facilitate control of environmental pollution aims at being pragmatic and the categorization should be applicable both for organizations that already work with these issues and organizations that have not started to work with it yet. The ambition is not to present an exhaustive investigation of efforts for source control and 'upstream work', but to exemplify how one can go about preventing and reducing environmental pollution in an effective way and what is needed to do so. The efforts/measures did not have to be particularly developed for the water and wastewater sector to be seen as useful and hence included, but the information derived should be relevant for the sector. On this basis, several efforts/measures were discarded as they were so comprehensive that it was challenging to understand what the practical benefits and specific contribution of using it would be.

First, the development of 'upstream work' is described. Second, the differences and similarities between the concepts source control and 'upstream work' and the ways in which source control can give insights into the development of 'upstream work' are discussed. Last, categorized approaches, methods and tools for controlling environmental pollution are presented.

2 THE DEVELOPMENT OF 'UPSTREAM WORK' IN SWEDEN

The first step towards what today is referred to as 'upstream work' in Sweden was initiated in Stockholm and Gothenburg in the beginning of the 1960s. These first attempts were not called 'upstream work' but 'industrial wastewater control' in Gothenburg [4] and 'measures for treatment plants' in Stockholm [3]. In both cities, industries was the target group and the main aim was to protect pipe-system, pump stations and treatment processes [3, 4]. In 1976, the water and wastewater company in Gothenburg formed a working group to evaluate how the industrial wastewater was handled in the region [4]. The group changed its name from 'industrial wastewater control' to 'wastewater quality' in the 1980s to not denounce industries. The name was changed to the 'upstream unit' in 2008. By that time, 'upstream work' was an established concept.

During the 1980s, sewage sludge was spread on agricultural land without much regulation. When Sweden entered the European Union in 1994, the EU sludge directive from 1986 was to be implemented. However, the Federation of Swedish Farmers did not see the directive as strict enough [13]. Therefore, the so-called sludge agreement between the Federation of Swedish Farmers, the Swedish Environmental Protection Agency and the Swedish Water and Wastewater Association was formed. With the 'sludge agreement' the water and wastewater sector had to be able to demonstrate the practical measures that was taken to increase sludge quality and this marks the beginning of the term 'upstream work' [5]. The agreement was broken by the Federation of Swedish Farmers in 1999 following a debate on flame retardants [13]. After the disagreement, the Swedish Water and Wastewater Association initiated a new project to increase sludge quality, which was later developed into a certification system named Revaq. Today Revaq is a cooperative effort between the Swedish Water and Wastewater Association, the Federation of Swedish Farmers, the food industry & Swedish FMCG association [14]. The first WWTP was certified in 2008 and in 2015, 42 municipalities were certified. This translates into that slightly more than half of the Swedish population is connected to a certified WWTP [15]. With the implementation of the Water Framework Directive in the 2000s, 'upstream work' for water quality has become increasingly important [16]. In 2013, the Swedish Environmental Protection Agency recommended the Swedish government to legislate on an obligation for all WWTPs to initiate 'upstream work' [6].

3 COMPARISON OF UPSTREAM WORK AND SOURCE CONTROL

Source control and upstream work have a difference in that 'upstream work' focuses on avoiding introduction of environmental pollution in the water cycle or cycle of nutrients (i.e. sewage sludge), while source control does not particularly target water. Although the term 'source control' appears in the pollution literature, definitions are rare. An exception is Larsen and Gujer [17], who define source control as "the interaction of the consumer with industry that provides the goods and services the consumer desires" (p. 310). In this definition, the relationship between the industry and the consumer is highlighted, while upstream work focuses on the relationship between the water and wastewater utilities and industry as well as consumers and water and wastewater utilities. There are, however, several similarities in the practical application of the two concepts. Larsen and Gujer [17] give examples of substituting chemicals, on-site treatment in industries and segregated waste streams. The Revaq guidelines exemplify 'active upstream work' as mapping industries, requesting lists of chemical usage from connected enterprises, measuring pollutants in sludge and carrying out information campaigns to the public [18]. The measures are rather similar to what Larsen & Gujer exemplify with, although they focus on the industries and not the water and wastewater sector. A water and wastewater company can make use of legislation to pressure industrial companies to substitute chemicals and install treatment, but the industries have to implement the measures.

Eriksson et al. [19] use source control broadly (and calls it emission control strategies) as strategies to manage releases in different parts of the system, from the point of release to the receiving waterway. This says something important on the effectiveness of measures for reducing environmental pollution. The source (as in point of release) might not be the most effective way to control the pollution. Instead, the best alternative, seen as total reduction of pollution, can differ dependent on pollution and source. Therefore, a view of the whole urban water cycle is needed in order to find suitable places to intervene.

There are some indications that the focus on wastewater and sludge in 'upstream work' is beginning to change. The slogan for the Revaq certificate was changed in 2016 from focusing on sludge as a 'recycled crop fertilizer' to also emphasizing the benefits 'upstream work' has for water quality [20]. Further, a Swedish drinking water producer has initiated 'upstream work' for their water reservoirs [7]. To facilitate this broadened perspective, some aspects of source control can be useful. 'Upstream work' as a concept and practice was developed from needs to handle harmful substances in wastewater, which is probably why it works well for waters with clear boundaries, i.e. piped water. However, 'upstream work' for stormwater and raw water is not as widespread, although there are some initiatives. What source control can offer is systematic efforts from a widened perspective, more applicable to stormwater, such as best management practices [21] or raw water, such as source control interventions [22]. Considering the responsibility of the water and wastewater sector for the whole water and wastewater system, such a widened perspective is called for. By considering the whole system, the most suitable places to intervene can increase the effectiveness of the measure. As many environmental toxins are of persistent character, some measures only move pollution from one medium to another [8]. A widened perspective reduces the risk of measures only moving pollution between the medium.

4 APPROACHES, METHODS AND TOOLS FOR SOURCE CONTROL AND UPSTREAM WORK

In the area of environmental management in general and water management in particular, there are several approaches, methods and tools developed to ease the work. However, it can be difficult to know when to use which tool, method or approach. For effective source control and upstream work, there are some process steps that have to be taken and some approaches, methods and tools are more suitable for some process steps than the other.

4.1 Defining approach, method and tool

The use of approaches, methods and tools is often central part of a strategy/effort/initiative. These terms are regularly used in connection with environmental management, but they are rarely defined. de Ridder et al. [23] include methods, procedures and techniques in the term 'tool'. For the purpose of exemplifying practical ways to go about preventing and reducing environmental pollution, such a broad definition does not seem sufficient. van der Vorst et al. [24] instead separates approach from technique and tool, but argue that the three are connected. A *tool* is something used to perform an action. A *technique* is a tool with an added purpose, e.g. to inform decision-making. An *approach* is a technique with an intention or philosophical framing, e.g. sustainable development. These definitions demonstrate the different abstraction levels where *tool* is the most applied and describes something to be used when executing a specific task while *approach* is the most abstract and serves a greater goal. The term *technique* can, in this context, be problematic. In the water and wastewater sector, there has been a large focus and faith in technical solutions, something that has been critiqued [25]. Source control efforts and 'upstream work' can include technical solutions but is not limited to that. To avoid the risk of technique being associated to technical aspects, I instead use the term *method* to describe a process, with a specific purpose, to obtain new information. Hence, based on van der Vorst et al. [24], the definitions are as follows:

- A *tool* is something that is used to execute a certain task.
- A *method* consists of one (or several) tool(s) with a specific purpose.
- An *approach* provides a trajectory or loose structure, identified by a need, but does not provide detailed practical advice. It can consist of several methods but form a whole.

4.2 Analytical scheme

A tool, method or approach can be used in different parts of the process of controlling environmental pollution and provide assistance with different challenges. There are some general lists of harmful substances provided by, for example, the Water Framework Directive, but some substances are more prevalent in some areas than other [26]. Hence, *identifying* problematic substances in different types of water and tracing the sources are important [18]. When the problematic substances in the local context and their sources are identified, a *quantification* of sources and flows give information on where the largest sources are [12]. This information can then be used to *prioritize* between substances and sources [1]. However, as pointed out by Hörsing and Ledin [26], a prioritization should not only be based on quantities but also how harmful the substance is. Some substances do not even have to be quantified to be of high priority. This concerns, for example, the so-called 'phase-out' substances that the Swedish Chemicals Agency has decided on [27]. The water and wastewater company has the right to request an action plan from connected industries on how they plan to phase out these substances. A quantification of flows and sources also result in different source control options [1]. However, this requires that there are some source *control* options identified. In summation, the analytical scheme is as follows:

- Does the tool/method/approach help *identify* substances, sources or abatement strategies?
- Does the tool/method/approach *quantify* substances?
- Does the tool/method/approach help *prioritize* substances, sources or abatement strategies?
- Does the tool/method/approach *control* sources?

4.3 Ways to facilitate the process steps of source control and upstream work

The process steps (identification, quantification, prioritization and control) for source control and upstream work and approaches, methods and tools that can facilitate each of the steps are shown in Fig. 1. Some of these efforts can be executed by the water and wastewater company, while for some the results can support source control or 'upstream' initiatives.

4.3.1 Identification

Identifying what substances are present where and the sources is a prerequisite to be able to control environmental pollution. Two approaches to identify substances that can be used by the water and wastewater sector, preferably on a WWTP or water resource level, are *water characterization* and *source mapping*. It is important to characterize raw water in a drinking water reservoir or influent to a WWTP in order to receive information on what substances that have to be handled. Source mapping include identifying activities in a catchment area of a drinking water reservoir and in-water activities that may have an impact on water quality. For wastewater, identifying connected enterprises and their chemical usage is one way of mapping potential sources. EnvoMap is a tool specifically developed for the water and wastewater sector with the aim to aid the task of mapping enterprises. The programme is map-based and can visualize the connected enterprises in relation to the pipe system. The enterprises can further report their chemical usage directly into the system, whereby the chemical usage is scanned against priority lists of harmful substances [28].

Although enterprises might use harmful substances in larger quantities, households can together be responsible for a significant amount of the pollution through used products [26] and these contributions can be more difficult to identify. *Life cycle assessments (LCA)* is a method to assess environmental impact of products, services or other activities [24]. The chemical footprint (ChF) tool developed by Hitchock et al. [29] is based on LCA and can be particularly relevant for the sector when identifying sources. The chemicals in each ingredient in a product is investigated and the ingredient is scored based on parameters such as if it

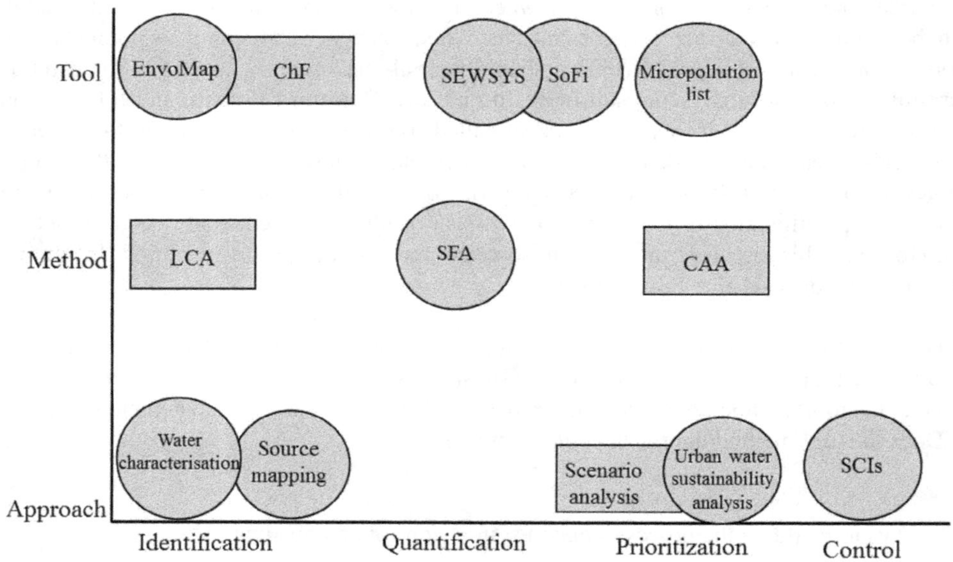

Figure 1: The efforts plotted as a result of type (y-axis) and which part of the source control facilitation process it operates on (x-axis). Circles indicate that it can be used by the water and wastewater company and rectangles indicate that the results can benefit the source control or upstream practice. Abbreviations: ChF (chemical footprint), LCA (life cycle assessment), SCIs (source control interventions), SFA (substance flow analysis), SoFi (Source Finder) and SEWSYS (Sewage System).

is carcinogen, toxic to aquatic organisms, and bioaccumulative, to make up the toxicity of the product. This information can be important to produce an overview of the potential threats to water or sludge quality from products used in households.

4.3.2 Quantification

After the substances and sources have been identified, the next step is to quantify them to get an understanding of significant flows. Substance flow analysis (SFA) is a widely used method for quantifying flows (i.e. sources, transport routes and recipients) of environmental pollution in a defined system [21]. It has been used to quantify, for example, silver flows [30], heavy metals from road traffic [31] and pharmaceuticals [32]. An SFA requires resources, both financial and in terms of competence, which can make it difficult for a small municipality to carry out SFAs themselves. Some general conclusions can be drawn from an SFA in one municipality to another, particularly concerning stocks, sources and involved actors, while measures to handle the flows is more difficult to make general conclusions about [33].

SEWSYS (SEWage SYStem) and *Source Finder* (*SoFi*) are tools based on SFA developed for the water and wastewater sector. SEWSYS simulates flows of some common substances in sewer systems [34]. Other factors that may influence quantity, such as corrosion rates and atmospheric deposition, are also included in the model [21]. SoFi is a calculation tool for heavy metals in influent water to WWTPs. This Excel-based tool uses a mixture of template- and measured data and can separate sources into household, stormwater (from combined systems), small and large enterprises, drainage water and leachate. A feature allows the user to test how certain efforts perform in reducing the overall quantity of the heavy metal pollution to the WWTP [35], which is why it is categorized as somewhere between the category quantification and prioritization.

4.3.3 Prioritization

Quantification of substances and flows allows for prioritization. Prioritization could be on a substance or sources level. It can also concern a prioritization among abatement strategies to get the most effective control.

On a substance level, there are several priority lists of substances that are perceived as particularly harmful. The *micro-pollution list* is a Swedish tool developed by Hörsing and Ledin [26] to help practitioners with their 'upstream work'. The list consists of organic micro-pollution present in sludge that poses the risk of being transferred to ground water or crops. The substances are based on the WFD-list, the SIN-list and the 'phase-out' substances decided by the Swedish Chemical Agency. The micro-pollution list can assist WWTPs when prioritizing what substances to focus on and develop control strategies for.

Chemical alternative assessment (CAA) is a method that can support water and wastewater companies with substitution [36]. A main aim of the method is to avoid the so-called 'regrettable substitution' where one toxic substance is substituted for an equally bad or worse one [37]. The assessment also considers if the substances are needed at all, such as with antimicrobial content in hand soap [36] or if the substances achieve what they are supposed to in a sufficient way, for example, if flame retards improve fire safety or increase escape time [37]. Although the water and wastewater company will not perform a CAA itself, insights from CAAs can be used by the water and wastewater company. This too, for example, prioritizes what substances to highlight in guideline, lobby for substitution of and substitute in their own facilities.

Scenario analysis is an approach that can be used to describe possible future developments [38]. By creating scenarios, different source control solutions and efforts can be tested. Scenarios for source control purposes generally need some type of input-data, and SFAs can be used as a basis for the scenario development (see, e.g. [21, 39]). The so-called prescriptive scenarios [40] are most relevant for practical source control efforts as they aim to give opportunity to plan for the future. Prescriptive scenarios are divided into 'forecasts' and 'what-if' [40]. The 'what-if'- type has been used for source control purposes in the EU project Score- PP (Source Control Options for Reducing Emissions of Priority Pollutants). The project used both technical (e.g. best available technology) and management-oriented scenarios (e.g. implementation of EU directives) in different medium, such as sewage sludge, groundwater, surface water, air and soil [19]. The study indicates that implementing EU directives should be prioritized. Carrying out a large-scale scenario analyses might be too time-consuming for an individual water and wastewater company, but the information derived from projects such as Score-PP can help prioritize between abatement strategies. When investigating different possible abatement strategies and prioritizing between them, there are several sustainability aspects to consider, such as available technology, economical feasibility, acceptance by the public and that the environmental benefits are secured. The urban water sustainability analysis is an approach aimed to address these issues, specifically developed for the water and wastewater sector. The analysis is based on a conceptual framework with three subsystems, namely technology, user and organization [41]. The framework also includes five criteria: hygiene, environment, sociocultural aspects, technical function and economy (for details, see [42]). This comprehensive approach will aid in choosing the most efficient and sustainable strategy.

4.3.4 Control

Controlling the identified, quantified and prioritized substances and sources is the last step of the process. Some general ways to control pollution is prohibition and substitution (legislative and voluntary) and de-centralized treatment in, for example, stormwater and by industries following best available technology (BAT) [19]. Prohibition can be on substance level such as with 'phase-out' substances or that certain activities (such as leachate from landfills) are not permitted to be connected to the WWTP [18]. However, there seems to be rather few examples of systematic approaches, methods and tools to be used in the control step.

Source control interventions (SCIs) is an example of a systematic approach. It is focusing on controlling pollution from farming practices to water reservoirs. SCIs are based on cooperation between farmers or land managers and the water and wastewater company. Four kinds of practical efforts are included in SCIs. 'Liaison', which is the most commonly used, consists of different types of collaboration such as data sharing. The other categories are 'lobbying', attempts to influence increased water protection, 'advice & support', such as investments on farms or compensation of land, and 'education' which include, for example, information brochures and communication efforts with the farmers. SCIs have mainly been used to handle pesticide problems as they pose a large problem to the drinking water quality and there is, at present, no feasible treatment option. Further, the approach has more often been used for surface water than for groundwater. When SCIs have been used for groundwater, it has been connected to nitrate abatement [22].

5 CONCLUSIONS

In this contribution, I have categorized approaches, methods and tools that can facilitate control of environmental pollution aimed at the water and wastewater sector. Different steps of the facilitation process require different efforts, which is why the approaches, methods and tools have been categorized based on if it assists in identification, quantification, prioritization or control. Dependent on where the organization is at with their own work, they can start at different levels. In addition, approaches, methods and tools that can be executed by the water and wastewater sector are separated from those where the results can inform the source control or 'upstream' practice. Moreover, source control and 'upstream work' as concepts and practices were compared and although there are several similarities, source control is a broader concept than 'upstream work'. Since 'upstream work' as a concept and practice is widening, there are some lessons to be learnt from source control, especially regarding stormwater and raw water. 'Upstream work' has a good foundation in the water and wastewater sector in Sweden. By expanding 'upstream work' to include more measures in the watercycle, important steps towards a more sustainable urban water management are taken.

ACKNOWLEDGEMENTS

Svenskt Vatten Utveckling has granted financial support for this research project.

REFERENCES

[1] Holten-Lützhøft, H.C. et al., A source classification framework supporting pollutant source mapping, pollutant release prediction, transport and load forecasting, and source control planning for urban environments. *Environmental Science and Pollution Research,* **19**, pp. 1119–1130, 2012. DOI: 10.1007/s11356-011-0627-9

[2] Uppströmsarbete och kretslopp; Svenskt Vatten Web Site http://www.svensktvatten.se/vattentjanster/avlopp-och-miljo/kretslopp-och-uppstromsarbete/ (accessed 05 December 2016).

[3] Brattberg, G. Personal communication, 02 March 2017, Former Vice President, Stockholm Vatten, Stockholm, Sweden.

[4] Nordén, L. Personal communication, 27 February 2017, Former Environmental Engineer, Gryaab AB, Gothenburg, Sweden.

[5] Hugmark, P. Personal communication, 09 March 2017, Former Director of the Industrial Unit, Käppalaförbundet, Stockholm, Sweden.

[6] Naturvårdsverket, *Hållbar återföring av fosfor: Naturvårdsverkets redovisning av ett uppdrag från regeringen*, 6580, 2013.

[7] Norrvatten initierar aktivt uppströmsarbete; Norrvatten Web Site https://www.norrvatten.se/Press/Nyhetsarkiv/Nyheter-2016/Norrvatten-initierar-aktivt-uppstromsarbete/ (accessed 11 November 2016).

[8] Revitt, D.M., Lundy, L., Eriksson, E. & Viavattene, C., Comparison of pollutant emission control strategies for cadmium and mercury in urban water systems using substance flow analysis. *Journal of Environmental Management*, **116**, pp. 172–180, 2013. DOI: 10.1016/j.jenvman.2012.12.007.

[9] Safer Products, cleaner cities!, NonHazCity, available at http://nonhazcity.eu/ (accessed 03-09 2017)

[10] Brack, W. et al., The SOLUTIONS project: Challenges and responses for present and future emerging pollutants in land and water resources management. *Science of the Total Environment*, **503–504**, pp. 22–31, 2015. DOI: 10.1016/j.scitotenv.2014.05.143.

[11] Guinée, J.B. et al., Human and Ecological Life Cycle Tools for the Integrated Assessment of Systems (HELIAS). *International Journal of Life Cycle Assessment*, Special Issue **1**, pp. 19–28, 2006. DOI: 10.1065/lca2006.04.008

[12] Amneklev, J., *Diffuse emissions from goods – influences on some societal end products.* PhD dissertation, Faculty of Health and Life Sciences, Linneaus University, Sweden, 2015.

[13] Fridström, L.M. & von Seth, S., Certifiering av avloppsslam för användning som växtnäring på åkermark. Master's thesis. Faculty of Engineering LTH, Lund University, Sweden.

[14] Aktivt uppströmsarbete med Revaq-certifiering; Svenskt Vatten Web Site http://www.svensktvatten.se/vattentjanster/avlopp-och-miljo/kretslopp-och-uppstromsarbete/revaq-certifiering/.(accessed 2017-02-03)

[15] Revaq, *Årsrapport 2015.* 2016.

[16] Flygare, S. Personal communication, 11 April. 2016, Environmental engineer, VA SYD, Malmö, Sweden.

[17] Larsen, T.A. & Gujer, W., Waste design and source control lead to flexibility in wastewater management. *Water Science and Technology*, **43(5)**, pp. 309–318, 2001.

[18] Revaq, *Regler för certifieringssystemet.* Utgåva 4.01, 2017.

[19] Eriksson E., Revitt, D.M., Ledin, A., Lundy, L., Holten Lützhøft, H.C., Wickman, T. & Mikkelsen, P.S., Water management in cities of the future using emission control strategies for priority hazardous substances. *Water Science & Technology*, **64(10)**, pp. 2109–2118, 2011. DOI: 10.2166/wst.2011.797

[20] Frenzel, M. Personal communication, 8 Mar. 2017, Environmental engineer, Käppaläförbundet, Stockholm, Sweden.

[21] Ahlman, S. Malm, A., Kant, H., Svensson, G. & Karlsson, P., Modelling non-structural Best Management Practices– focus on reductions in stormwater pollution. *Water Science & Technology,* **52(5)**, pp. 9–16, 2005.

[22] Spiller, M., McIntosh, B.S., Seaton, R.A.F. & Paul Jeffrey, P., Pollution source control by water utilities – characterization and implications for water management: Research results from England and Wales. *Water and Environment Journal*, **27**, pp. 177–186, 2013. DOI: 10.1016/S0304-4017(98)00226-X

[23] de Ridder, W., Turnpenny, J., Nilsson, M. & von Raggamby, A., A framework for tool selection and use in integrated assessment for sustainable development. *Journal of Environmental Assessment Policy and Management*, **9(4)**, pp. 423–441, 2007. DOI: 10.1007/978-94-010-0456-5

[24] van der Vorst, R., Grafé-Buckens, A. & Sheate, W.R., A systemic framework for environmental decision-making. *Journal of Environmental Assessment Policy and Management,* **1(1)**, pp. 1–26, 1999. DOI: 10.1080/02688867.1992.9726859

[25] Brown, R.R. & Farrelly, M.A., Delivering sustainable urban water management: A review of the hurdles we face. *Water Science & Technology,* **59(5)**, pp. 839–846, 2009. DOI: 10.2166/wst.2009.028

[26] Hörsing, M. & Ledin, A., *Identifiering av fokusämnen för slam – organiska miljögifter.* Svenskt Vatten Utveckling, 2016–08, 2016.

[27] Prioritisation levels; Swedish Chemicals Agency Web Site http://www.kemi.se/en/prio-start/before-starting/important-points-to-note/prioritisation-levels, 2015 (accessed 9 March 2017)

[28] Gemit Solutions AB, *EnvoMap ® Verksamhetsstöd för uppströmsarbete*. n.d.
[29] Hitchcock, K., Panko, J. & Scott, P., Incorporating chemical footprint reporting into social responsibility reporting. *Integrated Environmental Assessment and Management*, **8**, pp. 386–388, 2012. DOI: 10.1002/ieam.1288
[30] Amneklev, J., Bergbäck, B., Sörme, L., Lagerkvist, R., Kotsch, M. & Augustsson, M., *Strategier att reducera silver och vismut i urbant avloppsvatten*. Svenskt Vatten Utveckling, 2014-10, 2014.
[31] Hjortenkrans, D., *Road traffic metals – sources and emissions*. PhD dissertation. Faculty of Natural Sciences and Engineering, University of Kalmar, Sweden. 2008.
[32] Chevré, N. et al., Substance flow analysis as a tool for mitigating the impact of pharmaceuticals on the aquatic system. *Water Research*, **47**, pp. 2995–3005, 2013. DOI: 10.1016/j.watres.2013.03.004
[33] Lindqvist, A. & von Malmborg, F., What can we learn from local substance flow analyses? The review of cadmium flows in Swedish municipalities. *Journal of Cleaner Production,* **12**, pp. 909–918, 2004. DOI: 10.1016/j.jclepro.2004.02.012
[34] Ahlman, S. & Svensson, G., Modelling substance flows in urban sewer systems using MATLAB/Simulink. *Proceedings of the 9th International Conference on Urban Drainage*, eds. E.W. Strecker & W.C. Huber, September 8–13, Portland, OR, USA, 2002.
[35] Agduhr Eronen, S., *Substansflödesanalys av tungmetaller i avloppssystemet – Nytt verktyg testat på Sigtuna och Solna kommuner*. Master's thesis. Faculty of Science and Technology, Uppsala University, Sweden, 2010.
[36] Geiser, K. & Tickner, J., Edwards, S. & Rossi, M., The architecture of chemical alternative assessment. *Risk Analysis,* **35**(**12**), pp. 2152–2161, 2015. DOI: 10.1111/risa.12507
[37] Howard, G.J., Chemical alternatives assessment: The case of flame retardants. *Chemosphere*, **116**, pp. 112–117, 2014. DOI: 10.1016/j.chemosphere.2014.02.034
[38] Duinker, P. N. & Greig, L.A., Scenario analysis in environmental impact assessment: Improving explorations of the future. *Environmental Impact Assessment Review*, **27**, pp. 206–219, 2007. DOI:10.1016/j.eiar.2006.11.001
[39] Eriksson, E., Revitt, M., Holten-Lützhøft, H.C., Viavattene, C., Scholes, L. & Mikkelsen, P.S., Emission control strategies for short-chain chloroparaffins in two semi-hypothetical case cities, *Urban Environment: Proceedings of the 10th Urban Environment Symposium: Urban Futures for a Sustainable World*, eds. S. Rauch. & G.M. Morrison, Alliance for Global Sustainability Book series, Vol. 19. Springer: Dordrecht, pp. 213–223, 2012.
[40] Börjeson, L., Höjer, M., Dreborg, K.H., Ekvall, T. & Finnveden, G., Scenario types and techniques: Towards a user's guide. *Futures*, **38**, pp. 723–739, 2006. DOI:10.1016/j.futures.2005.12.002
[41] Carlsson, K. & Kärrman, E., *Beslutsstöd inför stora investeringar inom VA: Hållbarhetsanalyser och samhällsekonomiska bedömningar*. Svenskt Vatten Utveckling, 2014–13, 2014.
[42] Hellström, D., Jeppson, U. & Kärrman, E., A framework for systems analysis of sustainable urban water management. *Environmental Impact Assessment Review,* **20**, pp. 311–321, 2000. DOI: 10.1016/S0195-9255(00)00043-3

TREATMENT OF MANGANESE IN ACID AND NEUTRAL MINE DRAINAGE USING MODIFIED DOLOMITE

IULIANA LAURA CALUGARU[1], THOMAS GENTY[1,2] & CARMEN MIHAELA NECULITA[3]
[1]Technology Center for Industrial Waste (Centre Technologique des Résidus Industriels – CTRI), Canada
[2]Agnico Eagle Mines Limited, Canada
[3]Research Institute on Mines and Environment (RIME), University of Quebec in Abitibi-Temiscamingue, Canada

ABSTRACT

The treatment of manganese [Mn] in acid mine drainage (AMD) and neutral mine drainage (NMD) is prone to variation in performance, which depends on several factors, including pH, Mn and dissolved oxygen (DO) concentrations, catalysis by Mn-/iron [Fe]-solids, presence of ligands, and bacteria. Notwithstanding the recent technological and scientific advancements, Mn removal is notoriously difficult and still challenging because of its complex chemistry and high solubility. Precipitation of Mn(II), in AMD and NMD, occurs in the form of oxides, with prior oxidation to Mn(III) and Mn(IV), or as carbonates and sulphides. However, most treatment systems have limited efficiency in simultaneously treating Mn and other metals in AMD/NMD, necessitating secondary treatment for Mn removal. Modification of natural or residual materials was found efficient for the treatment of NMD. In this context, this study evaluated the performance of half-charred dolomite (1 h at 750°C) in batch testing (solid: liquid ratio of 3 g:400 mL) for Mn treatment in AMD and NMD. Results showed that modified dolomite treated more than 98% Mn in synthetic NMD (pH 6.1–6.3, up to 1 g/L Mn), at final pH of 9.7–10.6. Similar efficiency (98%) in Mn removal was also found for synthetic AMD (pH 3.6, up to 100 mg/L Mn and 1 g/L Fe, Mn:Fe molar ratio 1:10), while Fe was completely treated, at final pH of 9.7–9.8. In addition, Mn removal was 99.5% within the first 2 h (when pH increased to 8.0), while after 4 h, the efficiency was up to 99.9% (at final pH of 9.6). DO also decreased (from 8 to 2.2 mg/L), at initial Mn concentrations of 1 g/L, and dropped (from 8 to 0.7 mg/L), when Fe concentrations increased to 1 g/L (Mn:Fe molar ratio 1:10). Based on these findings, half-charred dolomite seems a promising option for the treatment of Mn in both AMD and NMD.

Keywords: acid mine drainage, Mn treatment, modified dolomite, neutral mine drainage.

1 INTRODUCTION

Mining operations can adversely affect the physicochemical and biological quality of the surrounding surface and ground waters in terms of pH (acidic/alkaline) and concentrations of toxic metals (e.g., aluminium [Al], cadmium [Cd], cobalt [Co], chromium [Cr], copper [Cu], Fe, mercury [Hg], Mn, molybdenum [Mo], nickel [Ni], lead [Pb], uranium [U], zinc [Zn]), metalloids (e.g., arsenic [As], antimony [Sb], selenium [Se]), as well as polyatomic anions [e.g., sulphate (SO_4^{2-}), cyanide (CN^-), thiocyanate (SCN^-), cyanate (OCN^-), nitrite (NO_2^-), nitrate (NO_3^-)] and cations [e.g., ammonium (NH_4^+)] [1–4].

Among the above-mentioned contaminants, Mn is notoriously difficult to remove from mine effluents prior to their discharge into the environment, mainly due to the high solubility of its minerals relative to other metals, inhibition of Mn precipitation at a molar ratio Fe/Mn > 4, reductive dissolution of Mn oxides by organic matter, sulphides and Fe(II), as well as the high pH (>8) required for its oxides formation and precipitation [5–8].

In acidic (pH < 6) and neutral (pH 6–8) mine drainage (AMD and NMD, respectively), the soluble Mn(II) form is present [9–11]. The oxidation of Mn to higher oxidation states (III and IV) to form slightly soluble oxides occurs at neutral pH in the presence of Mn-oxidizing bacteria, commonly present in nature [7, 8]. Thus, the oxidation requires a growth surface for the above-mentioned aerobic heterotrophic bacteria and sufficient DO, while the formation

DOI: 10.2495/EI-V1-N3-323-333

Table 1: Raw/modified materials utilized for Mn removal in various real/synthetic effluents.

Effluent characteristics	Material characteristics	Test characteristics	Performance	Ref.
Synthetic effluent, 19.39 mg/L Mn	Raw dolomite	Solid to liquid ratio 10 g:1 L ml,15 days	<19.3% removal, >15.6 mg/L residual conc., pH 7.2–7.3	[20]
	Charred dolomite (1 h at 600°C)		>95% removal, <0.97 mg/L residual conc., pH 9.3–9.7	
	Modified dolomite [charred/ aq. Mn(II)]		>89% removal, <1.99 mg/L residual conc., pH 7.7–8.1	
	Modified dolomite [charred/ aq. Mn(II) & $KMnO_4$]		>95% removal, <0.84 mg/L residual conc., pH 8.5–8.7	
Real AMD, pH=3.5, 6.2 mg/L Mn and 53.5 mg/L Fe	Wood ash	Solid to liquid ratio 0.5 g:1 L, 20 min	100% Fe removal, >56% Mn removal, 2.68 mg/L residual Mn conc., pH 8.3	[19]
	$Ca(OH)_2$	Solid to liquid ratio 0.2 g:1 L, 20 min	100% Fe removal, <30% Mn removal, 4.35 mg/L residual Mn conc., pH 8.26	
Synthetic effluent, pH = 5.2, 50 mg/L Mn or 50 mg/L Fe	Deciduous sawdust	Solid to liquid ratio 20 g:1 L, 10 min	30% Mn removal, 10% Fe removal	[21]
Real NMD, pH = 6.2, 20.5 mg/L Mn and 104 mg/L Fe	Zeolite (obtained by hydrothermal treatment) from AMD sludge, originated from coal fly ash	Solid to liquid ratio 10 g:1 L, 1 h	>99% Fe removal, 64% Mn removal, pH 8.3	[18]
Synthetic effluent, 10 mg/L Mn	Coal fly ash	Solid to liquid ratio 100 g:1 L, 5 min	>99.7% removal, <0.02 mg/L residual conc.	[14]
	Coal fly ash zeolite (from hydrothermal treatment)		>99.6% removal, <0.04 mg/L residual conc.	
Synthetic effluent, 100 mg/L Mn	Coal fly ash	Solid to liquid ratio 10 g:1 L, 2 h	3.4% removal, 96.6 mg/L residual conc.	[16]

	Coal fly ash zeolite (from hydrothermal treatment)		85.6% removal, 14.4 mg/L residual conc.	
Synthetic effluent, pH = 2 and 6, 10 mg/L Mn	Natural zeolite, clinoptilolite rich	Solid to liquid ratio 10 g:1 L	37.1% removal, for initial pH=2, 8 h, 97.3% removal, for initial pH=6, 1 h	[10]
Synthetic effluent, pH 6, 10 mg/L Mn and 10 mg/L Fe			38.8% Fe removal, 1 h, 12.9% Mn removal, 8 h	
Synthetic AMD, pH = 3.5, 20 mg/L Mn and 400 mg/L Fe	Natural zeolite, clinoptilolite	Solid to liquid ratio 37 g:1 L, 6 h	18.9% Mn removal, 59.9% Fe removal	[17]
Synthetic effluent, 20 mg/L Mn		Solid to liquid ratio 37 g:1 L, 6 h	95.2% Mn removal, 0.96 mg/L residual conc.	
Synthetic effluent, pH = 3, 10 mg/L Mn	Raw crab shell	Solid to liquid ratio 5 g:1 L	≥95% Mn removal, 72 h, pH 9.2–10.2	[5]
	Deproteinized crab shell		≥95% Mn removal, 48 h, pH 9.2–10.2	
	$CaCO_3$	Solid to liquid ratio 125 g:1 L	≤85% Mn removal, 72 h, pH 7.8–8.3	

and precipitation of Mn oxides are autocatalytic. Nevertheless, the major hurdle for an efficient Mn removal is represented by the dissolved Fe(II), which is adsorbed on the surface of Mn oxides and reduces them to Mn(II) soluble [7, 8]. Besides oxides, Mn(II) can be removed from mine effluents as sulphides and carbonates, under reducing and alkaline conditions [5, 12]. In general, the effective removal of Mn from mine drainage requires additional polishing steps in the design of the treatment systems, to priorly reduce Fe and Al concentrations [8, 13].

To limit the cost of Mn treatment, several natural and residual materials were evaluated, including the following: hydrated lime [$Ca(OH)_2$], limestone ($CaCO_3$), montmorillonite, bentonite, hydrotalcite, natural zeolite, coal fly ash and its hydrothermal synthetic zeolite, sludge originated zeolite, dolomite [$CaMg(CO_3)_2$] and activated dolomite, raw and deproteinized crab shells, Fe(III)-impregnated activated carbon, sawdust and wood ash [5, 8, 10, 11, 14–21]. However, some of these materials (Table 1), such as dolomite and activated dolomite, deciduous sawdust, coal fly ash and coal fly ash zeolite, natural zeolite, limestone, and raw and deproteinized crab shells, were evaluated for Mn removal in synthetic effluents, in the absence of Fe. In addition, except for the raw dolomite and deciduous sawdust, the referred to materials were found effective for Mn removal at concentrations less than 20 mg/L. Nonetheless, the fastest kinetics (5 min) were observed for the coal fly ash and its synthetic zeolite [14], whereas the slowest kinetics (15 days) were observed for the activated dolomite [20]. Conversely, whenever the Fe is present, in either real or synthetic effluents, Mn removal is adversely affected, as in the case of wood ash and hydrated limestone [19], sludge originated zeolite [18] or natural zeolite [10].

Among the materials evaluated for Mn removal, dolomite (a sedimentary carbonate mineral, worldwide spread and available at low cost) was found efficient for the treatment of several metals and metalloids (including in mine drainage), such as Pb, Cd, Cu, Ag, Co, and As [22–27]. Moreover, thermal activation considerably enhanced its efficiency for the removal of As, Co, Cr, Mn, Ni and Zn [20, 28–31]. Thermally activated dolomite was also found effective for the treatment of several metals (Ag, As, B, Ba, Co, Cr, Hg, Mn, Ni, Pb, Sn, Sr, Ti and V) in a shipyard wastewater [32].

In this context, the objective of this study was to comparatively evaluate the performance of raw and half-charred dolomite, for Mn treatment in NMD or in AMD, in the presence of Fe (Mn:Fe molar ratio of 1:10).

2 MATERIALS AND METHODS

2.1 Raw and activated dolomite

The dolomite rock used in this study was provided by Temiska Silice (Saint-Bruno-de-Guigues, Quebec, Canada). Its thermal activation was performed by charring in an oven (Thermolyne Furnace) for 1 h, at 750°C. These activation conditions were found optimal in a previous study where Ni and Zn were successfully treated in NMD [31].

Both raw and charred dolomite samples were first grinded. Then, the following physicochemical parameters were determined: particle size distribution, specific surface, pH, pH_{PZC}, elemental composition and mineralogy [31].

Thermal decomposition of dolomite produces [20, 28–30] either a mixture of $CaCO_3$ (calcite) and MgO (periclase, resulting from the dissociation of $MgCO_3$) (eqn (1)) or MgO and CaO, by further decomposition of $CaCO_3$ (eqn (2)). If the process is stopped at ≈800°C, a

Table 2: Conditions used in treatability testing.

Effluent/ dolomite type	Parameter	Metal	pH	Sampling (h)	Doses (mg/L)
NMD/raw	Uptake function of concentration	Mn	6.2–6.3	48	50, 100, 153, 202
NMD/charred			6.1–6.3		50, 100, 153, 202, 426, 1353, 1815, 2693
AMD/charred		Mn/Fe	3.6		58/443, 99/934 145/1470, 188/1860
NMD/raw NMD/charred	Uptake function of time	Mn	6.7	1, 2, 4, 8, 24, 48	77.5

solid rigid mixture of calcite and periclase is obtained. This new material has a porous structure, with over 50 times the specific surface area of the raw dolomite [29, 31–33]:

$$CaMg(CO_3)_2 \rightarrow CaCO_3 + MgO + CO_2 \tag{1}$$

$$CaCO_3 \rightarrow CaO + CO_2 \tag{2}$$

2.2 Synthetic effluents (solutions)

The performance of raw and half-charred dolomite was assessed using synthetic effluents (NMD and AMD) prepared with ACS grade of sulphate salts ($MnSO_4{\cdot}H_2O$ and $FeSO_4{\cdot}7H_2O$) and deionized water (Table 2). The pH of the AMD was adjusted with concentrated (98%) H_2SO_4. The ionic strength for both NMD and AMD types of effluents was adjusted with solid NaCl.

2.3 Batch testing conditions, sampling, analysis and data processing

The experiments were conducted in 50 mL propylene tubes containing a volume of 40 mL synthetic effluent, at a temperature of 25°C and at 300 rpm agitation speed. The solid to liquid ratio (3 g:400 mL) was selected according to the literature [24, 31]. The maximum Mn concentration in the synthetic NMD (pH 6.1–6.3) was 202 mg/L (for the raw dolomite) and 2,693 mg/L (for the half-charred dolomite) (Table 2). In the synthetic AMD (pH 3.6), the maximum Mn and Fe concentrations were 188 and 1,860 mg/L, respectively. Solely the half-charred dolomite was evaluated for the treatment of the synthetic AMD (Table 2). Nitrogen was purged for 2 min, within each propylene tube. At the end of the 48 h batch experiments, the supernatant was filtered, the pH and DO were measured, while the concentrations of Mn and Fe were analysed with an AA spectrometer. Additional batch testing was performed in order to evaluate the uptake of Mn in synthetic NMD (pH 6.7) as a function of time. The synthetic NMD contained 77.5 mg/L Mn, and the sampling was carried out at 1, 2, 4, 8, 24 and 48 h (Table 2). The pH, and the DO and Mn concentrations were measured for each filtered sample.

Metal removal (%) was calculated with eqn (3):

$$\% \text{ Removal} = [C_o - C_f].100/\, C_o \tag{3}$$

where C_0 and C_f, represent, respectively, the initial and final concentrations of metal in solution (mg/L).

3 RESULTS AND DISCUSSION

3.1 Materials characterization

The dolomite samples used in this study had high purity (87.1%). After 1 h of charring at 750°C, 7.2% of the solid consisted of dolomite, while two new mineral phases appeared, i.e. calcite (53.7%) and periclase (19.9%). The resulting charred dolomite (pH 11.6) showed strongly basic relative to the raw dolomite (pH 7.9), while the pH_{PZC} of raw dolomite increased from 9.6 to 11.1 after the charring. A sevenfold increase in specific surface (0.6–4.2 m^2/g) was also observed for the charred dolomite [31].

3.2 Batch testing results

Charred dolomite showed very effective for Mn removal in the synthetic NMD, as more than 89% of Mn was removed from solutions initially contaminated up to 1,353 mg/L (Table 3).

Minor residual concentrations (up to 3.3 mg/L) of Mn in the treated NMD were found only up to maximum initial concentration of 892 mg/L (Table 3). Effectiveness of the charred dolomite then decreased, when Mn concentration increased above 892 mg/L in the synthetic NMD, whereas raw dolomite seemed sparsely effective even at low (49.6 mg/L) Mn

Table 3: Performance of raw and charred dolomite for Mn removal in synthetic NMD as a function of the initial concentration, in batch testing.

C_0 (mg/L)	C_f (mg/L)	Removal (%)	pH	DO (mg/L)
		Dolomite		
49.6	43.7	11.9	7	8.7
99.9	91.5	8.4	6.8	8.6
153.3	148.4	3.2	6.7	8.3
202.3	196.1	3.1	6.6	8.2
		Charred dolomite		
49.6	1.4	97.2	10.6	7.9
99.9	1.5	98.5	10.5	7.7
153.3	3.1	98.0	10.4	6.9
202.3	3.3	98.4	10.3	6.3
426	0.02	99.9	10.2	4.1
892	0.04	99.9	9.7	2.3
1353	147	89.1	8.4	1.4
1815	592	67.4	7.8	1.4
2693	1350	49.9	7.5	1.7

Table 4: Performance of charred dolomite for Mn removal in the presence of Fe (1:10) in synthetic AMD, as a function of the initial concentration, in batch testing.

Mn			Fe			Mn:Fe Final ratio	pH	DO (mg/L)
C_0 (mg/L)	C_f (mg/L)	Removal (%)	C_0 (mg/L)	C_f (mg/L)	Removal (%)			
57.6	0.005	99.9	443	BDL	99.9	N/A	9.8	3.33
99	1.5	98.5	934	BDL	99.9	N/A	9.7	0.72
145	107	26.2	1470	145	90.1	1:1.3	6.9	0.24
188	162	13.8	1860	714	61.6	1:4.3	6.8	0.13

BDL: below detection limit; N/A: not applicable.

concentrations (Table 3). The initial pH of the uptake experiments was 6.1–6.3, whereas the final pH was 7–6.6 (for the raw dolomite) and 10.6–7.5 (for the charred dolomite).

Solely charred dolomite was evaluated for the treatment of Mn in the presence of Fe (1:10), in synthetic AMD (pH 3.6), and it was found effective up to 99 mg/L Mn and 934 mg/L Fe (Table 4).

It was observed that further increase of Fe concentration significantly decreased Mn removal, consistently with previously reported findings [10, 18, 19]. Fe removal was satisfactory (90%) up to 1,470 mg/L as initial concentration, but residual Fe was usually high (145 mg/L) in treated AMD (Table 4).

The pH varied from basic to neutral while DO concentration was significantly reduced. Oxidation of Fe(II) to Fe(III) followed by Fe(III) hydrolysis could explain the decrease of both DO and pH.

The results of batch testing on the uptake of Mn in synthetic NMD as a function of time (Table 5) were consistent with the previous results of this study. Hence, dolomite showed limited efficiency, whereas charred dolomite allowed significant reduction of Mn concentration within only 2 h of contact with the synthetic NMD. Noteworthy, 80% of Mn was removed within 1 h; however, residual Mn concentration was still high (15.6 mg/L). The initial Mn and DO concentrations were 77.5 and 8.1 mg/L, respectively, at a pH of 6.7. The DO recorded little variation for both raw and charred dolomite. The neutralizing effect of dolomite seemed slow compared to charred dolomite (that settled to 10.4 after 24 h). Consequently, to explain Mn removal, chemical sorption seems more probable than oxidation and precipitation, as Mn concentrations dropped from 77.5 to 0.37 mg/L within 2 h, whereas pH raised from 6.7 to 8 within the same interval of time.

3.3 Comparative performance of half-charred dolomite and some recently evaluated raw and modified materials for Mn treatment

The performance of some materials such as raw and activated dolomite, hydrated lime, limestone, wood ash and sawdust, natural zeolite, coal fly ash and coal fly ash zeolite, synthetic zeolite prepared from sludge resulting from the AMD treatment with coal fly ash, raw and deproteinized crab shell, recently evaluated for the treatment of Mn in different aqueous solutions, is presented in Table 1.

Table 5: Performance of charred dolomite for Mn removal in synthetic NMD as a function of time, in batch testing.

Time (h)	C_f (mg/L)	Removal (%)	pH	DO (mg/L)
		Dolomite		
0	77.5	0.0	6.7	8.1
1	72.5	6.5	6.8	8.0
4	70	9.7	6.8	7.0
24	70.5	9.0	6.9	8.0
48	69	11.0	8.2	8.0
		Charred dolomite		
0	77.5	0.0	6.7	8.1
1	15.6	79.9	7.2	7.5
2	0.37	99.5	8.0	6.8
4	0.01	99.9	9.6	7.22
8	0	100	9.8	7.49
24	0.1	99.9	10.4	7.0
48	0.01	99.9	10.4	7.2

Among natural materials, raw dolomite seems the least effective [20], natural zeolite the most effective [10, 17], whereas an important solid to liquid ratio is necessary to improve the effectiveness of the limestone [5] for Mn removal.

Concerning the waste materials, the least effective seems the sawdust of deciduous trees, and the most effective the coal fly ash and the raw crab shell. In the same time, Mn removal is faster by fly ash (>99.7% removal within 5 min) than by raw crab shell (≥95% Mn removal within 72 h).

Modification of natural and waste materials improves their performance in Mn withdrawal. For instance, modified dolomite [20] becomes very effective in terms of % removal (>95%) and residual Mn concentration (<0.84 mg/L) while maintaining a neutral pH of treated effluent (8.5–8.7). However, the initial concentration of Mn was lower (about 20 mg/L) compared to this study (892 mg/L), whereas the duration of batch testing was longer (15 vs 2 days). It is to observe that half-charred dolomite evaluated in this study decreased the Mn concentration from 77.5 to 0.37 mg/L within 2 h, while the pH of the treated effluent was 8.0. The solid to liquid ratio in this study was 7.5 g for 1 L, which is comparable to 10 g for 1 L, employed by Mamchenko et al. [20].

Coal fly ash seems effective for Mn removal at low initial concentrations [14] (10 mg/L, >99.7% removal), but less effective at higher initial concentrations (100 mg/L, 3.4% removal) [16]. Yet, its conversion into zeolite improves the performance for Mn removal (100 mg/L, 85.6%) [16].

Although very effective for Mn removal, natural zeolite seems to significantly lose its efficiency wherever Fe is present [10, 17]. However, synthetic zeolite prepared by hydrothermal treatment of sludge originated from AMD treatment with coal fly ash, removed 99% Fe and 64% Mn from a real NMD (pH 6.2, 20.5 mg/L Mn, and 104 mg/L Fe) while maintaining the pH of the final effluent at 8.3 [18].

Better performance was obtained by half-charred dolomite, within this study, as Mn and Fe were removed up to 98.5% (1.5 mg/L residual concentration) and 99.9% (BDL) respectively, from a synthetic AMD (pH 3.6, 99 mg/L Mn, and 934 mg/L Fe).

4 CONCLUSIONS

In this study, both raw and half-charred dolomite (1 h, at 750°C) were evaluated through batch testing for their performance in Mn removal from synthetic AMD and NMD. The results showed that charred dolomite treated more than 98% Mn in synthetic NMD (pH 6.1–6.3) when the initial Mn concentration was up to 892 mg/L. Moreover, Mn removal was very efficient (89%) when the initial Mn concentration was 1,353 mg/L. Nonetheless, the residual concentration was 147 mg/L Mn. The uptake kinetic was relatively fast, as Mn concentration dropped from 77.5 to 0.37 mg/L within 2 h. However, 79.9% Mn was removed within the first 1 h. Meanwhile, raw dolomite showed poor performance in the same testing conditions.

In synthetic AMD (pH 3.6, Mn:Fe molar ratio 1:10), better performance in Mn and Fe removal was achieved at maximum initial concentrations of 99 and 934 mg/L, respectively. Above 1,470 mg/L, Fe removal was still satisfactory (90.1%), despite the fact that residual Fe was high (145 mg/L), whereas Mn removal was poor (29.2% for 145 mg/L initial concentration).

Half-charring of dolomite strongly enhanced its alkaline character and neutralizing capacity. However, the pH increase seems rather slow to cause an eventual major removal of Mn by precipitation, as 99.5% of Mn was removed from 77.5 mg/L initial Mn concentration at a final pH of 8.0.

The most severe drop in DO was found during AMD treatment. As simultaneously decrease of pH was also observed, Fe(II) was probably partially oxidized to Fe(III), which then hydrolysed and precipitated. An important decrease of DO was also observed at higher concentrations of Mn in NMD (above 426 mg/L). In this last case, Mn(II) probably precipitated as oxidized Mn(III) or Mn(IV).

Based on these findings, half-charred dolomite could be considered as a promising option for the treatment of Mn in both AMD and NMD. Further research is warranted to evaluate the mechanisms of Mn removal.

ACKNOWLEDGEMENTS

This research was supported by the Natural Sciences and Engineering Research Council of Canada (NSERC), the College of Abitibi-Témiscamingue and the industrial partners of the Centre Technologique des Résidus Industriels (Technological Center for Industrial Waste): Regénération, Iamgold Corporation, Hecla Québec, Mine Canadian Malartic, Technosub and the Organisme Bassin Versant du Témiscamingue.

REFERENCES

[1] Acheampong, M.A., Meulepas, R.J.W. & Lens, P.N.L., Removal of heavy metals and cyanide from gold mine wastewater. *Journal of Chemical Technology and Biotechnology*, **85**(5), pp. 590–613, 2010. DOI: 10.1099/00221287-10-2-261.

[2] Nordstrom, D.K., Mine waters: Acidic to circumneutral. *Elements*, **7**(6), pp. 393–398, 2011. DOI: 10.2113/gselements.7.6.393.

[3] Jermakka, J., Wendling, L., Sohlberg, E., Heinonen, H. & Vikman, M., Potential technologies for the removal and recovery of nitrogen compounds from mine and quarry waters in subarctic conditions. *Critical Reviews in Environmental Science and Technology*, **45**(7), pp. 703–748, 2015. DOI: 10.1021/es060661h.

[4] Nordstrom, K., Blowes, D.W. & Ptacek, C.J., Hydrogeochemistry and microbiology of mine drainage: An update. *Applied Geochemistry*, **57**, pp. 3–16, 2015. DOI: 10.1016/j.apgeochem.2015.02.008.

[5] Robinson-Lora, M.A. & Brennan, R.A., Anaerobic precipitation of manganese and co-existing metals in mine impacted water treated with crab shell-associated minerals. *Applied Geochemistry*, **26**(5), pp. 853–862, 2011. DOI: 10.1016/j.apgeochem.2011.02.006.

[6] Song, H., Yim, G.-J., Ji, S.-W., Nam, I.H., Neculita, C.M. & Lee, G.Y., Performance of mixed organic substrates during the treatment of acidic and moderate mine drainage in column bioreactors. *Journal of Environmental Engineering*, ASCE **138**(10), pp. 1077–1084, 2012. DOI: 10.1061/(ASCE)EE.1943-7870.0000567.

[7] Lesley, B. Daniel, H. & Paul, Y., Iron and manganese removal in wetland treatment systems: Rates, processes and implications for management. *Science of the Total Environment*, **394**, pp. 1–8, 2008. DOI: 10.1016/j.scitotenv.2008.01.002.

[8] Skousen, J., et al., Review of passive systems for acid mine drainage treatment. *Mine, Water and the Environment*, **36**(1), pp. 133–153, 2017. DOI: 10.1007/s10230-016-0417-1.

[9] National Institute of Advanced Industrial Science and Technology Research Center for Deep Geological Environments Naoto TAKENO, *Atlas of Eh-pH diagrams, Intercomparison of thermodynamic databases*, Geological Survey of Japan Open File Report No. 419, p. 287, May 2005, available at http://www.eosremediation.com/download/Chemistry/Chemical%20Properties/Eh_pH_Diagrams.pdf.

[10] García-Mendieta, A., Solache-Ríos, M. & Olguín, M.T., Evaluation of the sorption properties of a Mexican clinoptilolite-rich tuff for iron, manganese and iron-manganese systems. *Microporous and Mesoporous Materials*, **118**(1–3), pp. 489–495, 2009. DOI: 10.1016/j.micromeso.2008.09.028.

[11] Goldani, E., Moro, C.C. & Maia, S.M., A study employing different clays for Fe and Mn removal in the treatment of acid mine drainage. *Water, Air and Soil Pollution*, **224**, p. 1401, 2013. DOI: 10.1080/12269328.2010.10541314.

[12] Neculita, C.M., Zagury, G.J. & Bussière, B., Effectiveness of sulfate-reducing passive bioreactors for treating highly contaminated acid mine drainage: II. Metal removal mechanisms and potential mobility. *Applied Geochemistry*, **23**, pp. 3545–3560, 2008. DOI: 10.1016/j.apgeochem.2008.08.014.

[13] Rakotonimaro, T.V., Neculita, C.M., Bussière, B. & Zagury, G.J., Effectiveness of various dispersed alkaline substrates for the pretreatment of ferriferous acid mine drainage. *Applied Geochemistry*, **73**, pp. 13–23, 2016. DOI: 10.1016/j.apgeochem.2016.07.014

[14] Belviso, C., et al., Removal of Mn from aqueous solution using fly ash and its hydrothermal synthetic zeolite. *Journal of Environmental Management*, **137**, pp. 16–22, 2014. DOI: 10.1016/j.jenvman.2014.01.040

[15] Mondal, P., Balomajumder, C. & Mohanty, B., A laboratory study for the treatment of arsenic, iron, and manganese bearing ground water using Fe^{3+} impregnated activated carbon: Effects of shaking time, pH and temperature. *Journal of Hazardous Materials*, **144**, pp. 420–426, 2007. DOI: 10.1016/j.jhazmat.2006.10.078.

[16] Nascimento, M., Soares, P.S.M. & de Souza, V.P. Adsorption of heavy metal cations using coal fly ash modified by hydrothermal method. *Fuel*, **88**, pp. 1714–1719, 2009. DOI: 10.1016/j.fuel.2009.01.007.

[17] Motsi, T., Rowson, N.A. & Simmons, M.J.H. Adsorption of heavy metals from acid mine drainage by natural zeolite. *International Journal of Mineral Processing*, **92**, pp. 42–48, 2009. DOI: 10.1016/j.minpro.2009.02.005.

[18] Vadapalli, V.R.K., Gitari, W.M., Ellendt, A., Petrik, L.F. & Balfour, G., Synthesis of zeolite-P from coal fly ash derivative and its utilisation in mine-water remediation. *South African Journal of Science*, **106**(5–6), pp. 1–7, 2010. DOI: 10.4102/sajs.v106i5/6.231.

[19] Heviánková, S., Bestová, I. & Kyncl, M., The application of wood ash as reagent in acid mine drainage treatment. *Minerals Engineering*, **56**, pp. 109–111, 2014. DOI: 10.1016/j.mineng.2013.10.032.

[20] Mamchenko, A.V., Kiy, N.N., Chernova, L.G. & Misochka, I.V., The investigation of the impact of the modification methods of natural dolomite on water demanganesation. *Journal of Water Chemistry and Technology*, **30**(4), pp. 191–196, 2008. DOI: 10.3103/S1063455X08040012.

[21] Božic, D., Stanković, V., Gorgievski, M., Bogdanović, G. & Kovačević, R., Adsorption of heavy metal ions by sawdust of deciduous trees. *Journal of Hazardous Materials*, **171**, pp. 684–692, 2009. DOI: 10.1016/j.jhazmat.2009.06.055.

[22] Kokaoba, S., Comparaison of Amberlite IR 120 and dolomite's performances for removal of heavy metals. *Journal of Hazardous Materials*, **147**, pp. 488–496, 2007. DOI: 10.1016/j.jhazmat.2007.01.037.

[23] Ayoub, G.M. & Mehawej, M. Adsorption of arsenate on untreated dolomite powder. *Journal of Hazardous Materials*, **148**(1–2), pp. 259–266, 2007. DOI: 10.1016/j.jhazmat.2007.02.011.

[24] Pehlivan, E., Özkan, A.M., Dinç, S. & Parlayici, Ş., Adsorption of Cu^{2+} and Pb^{2+} ion on dolomite powder, *Journal of Hazardous Materials*, **167**, pp. 1044–1049, 2009. DOI: 10.1016/j.jhazmat.2009.01.096.

[25] Salameh, Y., Al-Lagtah, N., Ahmad, M.N.M., Allen, S.J. & Walker, G.M., Kinetic and thermodynamic investigations on arsenic adsorption onto dolomitic sorbents. *Chemical Engineering Journal*, **160**, pp. 440–446, 2010. DOI: 10.1016/j.cej.2010.03.039.

[26] Irani, M., Amjadi, M. & Mousavian, M.A., Comparative study of lead sorption onto natural perlite, dolomite and diatomite. *Chemical Engineering Journal*, **178**, pp. 317–323, 2011. DOI: 10.1016/j.cej.2011.10.011.

[27] Ghaemi, A., Torab-Mostaedi, M., Shahhosseini, S. & Asadollahzadeh, M., Characterization of Ag(I), Co(II) and Cu(II) removal process from aqueous solutions using dolomite powder. *Korean Journal of Chemical Engineering*, **30**(1), pp. 172–180, 2013. DOI: 10.1021/ie0402620.

[28] Stefaniak, E., Dobrowolski, R. & Staszczuk, P., On the adsorption of chromium(VI) ions on dolomite and 'dolomitic sorbents'. *Adsorption Science & Technology*, **18**(2), pp. 107–115, 2000. DOI: 10.1177/026361749901700202.

[29] Salameh, Y., Albadarin, A.B., Allen, S.J., Walker, G. & Ahmad, M.N.M., Arsenic (III, V) adsorption onto charred dolomite: Charring optimisation and batch studies. *Chemical Engineering Journal*, **259**, pp. 663–671, 2015. DOI: 10.1016/j.cej.2014.08.038.

[30] Ivanets, A.I., Shashkova, I.L., Kitikova, N.V., & Drozdova, N.V., Extraction of Co(II) ions from aqueous solutions with thermally activated dolomite. *Russian Journal of Applied Chemistry*, **87**(3), pp. 270–275, 2014. DOI: 10.1134/S1070427214030045.

[31] Calugaru, I.L., Neculita, C.M., Genty, T., Bussière B., & Potvin, R., Performance of thermally activated dolomite for the treatment of Ni and Zn in contaminated neutral drainage. *Journal of Hazardous Materials*, **310**, pp. 48–55, 2016. DOI: 10.1016/j.jhazmat.2016.01.069.

[32] Walker, G.M., Hanna, J.-A., & Allen, S.J., Treatment of hazardous shipyard wastewater using dolomitic sorbents. *Water Research*, **39**(11), pp. 2422–2428, 2005. DOI: 10.1016/j.watres.2005.04.025.

[33] Duffy, A., Walker, G.M. & Allen, S.J., Investigations on the adsorption of acidic gases using activated dolomite. *Chemical Engineering Journal*, **117**(3), pp. 239–244, 2006. DOI: 10.1016/j.cej.2005.11.016.

HISTORICAL USE OF NITROGEN AND PHOSPHORUS FERTILIZERS IN RAIN-FED REGIONS AND THEIR CONSEQUENTIAL IMPLICATIONS ON WATER QUALITY IN THE PACIFIC NORTHWEST, USA

R.L. MAHLER[1] & M.E. BARBER[2]
[1]Department of Soil and Water Systems, University of Idaho, Moscow, ID, USA
[2]Department of Civil and Environmental Engineering, University of Utah, Salt Lake City, UT, USA

ABSTRACT
Nitrogen (N) and phosphorus (P) fertilizers are major sources of surface water and/or groundwater pollution in the Pacific Northwest. Soil scientists have tried to increase the efficiency of these nutrients in crop production within the region by refining both the rates of application, timing of application and placement of these nutrients within soils. The purpose of this article is to document changes in the use of these nutrients since 1971. To do this four mail-based grower surveys were conducted in 1981, 1996, 2011 and 2016 to document changing N and P fertilizer use and their methods of placement within soils. In addition, the surveys documented erosion control efforts, which would reduce surface water P pollution. In general, N application rates have increased by approximately 10% in the last 45 years. Even though more N fertilizer has been used, offsets due to increased yields, better timing of fertilizer additions and better application location within the soil profile (closer to the roots) have been observed. Conversely, P application rates have remained the same or slightly decreased in the last 45 years. In addition, implemented soil erosion control efforts have reduced the quantity of P moving off fields and into surface waters. Based on historical N and P application rates it is estimated that farms within the selected study area have reduced their N and P loads to surface waters by 15% and 28%, respectively.
Keywords: historical use, nitrogen fertilizer applications, nutrient loads in surface waters, phosphorus fertilizer applications

1 INTRODUCTION

Over 15,000,000 ha of cropland are found within the Columbia-Snake River Watershed of the Pacific Northwest. Approximately 5,000,000 ha of land are irrigated, where the other 10,000,000 ha are used in producing rain-fed crops. Within the rain-fed region there is some very productive land. One of these productive rain-fed regions is known locally as the Palouse. The Palouse is famous for consistently producing the highest rain-fed winter wheat yields in the world. However, to produce these consistently high yields large quantities of nitrogen and phosphorus fertilizers are applied to agricultural fields [1, 2]. A significant portion of the applied N and P leak into the surface waters of the region. Consequently, from historical, environmental, economic and practical standpoints it is important to know the application rates, application timing and placement of these fertilizers and tillage management practices on the farm to understand and prevent losses of these nutrients into the larger watershed.

2 BACKGROUND

Nitrogen is the major nutrient applied in winter wheat cropping systems in rain-fed areas of the Pacific Northwest [3–5]. Traditionally, most of the nitrogen has been applied in the fall prior to or at seeding. Most of the region's annual precipitation occurs between November and May, resulting in fertilizer N being in the soil when conditions are highly favourable for losses caused by leaching and denitrification. This timing of application technique

DOI: 10.2495/EI-V1-N3-334-343

results in relatively low nitrogen use efficiency (NUE) by the winter wheat crop. In drier areas (<475 mm annual precipitation), NUE usually ranges between 50% and 65%. However, in the wetter areas (>475 mm annual precipitation) NUE is much lower and often ranges between 30% and 50% [6]. The nitrogen loss is a result of leaching; however, it eventually moves laterally in the soil and is ultimately a surface loss to local rivers.

Water quality problems associated with P are generally confined to surface waters [1, 2, 7]. Many human activities contribute P to surface waters. Agricultural land enriched with P by fertilization can contribute substantial amounts of P to surface waters as a result of runoff and/or erosional processes. Activities associated with modern agriculture often significantly increase soil erosion, water runoff and transport of sediment into surface waters. Surface water pollution from P is controllable by reducing soil erosion and thus keeping the soil out of creeks, streams, rivers and lakes. Specific management practices that will reduce P leakage into surface waters include (1) soil erosion control; (2) using P fertilizer application rates based on research and soil sampling; (3) correctly placing the P fertilizer near the seed and away from the soil surface; and (4) using conservation tillage practices including reduced tillage, minimum tillage, no-till and/or direct seeding.

The purpose of this article is to document changes in the use of nitrogen and phosphorus in agricultural settings in eastern Washington, northern Idaho and northeastern Oregon since 1971. To do this four mail-based grower surveys were conducted in 1981, 1996, 2011 and 2016. We documented changing N and P fertilizer use and their methods of placement within soils. In addition, the surveys documented erosion control efforts that reduce surface water P pollution. This report is a summary of our findings.

3 METHODOLOGY

Survey instruments containing between 28 and 33 questions were developed to access nutrient and soil management practices used by farmers in the 350–650 mm precipitation zone in the rain-fed farming area producing winter wheat in the Idaho, Oregon and Washington portions of the Inland Pacific Northwest [8–10]. The actual surveys were conducted in 1981, 1996, 2011 and 2016. All surveys contained the following six questions which were evaluated in this article:

Compared to 10 years ago, how has nitrogen fertilizer use on your farm changed?
Answer choices: use significantly more N, use slightly more N, no change, use slightly less N, use significantly less N

Compared to 10 years ago, how has phosphorus fertilizer use on your farm changed?
Answer choices: use significantly more P, use slightly more P, no change, use slightly less P, use significantly less P

What is your typical application rate of nitrogen to winter wheat (in kg/ha)? *Answer choices: <50, 50 to 100, 100 to 125, 125 to 150, >150*

What is your typical application rate of phosphorus to winter wheat (in kg/ha)? *Answer choices: <15, 15 to 30, 30 to 60, >60*

What is your typical application method for nitrogen on winter wheat?
Answer choices: all pre-plant injected, all pre-plant, all applied at planting, all applied after planting, combination of pre-plant or at planting and a later application

What is the primary tillage method you use in winter wheat production?
Answer choices: conservation tillage, reduced/minimum tillage, no till/direct seeding

The survey target audience was a representative sample of 2,500 farms in the area studied. In addition to the questions above demographic information including annual precipitation, number of years of farming experience, formal education level of the farm operator and size of farm were collected from survey respondents.

In each survey year, a target of 600 completed questionnaires was chosen as the survey goal to result in a sampling error of 4–7% [11]. The survey process was designed to receive a completed survey return rate of more than 40%. Addresses were obtained from county extension offices and USDA-NRCS field office mailing lists and four mailings were planned to achieve the 40% return rate [12, 13]. The mailing strategy used was identical in all surveys that had been conducted in the region since 1970 [8–10].

It only took four mailings to achieve the target return rate of 40% in 1981, 1996, 2011 and 2016. The first mailing included the water issues survey form, a business reply envelope and a cover letter that (1) identified the survey's authors; (2) explained the purpose of the survey; (3) assured the respondent of anonymity; and (4) asked the respondents to fill out and return the survey via the business reply envelope. The second mailing (4 weeks later) consisted of a postcard that stressed the importance of the survey and remind the respondent to fill out and return the survey sent out in the first mailing. Five weeks later the third mailing was sent to residents who did not respond to the first or second mailing. This mailing included a reminder letter, another copy of the water issues survey and a business reply envelope. The fourth mailing consisted of a reminder postcard 6 weeks after the third mailing.

Survey answers were coded and entered into Microsoft Excel. Missing data were excluded from the analysis. The data were analysed at two levels using SAS [13]. The first level of analysis generated frequencies, while the second level evaluated the impacts of demographic factors. Significance ($P < 0.05$) to demographic factors was tested using a chi-square distribution [11–13]. Since similar response rates were observed in all four survey years, data analysis procedures were identical for each sampling.

4 RESULTS AND DISCUSSION

The survey methodology used in the study was not designed to be unique, but rather to be used as a tool to ascertain useful information. The survey methodology was designed to access management practices used by farmers on their land in eastern Washington, northeastern Oregon and northern Idaho over a 45-year period.

All four survey years achieved a survey return rate goals of more than 50% [11, 12]. The 1981, 1996, 2011 and 2016 survey response rates were 56.4%, 51.3%, 54.2% and 50.7%, respectively. Consequently, when coupled with the low sampling error of the survey, respondents can be equated to residents in the following discussion. The average survey respondent had farmed for 34 years and was 61 years old. Based on survey demographics the surveyed growers were representative of the actual demographics of farmers living in the region.

4.1 Nitrogen management of farms in the region

Farmers in this region have been applying constant rates of N to their winter wheat crops for the last 45 years (Table 1). Over one-third of the farmers indicated that they have not changed their N application rates, while between 34.0% and 46.5% claimed to have slightly increased N application rates, depending on the survey year. Conversely, few farmers have significantly increased or significantly decreased their N application rates on winter wheat over time.

Table 1: Nitrogen fertilizer rate application trends over the previous 10-year periods for winter wheat based on results from the 1981, 1996, 2011 and 2016 grower nutrient surveys.

Nitrogen application rate trend	Year (%)			
	1981	1996	2011	2016
Significantly higher	15.4	10.2	8.0	6.4
Slightly higher	46.5	40.4	38.1	34.0
No change	33.2	36.4	38.1	34.0
Slightly lower	4.0	8.7	12.3	10.3
Significantly lower	0.9	4.3	3.5	4.2

Table 2: Nitrogen fertilizer rate application trends over the last 45 years (comparing 1971 with 2016) for winter wheat based on historical annual precipitation data and grower surveys.

Annual precipitation zone (mm)	Nitrogen application trends
<425	No change
425–475	No change
475–525	Slightly higher
525–575	No change
>575	Slightly lower

When the four survey years are pooled and N application rates are separated by the amount of annual precipitation a farm received important trends can be observed (Table 2). First, N application rates on winter wheat have not changed in areas receiving less than 450 mm of annual precipitation. This is because water is the most limiting factor in wheat yields in these two precipitation zones. Consequently, further N additions would not increase wheat yields. Second, farmers in the 475–525 mm precipitation zone have slightly increased applications of N to winter wheat crops over the last 45 years. Here improved management practices may have resulted in both increased N application rates and higher winter wheat yield (Table 2).

Third, N applications have remained flat in the 525–575 mm precipitation area (Table 2). Again, better management practices may have allowed wheat yields to actually increase without a corresponding increase in N fertilizer use. Fourth, N application rates have actually decreased in the wettest wheat growing precipitation zone. This observation will be explained later in the article when N management is further explained.

Nitrogen fertilizer application rates on winter wheat over the last 45 years have ranged from 0 to 180 kg/ha. The average N application rate based on the four surveys was 95 kg N/ha. A majority of farmers have typically applied between 50 and 100 kg N/ha per winter wheat crop (Table 3). By 2016, 30% of growers were applying more than 100 kg N/ha per wheat crop. When evaluated more closely, the consistency of N application rates over time is

Table 3: Nitrogen fertilizer rate applications for winter wheat based on results from the 1981, 1996, 2011 and 2016 grower nutrient surveys.

Nitrogen application rate (kg/ha)	Year (%)			
	1981	1996	2011	2016
<50	15.2	16.8	13.4	12.6
50–100	60.2	60.8	55.8	57.4
100–125	16.0	13.0	14.3	15.2
125–150	4.1	4.2	14.2	10.7
>150	4.5	5.2	2.3	4.1

apparent. There were no statistical differences in N application rates that could be tied to time (survey year) over the 45-year study period.

4.2 Phosphorus management of farms in the region

Historical P application rate trends on winter wheat are shown in Table 4. A majority of survey respondents have not changed the quantity of P fertilizers applied to wheat over time. Back in 1981 over 45% of the farmers had increased their P application rates from the previous 10-year period (1971–1981).

Phosphorus application rate trends over the last 45 years are shown in Table 5. The trends shown indicate that annual precipitation zone had an effect on P fertilization in this rain-fed region. First, P application rates have not changed in areas receiving less than 475 mm of annual precipitation. This result mirrors the N fertilization trends in this rainfall regime. Again, P rates have probably remained the same in this zone because lack of water has been the major factor limiting winter wheat yield. Thus, changes in P use would only have a minimal impact on winter wheat yield potential.

Phosphorus fertilizer application rates were slightly higher in 2016 compared to 1981 in the 475–575 annual precipitation zones (Table 5). Here agronomic practices initiated in these rainfall zones in the last 25–35 years have increased wheat yield potentials. This in turn has resulted in a higher rate of P fertilizer applications. No change in fertilizer P application rates was observed in the zone where annual precipitation exceeded 575 mm.

Table 4: Phosphorus fertilizer rate application trends over the previous 10-year periods for winter wheat based on results from the 1981, 1996, 2011 and 2016 grower nutrient surveys.

Phosphorus application rate trend	Year (%)			
	1981	1996	2011	2016
Significantly higher	30.6	18.4	8.0	4.2
Slightly higher	15.3	18.0	28.0	20.4
No change	58.4	47.2	51.6	56.4
Slightly lower	4.1	13.4	9.0	14.5
Significantly lower	1.6	3.0	3.4	4.5

Table 5: Phosphorus fertilizer rate application trends over the last 45 years (comparing 1971 with 2016) for winter wheat based on historical annual precipitation data and grower surveys.

Annual precipitation zone (mm)	Phosphorus application trends
<425	No change
425–475	No change
475–525	Slightly higher
525–575	Slightly higher
>575	No change

Table 6: Phosphorus fertilizer rate applications for winter wheat based on results from the 1981, 1996, 2011 and 2016 grower nutrient surveys.

Phosphorus application rate (kg/ha)	Year (%)			
	1981	1996	2011	2016
<15	46.1	49.5	40.2	37.5
15–30	45.1	43.4	49.5	53.4
30–60	8.2	6.9	10.1	9.7
>60	0.0	0.2	0.2	0.4

Phosphorus fertilizer application rates over the last 45 years have ranged from 0 to 70 kg/ha. The average P application rate on winter wheat in this region based on grower surveys was 22 kg/ha. A majority of farmers have typically applied between 15 and 30 kg P/ha per wheat crop (Table 6). In 2016, 11.4% of growers were applying more than 30 kg P/ha per wheat crop. Conversely, 37.5% of growers were applying less than 15 kg P/ha. As with the N application data, the P data show the consistency of P application rates over the previous 45 years.

4.3 Tillage management of farms in the region

The primary tillage method used on winter wheat by farmers in the region is shown in Table 7. The primary type of tillage has changed over the last 45 years. In 1981, over 70% of farmers were using the traditional conventional tillage practice. In 2016, only 20% of the growers were still using conventional as their primary tillage practice. The use of reduced/minimum tillage increased from 28.3% of growers in 1981 to almost 50% by 2016 (Table 7). No-till/direct seeding was used by less than 2% of farmers in 1981, but had grown to be used by almost one-third of farmers in 2011 and 2016.

The demographic factors of grower experience, grower education level and survey year also significantly impacted the use of reduced/minimum tillage (Table 8). When survey years were pooled, growers with less than 10 years of farming experience were more likely to use reduced/minimum tillage than growers that had been farming for more than 25 years (68.3% vs 49.4%). It appears that the farmers with less experience were more likely to use reduced tillage practices to reduce soil erosion.

Table 7: Primary tillage method identified by growers in 1981, 1996, 2011 and 2016 surveys.

Tillage method	Year (%)			
	1981	1996	2011	2016
Conventional tillage	70.1	40.4	25.6	20.4
Reduced/minimum tillage	28.3	44.1	43.0	49.4
No-till/direct seeding	1.6	15.5	31.4	30.2

Table 8: The influence of demographic factors of grower experience, education level and survey year on the use of reduced tillage (reduced + minimum + no-till + direct seeding) in winter wheat cropping systems based on 1981, 1996, 2011 and 2016 grower surveys.

Demographic factor	Parameter	Using reduced tillage (%)	Significance
Grower experience	<10 years	68.3	**
	10–25 years	60.4	
	>25 years	49.4	
Education level	High school	52.9	**
	Some college	62.4	
	College graduate	73.8	
Survey year	1981	29.9	***
	1996	59.6	
	2011	74.4	
	2016	79.6	

Formal education level of farmers also impacted the type of tillage used on farms (Table 8). Farmers that were college graduates were more likely to use reduced/minimum tillage practices than farmers with fewer years of formal schooling. Conversely, farmers with only a high school diploma were less likely to use reduced/minimum tillage. In general, the greater the level of formal education the more likely the grower used reduced/minimum tillage.

Survey year also had a significant impact on the use of reduced/minimum tillage by farmers (Table 8). Back in 1981, less than 30% of growers used reduced/minimum tillage; however, by 2016 almost 80% employed this practice.

Tillage management is a very important practice that can be used to reduce N and P leakage from fields into surface water bodies in the region. For instance, soil erosion rates of 15–35, 3–12 and <3 mt/ha are common under wheat management conditions of conventional tillage, minimum/reduced tillage and no-till/direct seeding, respectively.

Since most P loss from soils is through surface erosion, changing tillage practices to minimum or no-till will greatly reduce P emissions from farmland to surface waters. Even though P fertilizer application rates have not been reduced in the last 45 years, the shift to tillage practices that greatly reduce soil erosion substantially reduces the P load from agriculture.

The use of reduced tillage practices to prevent soil erosion probably does not have a big impact on N leakage into the surface waters of the region. This is because nitrogen fertilizer is mobile in the soil and likely leaches into the soil below the active erosive surface of the soil profile. Consequently, erosion control has little impact on reducing N in surface waters.

4.4 Placement and timing of nitrogen applications

The primary application method of N fertilizers on winter wheat is shown in Table 9. Based on collected survey data farmers have employed six basic methods of N application to winter wheat. Back in 1981, 100% pre-plant injected (34.6%), 100% applied at planting (24.0%) and a combination of pre-plant or at planting with a partial spring top-dress application (20.4%) were the three most common N application methods. In 2016, the three methods listed earlier were still the most common methods used to apply N; however, their order of use had changed (Table 9). In 2016, the combination of all pre-plants or at planting with a spring top-dress application was the most common method (41.4%). The growers that used this method indicated that their total N application rate did not change, but rather, they were applying a higher portion of their total N in the spring. Split applications such as these have been shown to improve NUE in crops, resulting in less leakage into the environment.

This observation is made more credible when combined with the data shown in Table 10. Here the percentage of farmers applying 25–50% of their total N to winter wheat in the spring increased from 12.4% in 1981 to 29.3% in 2016. There could be many reasons for the increased use of split applications; however, there are two obvious explanations. First, split applications improve NUE which is important for the producers' economic bottom line.

Table 9: Primary application method of nitrogen fertilizer on winter wheat identified by farmers in 1981, 1996, 2011 and 2016 surveys.

Nitrogen application method	Year (%)			
	1981	1996	2011	2016
All pre-plant injected	34.6	32.1	30.7	28.4
All pre-plant	18.6	18.2	14.8	8.2
All applied at planting	24.0	19.6	22.3	20.3
All applied after planting	2.4	3.0	3.1	1.7
Combo-pre-plant/at plant + spring	20.4	27.1	36.0	41.4

Table 10: The percentage of fields receiving spring top-dress applications of nitrogen based on 1981, 1996, 2011 and 2016 grower surveys.

Percent of the nitrogen applied in spring	Year (%)			
	1981	1996	2011	2016
<25	86.2	80.0	62.3	59.5
25–50	12.4	16.6	24.2	29.3
50–75	1.0	2.7	6.9	6.0
>75	0.4	0.7	6.6	5.2

Table 11: The percent of nitrogen applied to winter wheat the spring after planting based on annual precipitation zone.

Annual precipitation (mm)	Year (%)			
	1981	1996	2011	2016
>425	5.0	6.2	4.3	2.9
425–475	10.7	10.6	12.7	13.2
475–525	12.3	12.4	17.2	18.9
525–575	19.6	24.5	29.4	32.8
>575	23.4	25.9	35.2	34.9

Second, N fertilizer costs have escalated in the last 20 years to the point it is applied when needed instead of just as insurance to produce the desired yield. Both explanations tie into simple economics. The reduced N load in local surface waters is an important side benefit.

Split applications (part in the fall, part in the spring) were most commonly adapted in the wetter precipitation areas, where NUE values are the lowest. Split application N management increased between 1981 and 2016 in the 475–525, 525–575 and >575 annual precipitation zones (Table 11). In 2016, over 30% of the N applied by growers in the wettest region was being applied in the spring.

5 CONCLUSIONS AND RECOMMENDATIONS

N, P and tillage data collected over the last 45 years in the region suggest that N and P leakage from farmland into the region's rivers has been substantially reduced. Specific findings include:

- Nitrogen fertilizer use has increased by about 10% in the last 45 years. Application rates over the last 45 years have ranged from 0 to 180 kg/ha, with the average application rate of 95 kg N/ha. A majority of farmers have typically applied between 50 and 100 kg N/ha per winter wheat crop.
- Phosphorus fertilizer use has remained constant over the last 45 years. Application rates over the last 45 years have ranged from 0 to 70 kg/ha, with an average application rate of 22 kg/ha. A majority of farmers have typically applied between 15 and 30 kg P/ha per wheat crop.
- The primary type of tillage has changed over the last 45 years. In 1981, over 70% of farmers were using the traditional conventional tillage practice. The use of reduced/minimum tillage increased from 28.3% of growers in 1981 to almost 50% by 2016.
- Since most P loss from soils is through surface erosion, changing tillage practices to minimum or no-till greatly reduced leakage into surface waters.
- Split applications of N fertilizers have been shown to improve NUE, resulting in less N leakage into surface waters.

REFERENCES

[1] Mahler, R.L., *Northern Idaho Fertilizer Guide: Winter Wheat*, University of Idaho College of Agriculture, Moscow, ID, Current Information Series No. 453 (revision), 2007.

[2] McDole, R.E., & Mahler, R.L., *Northern Idaho Fertilizer Guide: Winter Wheat*, University of Idaho College of Agriculture, Moscow, ID, Current Information Series No. 453, 1985.

[3] Smith, R.A., Alexander, R.B., & Wolman, M.G., Water quality trends in the nation's rivers, *Science*, **235**, pp. 1607–1613, 1987. DOI: 10.1126/science.235.4796.1607.

[4] Alexamder, R.B., Smith, R.A., Schwarz, G.E., Boyer, E.W., Nolan, J.V., & Brakebill, J.W., Differences in phosphorus and nitrogen delivery to the Gulf of Mexico from the Mississippi River Basin. *Environmental Science and Technology*, **42**(3), pp. 822–830, 2008. DOI: 10:1021/es0716103.

[5] Huang, W., Heifner, R.G., Taylor, H., & Uri, N.D., Timing nitrogen fertilizer application to reduce nitrogen losses to the environment. *Water Resources Management*, **14**, pp. 35–58, 2000. DOI: 10.1023/A:1008137011813.

[6] Lutcher, L.K., & Mahler, R.L., Sources and timing of spring top-dress nitrogen on winter wheat in Idaho. *Agronomy Journal*, **80**, pp. 648–654, 1988. DOI: 10.2134/agronj1988.00021962008000040020x.

[7] Sharpley, A.N., Chapra, S.C., Wedepohl, R., Sims, J.T., Daniel, T.C., & Reddy, K.R., Managing agricultural phosphorus for protection of surface waters: Issues and options. *Journal of Environmental Quality*, **23**(3), 437–451, 1993. DOI: 10.2134/jeq1994.00472425002300030006x.

[8] Mahler, R.L., Simmons, R., Sorensen, F. & Miner, J.R., Priority water issues in the Pacific Northwest. *Journal of Extension*, **42**(5), Article 5RIB3, available at http://www.joe.org/joe/2004october/rb3.php, 2004 (accessed 5 June 2014).

[9] Mahler, R.L., Gamroth, M., Pearson, P., Sorenson, F., Barber, M.E. & Simmons, R., Information sources, learning opportunities and priority water issues in the Pacific Northwest. *Journal of Extension*, **48**(2), Article 2RIB2, available at http://www.joe.org/joe/2010april/rb2.php, 2010 (accessed 5 June 2014).

[10] Mahler, R.L., Simmons, R. & Sorensen, F., Drinking water issues in the Pacific Northwest. *Journal of Extension*, **43**(6), 6RIB6, available at http://www.joe.org/joe/2005december/rb6.php, 2005 (accessed 5 June 2014).

[11] Salent, P. & Dillman, D., *How to Conduct Your Own Survey*, John Wiley and Sons, Inc.: New York, NY, 1994.

[12] Dillman, D., *Mail and Internet Surveys: The Tailored Design Method*. John Wiley and Sons, Inc.: New York, NY, 2000.

[13] SAS Institute Inc., *SAS Online Document 9.1.3.*, SAS Institute Inc.: Cary, NC, 2004.

HOW DOES TARGETED GRASSING OF ARABLE LAND INFLUENCE DRAINAGE WATER QUALITY AND FARM ECONOMIC INDICATORS?

ANTONÍN ZAJÍČEK, PETR FUČÍK, RENATA DUFFKOVÁ & JANA MAXOVÁ
Research Institute for Soil and Water Conservation, Prague, Czech Republic.

ABSTRACT

A 7-year experiment with grassing of arable land in the catchment recharge zone was conducted in a small (60 ha) agricultural tile-drained catchment (Bohemian – Moravian Highlands, Czech Republic). Together with water quality, the effect of increasing grassland area on selected farm economic indicators was evaluated. The area for grassing was delimitated by the method of relative soil infiltration vulnerability based on the analysis of five-digit valuated soil ecological unit code, available as .shp file at the scale 1:5000. Effect of grassing on farm budget was estimated by comparison of revenues (crop sales and subsidies) with technology costs of three different grassland area proportion within the agricultural land managed by the farm (1 200 ha). It was proved that nitrate concentrations in drainage water were influenced predominantly by the land use of the recharge zones within the drainage subcatchment. The grassing of arable land focused into proper catchment area (recharge zone) demonstrated a significant decrease in both NO_3 concentrations and N loads by 35% and 25%, respectively. On the other hand, increasing areas of grasslands would lead to decrease of landscape productive service, farm turnover and profit and the bigger dependence on subsidies. That´s why it is necessary to consider the non-productive functions of grasslands also as public service, taking into account the savings in water cleaning costs and the price of increased water retention. Presented approach come forward when setting the subsidies dealing with soil and water protective measures in agrarian landscape.

Keywords: nitrate concentration subsidies, recharge area, targeted grassing, technological costs, tile drainage

1 INTRODUCTION

Nitrates represent one of the main products of non-point water pollution, particularly in association with a high extent of agricultural land [1]. Nitrates enter surface water mainly by subsurface runoff, which is often represented by water from agricultural drainage. In small agricultural catchments in area of the Bohemian-Moravian Highlands crystallinum, the proportion of stream flow and nutrient load coming from tile drainage outlets was estimated to be 65–75% [2]. Drainage systems in this area are characterized by their locations in slopes, in transient zones, or at interfaces of transient and discharge zones [3]. In such cases, the entire catchment must be taken into account for water balance studies, since a considerable proportion of the drainage runoff can originate outside the drained area [4, 5]. These areas, called recharge or source areas, situated in the upper parts of slopes with shallow, highly permeable soil may represent, in intensively managed agricultural land, a significant source of leaching nutrients (N, P) and other pollutants (pesticides), especially when these localities are connected to drainage systems capturing both springs and shallow subsurface flow.

Grassing of arable land belongs to the preventive nitrogen remedial measures aimed at reducing nitrogen leaching from the soil. This important non-productive function of grassland is induced mainly by the fact that grassland can absorb and use bigger amount of nitrogen in comparison to field crops and this ability remains for longer period within the year. Permanent grasslands (PGs) cover the soil year round and employ large active subsurface biomass in the root system, which can utilize a significant amount of soil nitrogen. Moreover, grassland increases activity of soil microbes, which is much higher than under field crops.

DOI: 10.2495/EI-V1-N3-344-352

This contributes to the retention of nitrates in the soil by their immobilization [6]. The regulatory ecosystem function of PG allows fertilization with high nitrogen doses without negative impacts on water quality (up to 200 kg/ha/year [7]). Besides nitrogen remedial ability, grasslands have also other supporting and regulatory functions such as carbon sequestration, reduction of soil erosion, increase of water retention in the landscape, and optionally reduction of pesticide leaching [8, 9]. On the other hand, the excessive conversion of arable land to grassland has also negative aspects, such as decrease in provisioning ecosystem services, namely reduction in the area for cultivation of commercial field crops [10].

Main goal of this article was to evaluate the effect of grassing on both the quality of drainage water and also on the selected economic indicators of a farming company (agricultural cooperative).

2 MATERIALS AND METHODS

The grassing experiment was conducted in small agricultural catchment, Dehtáře (Fig. 1), situated in the Bohemian-Moravian Highlands, Czech Republic. It is a locally typical small agricultural catchment, where the tile drainage acts as the only permanent runoff from the catchment and the drainage system was built in the slope. Its area is 58 ha, with tile-drained areas occupying 19 ha (32% of the catchment). It is used mainly as agricultural land, with low forest representation. The agricultural land is mostly exploited as arable, with PG in the lower part of the catchment. The altitude varies between 550 and 497 m asl. Average year precipitation is 666 mm. The representation of soils (according to the World Reference Base for Soil Resources 2006) is variable, with Gleyed Cambisols, Gleysols and sporadically Histosols. In the recharge area, the soil cover is more homogenous, with prevailing Modal, Ranker and Arenic Cambisols. More detailed catchment description is given, e.g. in [11]. The experiment itself consisted in delimitation of the source area using the method defined by Janglova et al. [12]. This method is based on the analysis of soil conditions characterized by five-digit code of valuated soil ecological units (VSEU). The VSEU code characterizes main soil units, slope, exposure, skeletal character and soil depth. Based on the categorization of these criteria, soil is classified into five relative groups according to its significance for the infiltration process, with category 1 corresponding to the maximum infiltration capacity. Part of catchment recharge area defined by this method (4.6 ha) was grassed in 2006. During further routine management, we monitored changes in nitrate concentrations in drainage water at several drainage subsystems located in places with differing land use in the source (upper slope parts near the drainage divide) and outlet (lower slope parts) areas. To evaluate the effects of grassing, the whole monitored time was divided into two periods, period 1 – before grassing (2003–2006) and period 2 – after grassing (2007–2013). All grasslands in the catchment were fertilized by approximately 100 kg N/ha per season (mostly by urea and pig slurry digestate). The arable land in the catchment was fertilized according to crop rotation (cereals, potatoes and oilseed rape) in the amount of ca. 120–170 kg N/ha per season.

There were four monitored sites on drainage system in experimental field with different land use recharge and discharge zones of their subcatchments (Fig. 1). Site K1 (PG in the outlet area, grassing of the recharge area in period 2), K5 (arable land in the entire subcatchment during the entire experiment) and K4 (PG in the outlet area and arable land in the recharge area during the entire experiment). Drainage water from all these drainage subsystems was monitored in site KL, which is the closing profile of the entire drainage group. The experimental methodology was described in detail by [13, 14].

The economic evaluation of arable land grassing included calculation of technological costs for cultivation of particular crops according to the typical crop rotation in studied region, and their comparison with the income composed of the value of the particular crop production

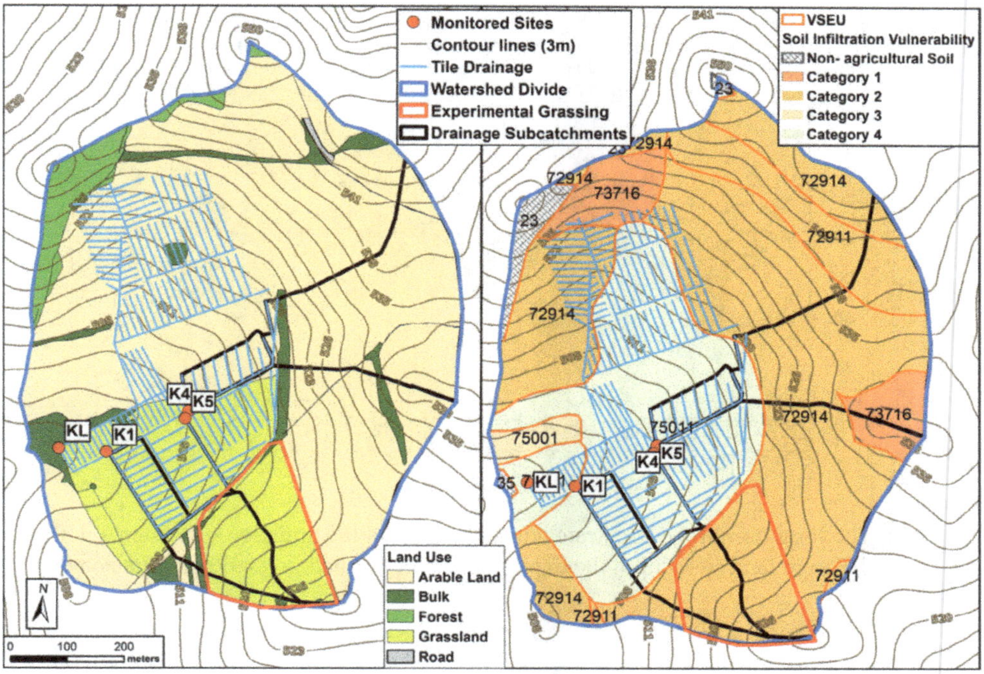

Figure 1: An overview map of the Dehtáře experimental catchment.

and the level of subsidies allocated to the field with this crop. The costs and income were both estimated for the year 2015. The data obtained from the Czech Statistical Office [15], various agricultural norms [16] and the level of subsidies were derived from the State Agricultural Intervention Fund and the Ministry of Agriculture of the Czech Republic [17]. These data were also consulted with farmers operating in the studied locality and its close neighbourhood. The costs and income were related to 1 ha and year and then to the 7-year crop rotation, which is mostly used in the studied locality (clover, winter wheat, spring barley/corn, winter oil seed rape, winter wheat, potatoes/corn, spring barley with clover undersowing). Evaluation of the experiment itself was done for 1 year and the area of 4.6 ha (the extent of grassing). The increase in the proportion of grassing was evaluated for three different variants of arable land/grassland proportions for a hypothetical farm with a typical size of agricultural land in the region. Variant A, considered as initial state, represents 1,500 ha of arable land and 300 ha of PG; variant B represents doubling of PG area at the expense of arable land, i.e., 1,200 ha of arable land and 600 ha of PG; and variant C represents equal shares of arable land and PG, i.e., 900 ha of arable land and 900 ha of PG. All incomes and costs are expressed in Czech Crows (CZK). The average exchange rate is 27 CZK for 1 EURO.

3 RESULTS

3.1 The effect of grassing on the drainage water quality

The nitrate concentrations in drainage waters were strongly variable during season (dependence on the discharge levels) and also in comparison of particular seasons (different precipitation course in particular season and by crops rotation). The measured NO_3

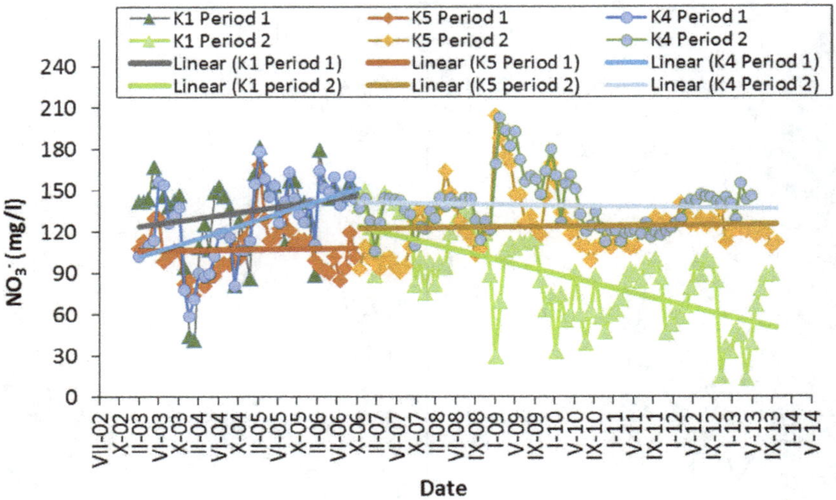

Figure 2: The long-term trend of nitrate concentrations in the drainage subsystems with different land use within the experiment. Site K1 had grassed recharge zone in period 2.

concentrations varied from 18 to 253 mg/l; the monthly flow-weighted concentrations (Cfw) varied from 13 to 197 mg/l. In period 1 (before grassing the recharge zone), Cfw values were surprisingly higher in sites K1, and K4 with the PG in drained area (discharge zone) than in site under arable land (K5) (Fig. 2). After grassing the K1 subsystem recharge area, some changes occurred. Approximately 1 year after grassing, the long-term course of NO_3 concentrations changed direction and became decreasing in the site with grassed recharge zone. In sites without land-use change (K5, K4), the nitrate Cfw trend remained increasing or the stagnation was found. Results of Kruskal–Wallis test, which evaluated the significance of nitrate concentration change, showed a statistically significant decrease happened in sites with the grassed recharge area. Further analysis of nitrate concentrations during both monitored periods (Fig. 3) showed the clearly demonstrable decrease by 26–32% at drainage subsystem K1, whose recharge area was completely grassed. The fall in concentrations was observed despite the fact that fertilization of the PG continued. The box plots (Fig. 3) also show increase of nitrate concentrations at drainage subsystem K5 with arable land in both discharge and recharge areas, and stagnation (or mild increase) of these concentrations at drainage subsystem K4 with PG in the discharge area and arable land in the recharge area. The entire drainage group KL (20% of recharge area grassed) showed a decrease in nitrate concentrations by 11%. Along with the decreased nitrate concentrations, there was also detected a drop in nitrogen load by 23% (from 38.4 kg/ha/year to 31.2 kg/ha/year) from the whole KL drainage group subcatchment whose source area was partly grassed, and by 47% (from 57.0 kg/ha/year to 39.6 kg/ha/year) at drainage subsystem K1 whose source area was grassed completely. In the same period, the drainage subsystem K5, where no change in land use occurred (arable land), displayed increase in nitrogen loss by 17%.

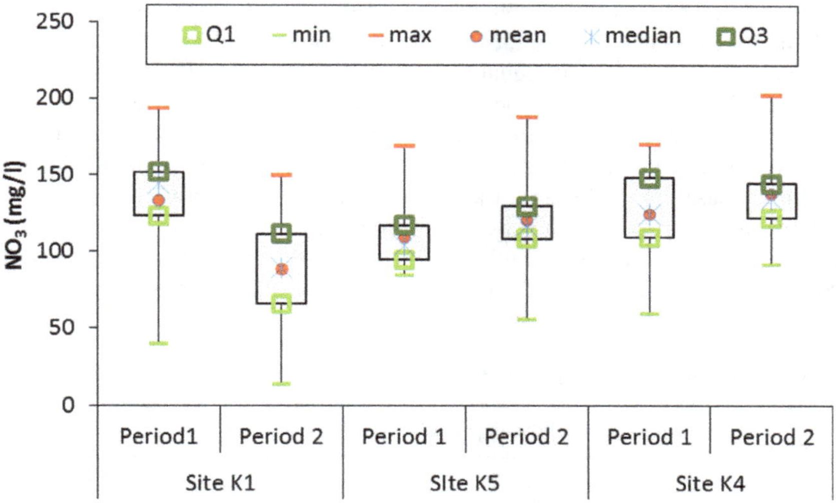

Figure 3: Nitrate concentrations in drainage waters of subsystem K1 (grasssed recharge zone in period 2), subsystem K5 (arable land in the whole area), and subsystem K4 (grassland in discharge zone, arable land in recharge area).

3.2 Financial assessment of grassing

The income from the main crops according to the crop rotation in the pilot area is shown in Table 1. The income value comprises the value of products from plant production and the value of subsidies per hectare of agricultural area depending on the particular crop. The highest income from the main crops was generated by potatoes, followed by oilseed rape and wheat; in contrast, the income from clover and grassland was the lowest. A significant part of the plant production income is currently represented by the subsidies. The subsidies level in 2015 for the majority of crops cultivated on arable land of the pilot locality was about 5 400 CZK/ha (SAPS – direct payment per 1 ha and Greening). This amount referred to clover, potatoes, maize, wheat and oilseed rape. The subsidies for potato cultivation was increased in 2015 by 'voluntary coupled support', and the total support for ware potatoes thus reached 18,605 CZK/ha in the pilot area. In the case of PG, total level of subsidies in the studied locality may reach approximately 10,300 CZK/ha (LFA and AEKO, amounting to approximately 1,240 and 2,592 CZK/ha, respectively). The proportion of subsidies in the income per hectare represents 75.5% for PG and an in an average of 17.2% for arable land.

Technological costs for the main assessed crops are given in Table 2. Again, the highest costs were associated with ware potatoes; the second position, far behind, was held by oilseed rape. The technological costs for cereals were similar. Logically, the lowest costs were associated with PG, where in most cases there are additional savings on seeds and chemicals in the following years. However, the relatively high costs (with regard to the income) of tractor work still remain.

The results of a simple analysis of income and costs of plant production for a 7-year crop rotation with differing ratio of arable land and PG are given in Table 3. The decrease in both income and technological costs for both assessed variants (B, C) displays a parallel slope. In case of doubling the PG share (variant B), the income decreased by 12.5% and the technological costs by 13.5%. In case of equal shares of arable land and PG (variant C; 900 ha of arable land and 900 ha of PG), the income decreased by 25.0% and technological costs by 26.7% compared to the initial situation. The difference between the income and the technological costs decreased slightly slower; in the case of 600 ha of PG (variant B) it dropped by 9.5% and in the case of equal share of arable land and PG (variant C) by 18.8%. This was caused by the higher proportion of subsidies for grassing included into the income, which in the studied variants grew from the initial 21.2% to 26.9% in the case of 1 200 ha of arable land/900 ha of PG and to 32.1% in the case of equal share of arable land and PG.

Table 1: Incomes from crop production (main product value and subsidies level).

Income from Crop Production by 1 year and 1 ha				
Crop	Crop Value (Thousands of CZK/ha)	Subsidies (Thousands of CZK/ha)	Income (Thousands of CZK/ha)	Subsidies Share (%)
Grassland	3.4	10.3	13.7	75.5
Clover	13.7	5.4	19.1	28.3
Winter wheat	28.6	5.4	34.0	15.9
Spring barley	18.7	5.4	24.1	22.5
Oilseed rape	39.5	5.4	44.9	12.0
Potatoes	84.0	17.8	101.8	17.5
Silage maize	27.3	5.4	32.7	16.5
Income from 7 year crop rotation and 4.6 ha experimental area				
Crop rotation	1,105.6	230.9	1,336.5	17.3
Grassland	107.9	331.7	439.5	75.5

Table 2: Technological costs for growing main crops in region in interest.

Technological Costs for Agricultural Production by 1 Year and 1 ha (thousands of CZK)								
Crop	Fuel	Work	Machinery	Seeds	Fertilizers	Chems.	Fixed	Total
Grassland	0.9	0.5	4.0	0.6	0.3	0.0	3.0	9.3
Clover	1.9	0.8	6.0	2.3	0.0	0.5	4.0	15.4
Winter wheat	1.8	0.8	4.7	2.2	5.3	4.0	4.8	23.7
Spring barley	1.9	0.6	4.3	2.0	6.4	1.5	5.0	21.6
Oil seed rape	1.9	0.8	4.3	2.0	10.5	6.0	4.6	30.1
Potatoes	3.9	4.0	17.6	21.1	5.2	8.9	20.0	80.7
Silage maize	3.0	1.0	6.5	2.0	5.8	1.4	7.3	27.0
Technological Costs for Agricultural Production by 7 year Crop Rotation and 4.6 ha Experimental Area (Thousands of CZK)								
Crop rotation	74.4	40.6	220.8	155.8	177.2	121.3	232.1	1,021.9
Grassland	29.0	14.8	128.9	0.6	9.7	0.0	96.6	299.1

Table 3: Comparison of Costs and Incomes of an agricultural company with changing proportion of grasslands and arable land.

Variant	Incomes (Millions of CZK)	Subsidies Share (%)	Technological Costs (Millions of CZK)	Difference (Millions of CZK)
A (1,500 ha arable/300 ha PG)	456,4	21,2	352,7	103,7
B (1,200 ha arable/600 ha PG)	399,5	25,9	305,6	93,9
C (900 ha arable/900 ha PG)	342,6	32,1	258,5	84,2

4 DISCUSSION

The above-mentioned results provide evidence for the positive effect of grassing on the drainage water quality. While PG situated in the outlet area directly above the drainage system (in the lowest part of the drained catchment) played no role in the nitrate concentrations in the drainage water, grassing of source areas resulted in their permanent, statistically significant reduction. To ensure effectiveness of grassing as an approach for reducing the nitrogen burden of water, it has to be applied to appropriately delimitated source (vulnerable) areas with shallowest soil, where most of the runoff from the catchment originates. For this, a number of certified methodologies developed by Research Institute for Soil and Water Conservation can be employed. Grassing is also advantageous by the possibility to use liquid manure, whose application to the grass cover is not associated with the risk of increased nitrate leaching [18].

With regard to economy, the increase in grassed areas undoubtedly results in reduction of farm turnover. As well, the drop in the costs and income will lead to reduction of the profit, which will be partly compensated for by higher PG support. Higher dependence on the support, however, is disputable in view of the long-term stability of a farm because the amount and focus of the support may change year after year. It should also be noted that the production value of PG and clover, and partly of corn as well, is relevant only when the agricultural subject also maintains animal production, or optionally a biogas station, while considering the amount of grass mass that the agricultural subject is able to process.

Besides the positive effect on water quality and extra-productive landscape function, grassing of arable land can also be perceived as a public service that will increase the value of other landscape ecosystem functions, such as regulatory and aesthetic functions. Monetary evaluation of these ecosystem services is only meaningful when grassing is actually perceived as a public service. It means not only the increase of ecosystem landscape functions and improvement of water retention in agricultural soil, namely in cases of intensive rainfall-runoff events, but also saving of the costs of removal of nitrates (and also of pesticides) from water in water processing plants. The reported value of 1 m^3 water retention in the landscape represents approximately CZK 500 per year [9]. The estimated costs of water processing plants (regeneration salt and elimination of wastewater) for decreasing the nitrate concentrations from 50 mg/l to 15 mg/l range around CZK 3.2 and for decreasing the nitrate concentrations from 100 mg/l to 43 mg/l even around CZK 6.3 per 1 m^3 (according to data from water management companies). These data should be taken into account when determining the amount of support allocated to public service provided by farm management, aimed at improvement of landscape functions.

5 CONLUSIONS

The experiment proved that appropriate management of properly delimitated recharge area can improve water quality in the catchment. While PG placed directly in the drained area (catchment discharge zone) did not show any effect, the grassing focused on the catchment recharge area demonstrated a significant decrease in NO_3 concentrations. Hence, the grassing focused to the proper catchment area (recharge zone) can be a useful, effective and relatively cheap measure for improvement of shallow groundwater quality, or optionally the quality of local drinking water sources.. On the other hand, increasing areas of grasslands would lead to decrease of landscape productive service, farm turnover and profit and the bigger dependence on subsidies. That´s why the grassing should be applied in small, precisely defined parts of the catchment and it is necessary to consider the non-productive functions of grasslands also as public service, taking into account the saves in water cleaning costs and the price of increased water retention. Presented approach come forward when setting the subsidies dealing with soil and water protective measures in agrarian landscapes.

ACKNOWLEDGEMENTS

This article was prepared under the support of project no. 633945 FATIMA Farming Tools for external Inputs and water Management (The EU Framework Programme for Research and Innovation, Horizon 2020). Authors wish to thank the Kojčice Agricultural Cooperative for research support and data providing.

REFERENCES

[1] Fučík, P., Kvítek, T., Lexa, M., Novák, P. & Bílková, A., Assessing the stream water quality dynamics in connection with land use in agricultural catchments of different scales. *Soil and Water Research* **2008**(**3**), pp. 98–112, 2008.

[2] Doležal, F., Kulhavý, Z., Soukup, M. & Kodešová, R., Hydrology of tile drainage Run off. *Physics and Chemistry of the Earth* (B), **26**(**7–8**), pp. 623–627, 2001. DOI: 10.1016/S1464-1909(01)00059-4.

[3] Doležal, F. & Kvítek T., The role of recharge zones, discharge zones, springs and tile drainage systems in peneplains of Central European Highlands with regard to water quality generation processes. *Physics and Chemistry of the Earth*, **2004**(**29**), pp. 775–785, 2004. DOI: 10.1016/j.pce.2004.05.005

[4] Herrmann, A. & Duncker, D., Runoff formation in a tile-drained agricultural basin of the Harz Mountain Foreland, Northern Germany. *Soil and Water Research*, **3**(**3**), pp. 83–97, 2008.

[5] Zajíček, A., Kvítek, T., Kaplická, M., Doležal, F., Kulhavý, Z., Bystřický, V. & Žlábek, P., Drainage water temperature as a basis for verifying drainage runoff composition on slopes. *Hydrological Processes*, **2011**(**25**), pp. 3204–3215, 2011. DOI 10.1002/hyp.8039.

[6] Griffiths, B.S., Hallett, P.D., Kuan, H.L., Gregory, A.S., Watts, C.W. & Whitmore, A.P., Functional resilience of soil microbial communities depends on both soil structure and microbial community composition. *Biology and Fertility of Soils*, **44**, pp. 745–754, 2008. DOI: 10.1007/s00374-007-0257-z.

[7] Fiala, J., Recent management systems of grasslands – (1st part). Úroda, 6, pp. 9–11, 2002. (In Czech)

[8] Hejduk, S., *Grasslands Ecosystem, Perennial Grassland as Landscape Forming Tool*. MZLU: Brno, pp. 45–50, 2007. ISBN 978–80–7375–045–9. in Slovak)

[9] Hönigová, I., Vačkář, D., Lorencová, E., Melichar, J., Götzl, M., Sonderegger, G., Oušková, V., Hošek, M. & Chobot, K., Survey on grassland ecosystem services – Report of the European Topic Centre on Biological Diversity. Nature conservation agency of the Czech Republic, 78 p., 2012.

[10] Hauck, J., Schleyer, CH., Winkler, K. J. & Maes, J., Shades of greening: reviewing the impact of the new EU agricultural policy on ecosystem services. *Change Adaptation Socioecological Systems*, **1**, pp. 51–62, 2014. DOI: 10.2478/cass-2014-0006

[11] Nováková, E., Karous, M., Zajíček, A. & Karousová, M., Evaluation of ground penetrating radar and vertical electrical sounding methods to determine soil horizons and bedrock at the locality Dehtáře. *Soil and Water Research*, **8(3)**, pp. 105–112, 2013.

[12] Janglová, R., Kvítek, T. & Novák, P., Categorisation of soil infiltration capacity based on GIS proceeding of soil survey data (In Czech). *Soil and Water Scientific Studies*, **2**, pp. 61–81, 2003.

[13] Zajíček, A. & Kvítek, T., The effect of grassland focused into recharge areas on nitrate concentration in drainage waters. *Sovak*, **22(9)**, pp. 14–17, 2013. (in Czech)

[14] Fučík, P., Zajíček, A., Duffková, R. & Kvítek, T., Chapter 11: Water quality of agricultural drainage systems in the Czech Republic – Options for its improvement. *Research and Practices in Water Quality*, ed. T.S. Lee, InTech: Rijeka, pp. 241–262, 2015. DOI: 10.5772/59298.

[15] Czech statistical office (ČSÚ), www.czso.cz

[16] Normatives for crop and food production (AGC), available at www.agronormativy.cz (in Czech)

[17] eAGRI Czech ministry of Agriculture (MZe), available at http://eagri.cz

[18] Duffková, R. & Zajíček, A., Evaluation of groundwater quality after mulching and using slurry on grassland. *Vodní hospodářství*, **61(7)**, pp. 290–293, 2011. (in Czech)

HYDROLOGIC ASSESSMENT OF WOODY BIOMASS REMOVAL FOR BIOFUEL PRODUCTION

C. TAYLOR SMITH[1], MICHAEL BARBER[1] & ROBERT MAHLER[2]
[1]Civil and Environmental Engineering, University of Utah, USA
[2]Soil Science Division, University of Idaho, Moscow, ID, USA

ABSTRACT
Wood-based energy markets have been proposed as a means to ensure sustainable forests, enhance energy security, promote environmental quality and realize social benefits. An important issue among stakeholders is that collecting small-diameter woody biomass may significantly alter the water budget of post-timber harvesting landscapes. However, little is actually known about the hydrologic impacts that additional biomass removal and post-harvest land treatments may have on the water budget. Climate data and volumetric water content measurements at depths of 10, 20, 30 and 100 cm were collected from 28 one-acre plots near Eugene, OR, USA, subject to seven different land treatments. This information was analysed and used to calibrate and validate a site-specific water balance model (UNSAT-H) to evaluate a null hypothesis that changes in biomass removal do not impact subsurface environment. Results showed a positive correlation between post-harvest land treatments involving soil compaction and evaporation with compacted sites producing over 30% more evaporation than non-compacted, harvested sites and 20% more than non-compacted, non-harvested sites. Furthermore, non-compacted sites subject to increased biomass removal had higher infiltration rates than both unharvested sites and harvested, compacted sites. In terms of changes to runoff and sediment production, maximum impact coincides with the period immediately after track construction and harvesting. However, these effects decrease significantly over the 5-year time frame, well within the inter-logging cutting cycle of 30–40 years in this region.
Keywords: evapotranspiration, sustainable fuels, water balance

1 INTRODUCTION

Sustainable production of bioenergy is necessary to meet future worldwide energy demands while helping to offset the global impacts of increased carbon dioxide from traditional fossil fuels [1–4]. Sustainability, the ability to meet the present needs without compromising the ability to meet the needs of the future, continues to gain attention as human population growth creates ever-greater pressure on diminishing natural resources [5]. According to the 2016 Federal Activities Report on the Bioeconomy released on February 2016, the Biomass R&D Technical Advisory Committee has recommended 'targeting a potential 30% penetration of biomass carbon into the U.S. transportation market by 2030'. Recently, scientists have been researching ways to produce bioenergy without adversely impacting food, land and other environmental resources. The Northwest Advanced Renewables Alliance (NARA), a broad alliance of private industry and educational institutions, takes a holistic approach to building a supply chain within WA, OR, ID and MT based on using forest residuals to make aviation biofuel. NARA's objective is to increase efficiency for each supply chain step from forestry operations to conversion processes, creating new bio-based products; providing economic, environmental and social sustainability analyses; engaging stakeholder groups; and improving bioenergy literacy for students, educators, professionals and the general public. Forest residuals from logging operations will be used as feedstock to fulfil the project aims of creating a sustainable industry to produce aviation biofuels and important co-products.

This portion of the investigation is responsible for assessing the hydrologic concerns of the water balance and is part of a larger effort by the sustainability measurement. Our study

DOI: 10.2495/EI-V1-N3-353-363

is aimed at examining the potential hydrologic repercussions of incremental woody biomass removal associated with various degrees of residual ground cover (or biomass) removal in the production of biojet fuel in the Pacific Northwest. The specific region of interest for this project is the vadose zone of the subsurface as net changes to groundwater recharge could create issues with critical base flow during drier parts of the year. To achieve the project goal, completion of three objectives was required:

1) Collect and process data measured at the test plots and develop a predictive water quantity model to evaluate site-scale regional impacts of small-scale biomass removal.
2) Evaluate if there is a correlation between differences in land cover as a function of unique soil characteristics, such as cation exchange capacity (CEC), clay content percentage, soil carbon percentage, slope of the land, microbial population, pH and electric conductivity.
3) Determine the capability of extending site-scale regional impacts of small-scale biomass removal to watershed-scale semiarid regional impacts of large-scale biomass removal.

These objectives helped quantify the effects of woody biomass removal on the soil and water balance of the study site, and therefore demonstrate the sustainability of harvesting woody biomass forest residuals as a source of biomass for the NARA bioenergy feedstock. The vadose zone groundwater model was calibrated using on-site measured values of volumetric water content (VWC) over time, and calculated for the infiltration/evaporation water budget values.

2 BACKGROUND

It has been previously stated that very severe compaction at high moisture contents causes soil deformation, thus decreasing soil water potential [6]. Furthermore, it's been studied previously that compaction impairs the conditions of the soil by increased soil bulk density, increased nitrification rates, reduced porosity and water infiltration rates, damaged soil structure and aggregate stability [5]. In some soils, even small deformations can cause large decreases in the saturated hydraulic conductivity of that particular soil. Therefore, the three study objectives described earlier will be evaluated using the following null hypotheses:

i) H_0: Increased biomass removal from the Long-Term Soil Productivity (LTSP) site will have no impact on infiltration rates or the water budget of the subsurface.
ii) H_0: There is no distinction between differences in land cover as a function of unique soil characteristics such as slope, saturated conductivity, soil temperature, CEC and microbial populations present.
iii) H_0: Data from the site-scale regional impacts can be applied to watershed-scale regional impacts of large-scale biomass removal through the Pacific Northwest.

As a part of Weyerhaeuser's effort to sustainably manage its more than 6 million acres of forested timberland in the United States, it continues to conduct, evaluate and support research associated with the North American LTSP programme. Data for this investigation were collected from Weyerhaeuser's LTSP site in the southern Willamette Valley near Springfield, Oregon, in the Pacific Northwestern United States. The site was created to support the NARA project. A total of 28 one-acre plots were selected to aid in this investigation and round out an existing regional study, to extend into warmer and drier parts of the Douglas-fir ranges,

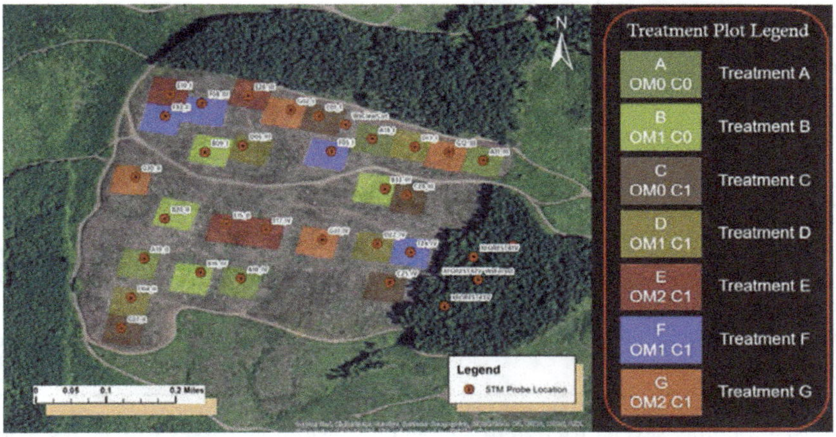

Figure 1: ArcMap model of LTSP study plots and location of soil moisture probes [7].

as to contribute more understanding into the broader LTSP network. Treatments were randomly assigned and laid out in such a way that any plot could feasibly receive that particular random assignment (see Fig. 1). These plots are part of a larger interconnected network for hydrologic analysis. The site elevation is between 600 and 670 m above the mean sea level on gentle slopes of 2–20%. Soil information was collected from Weyerhaeuser's NARA LTSP plot summary analysis. The majority of the information used to characterize the soil class were the percent silt, percent sand and percent clay values, CEC, carbon organic matter amount, pH and electric conductivity.

The average annual precipitation at the study site is 129.3 cm (50.9″). Summers tend to be dry with less than one-third that of the wettest winter month, and with less than 3.9 cm (1.54″) of precipitation in a summer month. The month with the most precipitation on average is November with 21.4 cm (8.44″) of precipitation. The month with the least precipitation on average is July with 1.6 cm (0.64″) of precipitation. The warmest month, on average, is August with an average temperature of 27.8°C (82°F) and the coolest month on average is December, with an average temperature of 7.8°C (46°F).

Multiple studies provide evidence that forest-harvesting practices undoubtedly have a significant effect upon surface soil properties, and in semi-arid regions where the clearing of land promotes desertification [8]. In rain-dominated portions of the Pacific Northwestern United States, annual water yield may be enhanced by the removal of forest vegetation from small upland watersheds, yet questions still remain regarding the effects of logging operations on infiltration under the variety of climatic, physiographic and vegetative conditions of this region. The objective of this study was to examine the ecological environment through the measurement of soil moisture at the site and develop a predictive water quantity and quality models based on each different types of removal (A–G).

3 METHODOLOGY

The plots shown in Fig. 1 represent the General LTSP 'Core' treatments consisting of a 3 × 2 factorial combination of compaction (C0, none; C1, moderate) and aboveground OM removal (OM0, bole/trunk only; OM1, whole tree; OM2, whole tree plus forest floor removal). Three levels of organic matter removal and two levels of compaction in a 3 × 2 complete factorial

Table 1: Tabular form of treatment removal technologies.

Compaction	C0	C1
Harvest	No compaction	Moderate compaction
OM0 – bole only	OM0 C0 – boles removed, no compaction A	OM0 C1 – boles removed, moderate compaction C
OM1 – boles and crowns removed 'Total Tree'	OM1 C0 – boles and crowns removed/no compaction B	OM1 C1 – boles and crowns removed/moderate compaction D/F
OM2 – boles, crowns, forest floor removed		OM2 C1 – boles, crowns, forest floor removed/moderate compaction E/G

design exist – totalling seven different treatment plots. Multiple passes with heavy machinery were used to compact soils. Seven different treatment combinations have been applied to 28 study plots, four replicate plots of each treatment. The different treatment types based on two different levels of compaction and three different levels of biomass land cover are categorized as follows, and presented via tabular form in Table 1, with their designated treatment type denoted by the letter in the top right corner. Additionally, Fig. 1 depicts each LTSP study plot location on the ArcMap model, including the location of the soil moisture sensor probes and weather stations.

The following is an explanation of the terminology used in Table 1:

C0 – No compaction – no ground trafficking on plot.

C1 – Compaction – fixed traffic lanes where plot is levelled with forestry machinery.

OM0 – Boles only – boles/trunks only, meaning harvest consists in removing saw log top (5″ minimum diameter), while all other limbs and tops remain on the site.

OM1 – Total Tree – whole-tree-type harvest where approximately 75+% of limb/top material is removed along with boles/trunks. Remaining material will be dispersed equally across plots.

OM2 – Total Tree + forest floor – whole-tree-type harvest where ~90–95% of limb/top material is removed along with the bole/trunk. Forest floor and legacy woody debris also removed.

A typical example of the LTSP plots after woody biomass removal is shown in Fig. 2. Note the clearly visible compaction tracks caused by harvesting activities. These tracks are quickly obliterated by the time the saplings were planted a year later.

The objectives of this study were completed by collecting the necessary weather and soil characteristics data. The mean absolute error (MAE) was calculated for each plot for each season of the year over the course of 2 years between its measured and simulated values. UNSAT-H Version 3.0: Unsaturated Soil Water and Heat Flow Model, developed by the Pacific Northwest National Laboratory for the United States Department of Energy, will be used to develop a predictive water balance model that correlates with data collected from the soil moisture probes installed in the field. The kind of information that will then be predicted from the model will be quantified amounts of runoff, infiltration, evaporation and overall water balance.

Figure 2: LTSP site for treatment G: (OM2 C1) compaction total tree + FF [7].

Approximately four soil moisture measurement probes were installed at each plot location on the map, respectively, totalling to approximately 112 probes for the whole LTSP site. The 5TM Soil Moisture & Temperature Sensor by Decagon Devices® was the chosen soil moisture and temperature probe selected for installation. Weyerhaeuser conducted the seven different land alterations at each plot, and installed four separate sensors at 10, 20, 30 and 100 cm depths, respectively. For the 10, 20 and 30 cm depths, a trench wall was dug and the sensors were installed with a horizontal orientation to the surface. For the 100 cm depth, an auger hole was drilled into the subsurface and the sensors were installed with a vertical orientation to the surface. In three locations, the 100 cm depth was too grave to install a 5TM sensor, therefore not collecting data at the 100 cm depth. Additionally, there are two weather stations installed at the sites to measure precipitation, air temperature, wind speed, relative humidity and solar radiation at every hour. One weather station is in the clear-cut section, and the other weather station is located in the forested section. An example of the data generated is shown in Fig. 3.

An unsaturated flow recharge model (UNSAT-H) was used to estimate infiltration and evaporation of the groundwater and also helped predict water drainage rates for the simulated conditions. UNSAT-H is a one-dimensional vertical or horizontal model which uses the finite difference method to solve for Richard's equation [9]. There are many soil water retention relationships that have been determined, such as the linked polynomials, the Haverkamp function, the Brooks and Corey function, the van Genuchten function and several special functions that account for water retention of very dry soils [10]. Therefore, eqn (1) illustrates how water content and hydraulic conductivity are functions of suction head according to the van Genuchten function and Mualem hydraulic conductivity model and has the form

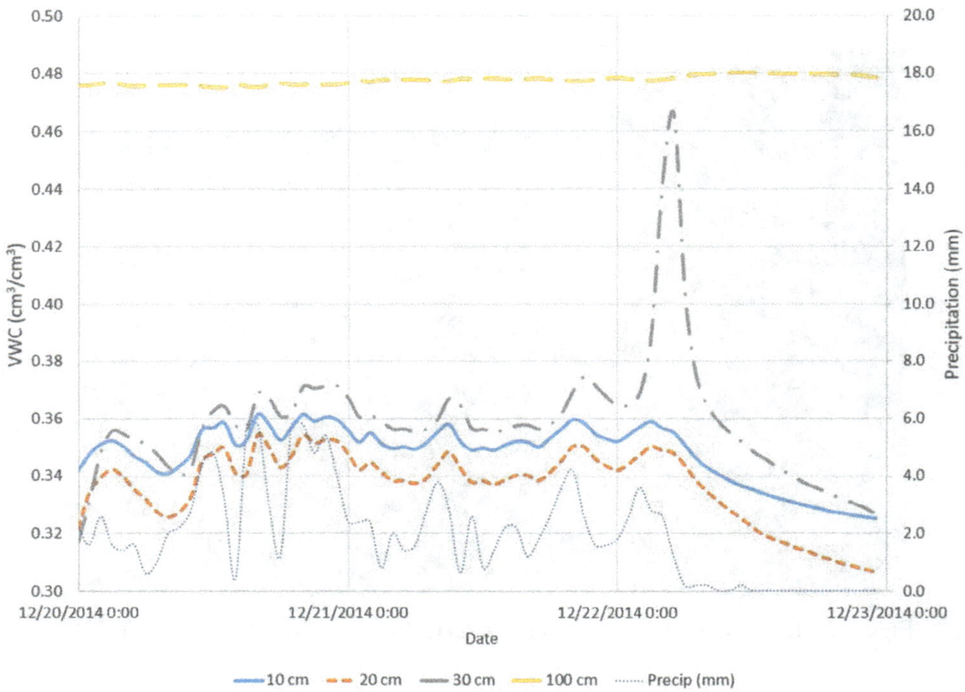

Figure 3: Treatment plot F08III VWC during peak rainfall event in December 2014 [7].

$$\theta = \theta_r + (\theta_s - \theta_r)\,[1 + (\alpha\, h)^n]^{-m} \tag{1}$$

where the parameters α, m and n are curve-fitting parameters based on soil type, and it is assumed that $m = 1 - 1/n$. θ_r is the residual water content and θ_s is the saturated water content. The conductivity function is based on the Mualem conductivity model [11] shown in eqn (2):

$$K_L = K_s\,\frac{(1-(\alpha\, h)^{n-2}\,[1+(\alpha\, h)^n]^{-m})^2}{[1+(\alpha\, h)^n]^{-m})^2} \tag{2}$$

where K_L is the unsaturated hydraulic conductivity and K_S is the saturated hydraulic conductivity.

Both statistical and graphical model techniques were reviewed for the calibration of the UNSAT-H simulations. For graphical techniques, curve fitting to simulate the measured data at the LTSP site for each respective plot as closely as possible for the changes in water content over time was used. For statistical model techniques, the MAE and percent bias (PBIAS) values were calculated between the observed and simulated values of water content at each given day. The constraining statistic model technique used for this study was PBIAS, and this measures the average tendency of the simulated data to be larger or smaller than their observed water content counterparts. The optimal value of PBIAS is 0.0, with low-magnitude values indicating accurate model simulation. Positive values indicate model underestimation bias, and negative values indicate model overestimation bias [12]. The reported performance ratings and corresponding values developed were adapted from the work of Moriasi [13] and include ranges of values used in establishing general performance which appear in Table 2.

Table 2: General performance ratings for recommended statistics.

Performance rating	PBIAS (%)
Very good	PBIAS < ±10
Good	±10 ≤ PBIAS < ±15
Satisfactory	±15 ≤ PBIAS < ±25
Unsatisfactory	PBIAS ≥ ±25

Performance was evaluated on the results of the simulation by calculating the MAE and PBIAS values and can be found in Table 2. A 'satisfactory' rating between ±15% and ±25% was preferred; therefore, multiple iterations of parameter changes were made until PBIAS values were under ±25%.

4 RESULTS AND DISCUSSION

Two-year simulations with the UNSAT-H model were completed for each of the 28 treatment plots. Time-series graphs for two typical simulations with respective VWC measurements (θ) are shown later. Figures 4 and 5 illustrate both variations in depth and seasonal variation in moisture content. Summer drainage and evaporation are clear in the near-surface model predictions.

The total basal liquid flux (drainage) was defined as the amount of water that penetrates the water table. Once the water has travelled this vertical distance below the subsurface, it is assumed to have infiltrated the water table and thus impact the water budget. Total infiltration was defined as the total amount of water that penetrates the water table located at 150 cm. Any amount of water that resides above the water table, which includes the current water moisture in the 0–150 cm soil profile, is considered as 'total infiltration'.

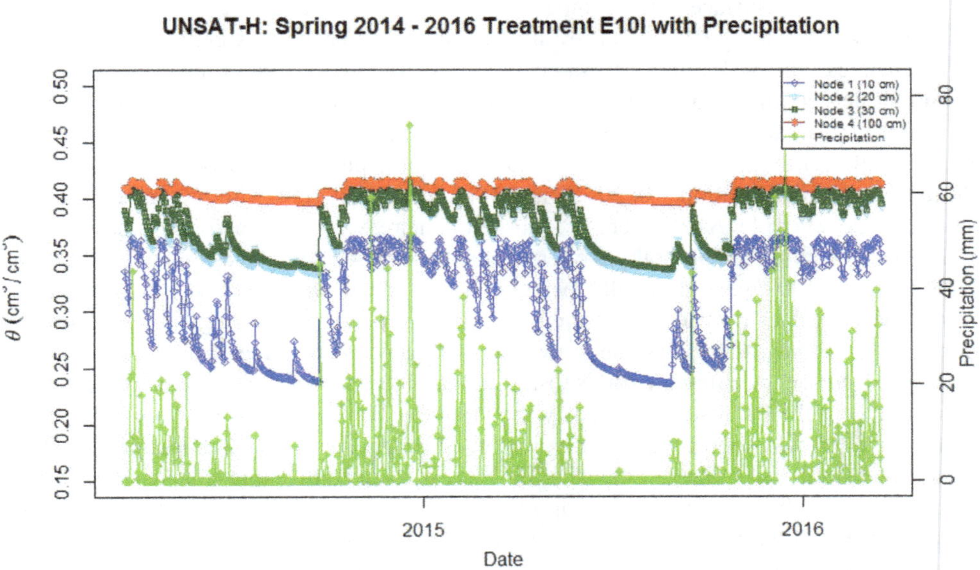

Figure 4: Volumetric water content and precipitation for treatment plot E10I.

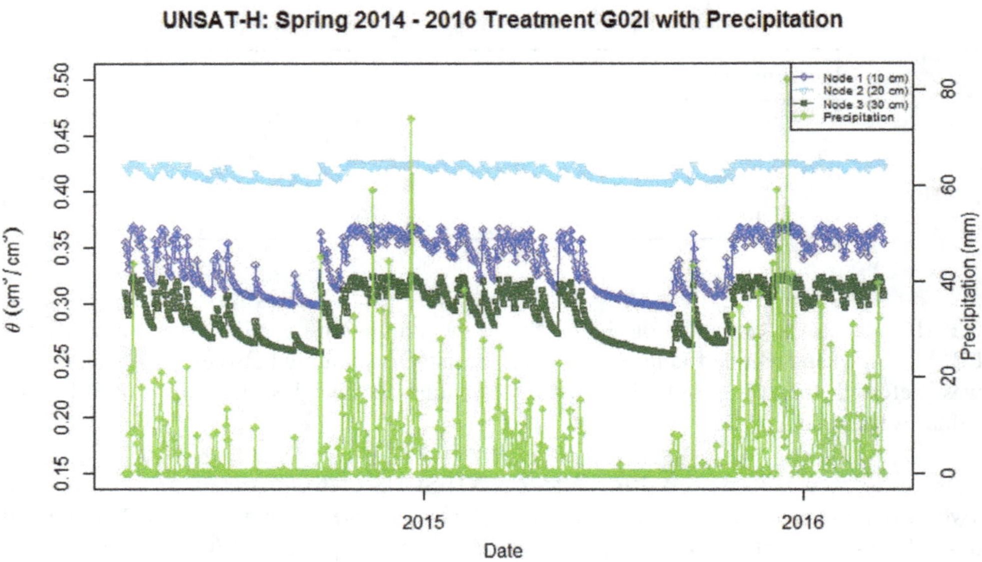

Figure 5: Volumetric water content and precipitation for treatment plot G02I.

After the analysis of the amount of evaporation from each treatment type A–G, it was observed that there were statistically higher evaporation rates in treatments C–F when compared with treatments A–B, ranging from 12% to 32% increase. Treatments D–F fall in the C1 compaction category, as well as the OM1 and OM2 harvest categories. To investigate why the compacted sites had more evaporation, an infiltration analysis was conducted between a non-compacted plot, A19II, with a compacted plot, F08III. These two plots had similar soil types in the subsurface when compared with one another, and therefore chosen for this infiltration analysis. A peak rainfall event was chosen to ensure a significant response in VWC changes. The peak rainfall event occurred on 20 December 2014 for a total of 7.4 cm of rainfall and continued on 21 December 2014 for a total of 5.14 cm of rainfall. The last day observed was 22 December 2014 with a total rainfall amount of 1.54 cm, bringing the total rainfall over the course of 3 days to 14.08 cm of precipitation. Hourly VWC values were plotted at each port location, 10, 20, 30 and 100 cm, respectively, with precipitation on the secondary axis. For plot A19II, there is a substantial increase in VWC before the main portion of the storm occurs. Possible sources of error could have risen from measurement error of the 5TM sensor at 100 cm. Additionally, percent change values were calculated from the daily average VWC values for each port location. A positive percentage change in VWC denotes an increase, meaning the soil is gaining water, while a negative percentage change in VWC denotes a decrease, meaning the soil is losing water. Figure 6 shows the results for A19II plot compared to the F08III results previously shown in Fig. 3.

Statistically it was shown that our first hypothesis suggesting that land treatments did not impact evaporation was false. Removal of biomass without compacting soils resulted in considerably more infiltration and less evaporation than compacted soils. However, it was not possible to reach a definitive conclusion for our second hypothesis. Three regression analyses were done for different hydrologic parameters for the three soil profiles conducted by Weyehaeuser during the 5TM sensor installation and soil analysis: 0–15, 15–30 and 30–100 cm.

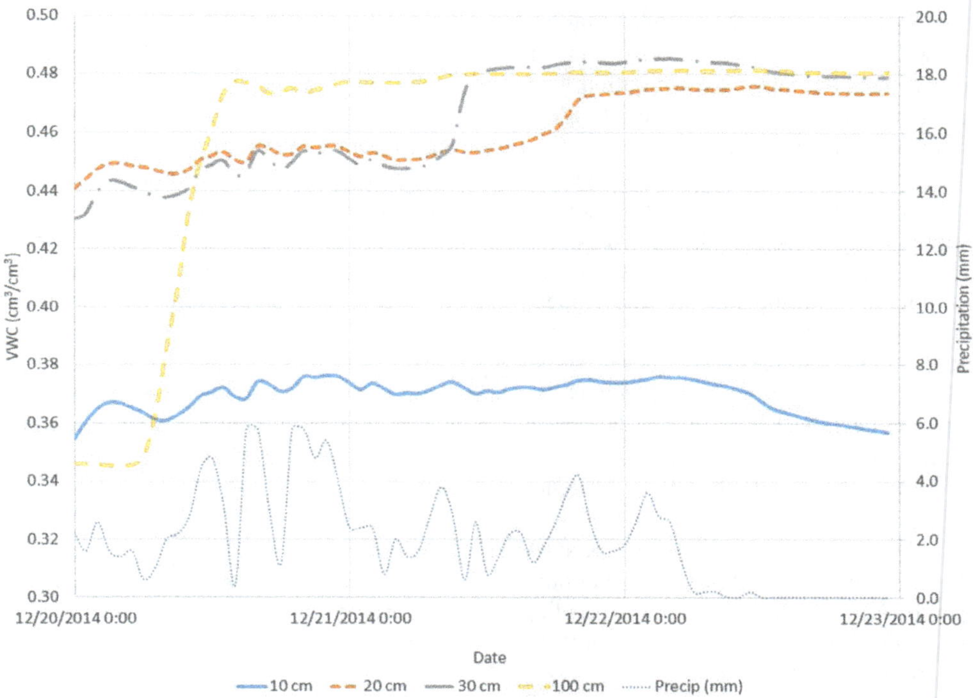

Figure 6: Water content for peak rainfall event for treatment plot A19II.

When the bin depth went across the 30 cm 5TM sensor, the two different values were averaged together to receive the 30 cm designated value. The statistical results for the analysis of each soil profile for total basal liquid flux drainage simulated values versus percent clay content percentage, pH, electrical conductivity (EC), CEC, soil carbon content in megagrams per hectare (SoilC), mean slope, saturated hydraulic conductivity, average temperature, the five major genus of bacteria present (*Bacillus*, *Clostridium*, *Shewanella*, *Thermoanaerobacter* and *Lactobacillus*) were calculated. There was no clear defining distinction between the land cover applications when comparing it with the plot's unique soil characteristics. While it is true that most of the sites contained loams, clay loams and clays, each plot had a different soil composition at each respective subsurface depth.

Our third hypothesis (data from the site-scale regional impacts can be applied to watershed-scale regional impacts of large-scale biomass removal through the Pacific Northwest) proved equally difficult to statistically prove. The data varied too much from treatment to treatment so the logical conclusion was that site critical information would be needed for each location where harvesting was proposed.

5 CONCLUSIONS AND RECOMMENDATIONS

Soil water evaporation coupled with infiltration and drainage can significantly affect soil water contents and moisture redistributions. UNSAT-H was used to model soil conditions and ground cover options for each of the 28 one-acre sites in the Oregon LTSP. Model results were compared to field measurements of soil moisture taken at four depths. Changes to surface runoff and sediment production due to erosion coincided with the period immediately after track construction and harvesting but these effects quickly decreased during our study and were not seen as a significant factor in the water budget analyses. Overall, statistically significant differences in evaporation were observed when compacted soil sites were compared with non-compacted sites. Evidence from the UNSAT-H modelling indicated that compacted sites will exhibit more evaporation from the surface. Furthermore, sites without compaction and with crowns removed exhibited the least amount of evaporation. It is recommended that the woody biomass removal includes crowns to ensure the least amount of evaporation occurs as to disrupt the water budget of the site the least. A statistically significant relationship between soil hydrologic parameters and drainage values to the water table could not be determined. Harvesting operations for biofuel production are therefore advised not to compact the sites but to remove the slash and crowns present at the plot to have the least amount of impact on the water budget of the site.

Future studies on this project are possible, and include plant modelling within the UNSAT-H on the LTSP site to see if and how regrowth of the Douglas Fir saplings affect the water budget, and researching more into drainage rates to other treatment technologies and procedures other than solely the soil hydrologic parameters.

ACKNOWLEDGEMENTS

This work was supported by the National Institute of Food and Agriculture, U.S. Department of Agriculture. Any opinions, findings, conclusions or recommendations expressed in this publication are those of the author(s) and do not necessarily reflect the view of the U.S. Department of Agriculture.

REFERENCES

[1] Berndes, G., Bioenergy and water—the implications of large-scale bioenergy production for water use and supply. *Global Environmental Change*, **12**(4), pp. 253–271, 2002. DOI: 10.1016/S0959-3780(02)00040-7.

[2] Johansson, D.J.A. & Azar, C., A scenario based analysis of land competition between food and bioenergy production in the US. *Climatic Change*, **82**, pp. 267–291, 2007. DOI: 10.1007/978-94-009-0129-2_9.

[3] Beringer, T., Lucht, W. & Schaphoff, S., Bioenergy production potential of global biomass plantations under environmental and agricultural constraints. *GCB Bioenergy*, **3**, pp. 299–312, 2011. DOI: 10.1016/j.enpol.2009.05.029.

[4] Aransiola, E.F., Ojumu, T.V., Oyekola, O.O., Madzimbamuto, T.F. & Ikhu-Omoregbe, D.I.O., A review of current technology for biodiesel production: State of the art. *Biomass and Bioenergy*, **61**(2), pp. 276–297, 2014. DOI: 10.1016/j.biombioe.2013.11.014.

[5] Mann, L. & Tolbert, V., Soil sustainability in renewable biomass plantings. *AMBIO: A Journal of the Human Environment*, **29**(8), p. 492, 2000. DOI: 10.1579/0044-7447-29.8.492.

[6] Soane, B. D. & Ouwerkerk, C.V., Soil compaction in crop production. Elsevier: Amsterdam, 1994.

[7] Smith, C.T., Evaluating the hydrological impact of removing woody biomass for biofuel production through unsaturated zone modeling. Master of Science Thesis, University of Utah: Salt Lake City, UT, 2017.

[8] Bonan, G.B., Frost followed the plow: Impacts of deforestation on the climate of the United States. *Ecological Applications*, **9**(4), p. 1305, 1999. DOI: 10.1029/95JD02169.

[9] Benson, C., Modeling unsaturated flow and atmospheric interactions. *Theoretical and Numerical Unsaturated Soil Mechanics*, ed. T. Schanz, Springer: Berlin, pp. 187–202, 2007.

[10] Fayer, M., *UNSAT-H Version 3.0: Unsaturated Soil Water and Heat Flow Model*. Pacific Northwest Laboratories: Richland, WA, 2000.
[11] Mualem, Y., A new model for predicting the hydraulic conductivity of unsaturated porous media. *Water Resources Research*, **12**, pp. 513–522, 1976. DOI: 10.1029/WR012i003p00513.
[12] Gupta, H.V., Sorooshian, S. & Yapo, P.O., Status of automatic calibration for hydrologic models: Comparison with multilevel expert calibration. *Journal of Hydrologic Engineering*, **4**(2), pp. 135–143, 1999. DOI: 10.1061/(ASCE)1084-0699(1999)4:2(135).
[13] Moriasi, D.N., Arnold, J.G., Van Liew, M.W., Bingner, R.L., Harmel, R.D. & Veith, T.L., Model evaluation guidelines for systematic quantification of accuracy in watershed simulations. *Transactions of the ASABE*, **50**(3), pp. 885–900, 2007. DOI: 10.13031/2013.23153.

RETENTION SOIL FILTER FOR TREATMENT OF COMBINED SEWER OVERFLOW AND WASTEWATER TREATMENT PLANT EFFLUENT

KATHARINA KNORZ, FRANZ-MICHAEL MERTENS, ANDREA F. BRUNSCH
& EKKEHARD CHRISTOFFELS
Erftverband, Bergheim, Germany

ABSTRACT

Point sources such as wastewater treatment plants (WWTP), combined sewer overflows (CSO) and separate sewer outlets (SSO) are acknowledged to be major pathways for pollutants to enter surface water. As response, retention soil filters (RSF) have been installed among others in the Erft river catchment area over the past two decades to treat outflows of combined sewer systems and rainwater from separate sewer systems. Studies conducted on several of these RSFs have shown that besides reducing solids by filtration, RSFs have good chemical and biological cleaning capacities for a variety of parameters. Concentrations of both nutrients and heavy metals are reduced. Numerous micropollutants such as pharmaceuticals, pesticides and industrial compounds as well as faecal bacteria are eliminated at high rates. With this background a new type of RSF (RSF^{plus}) has been designed for the flexible treatment of WWTP effluent and CSO. The RSF^{plus} treats CSO during rain events. During dry weather periods, the RSF^{plus} is used for polishing WWTP effluent. The RSF^{plus} is divided into three segments which are fed alternately to guarantee aerobic conditions and is equipped with innovative distribution channels to ensure homogeneous feeding. Granulated activated carbon is used as an additive for the filter medium for advanced treatment. The effectiveness of WWTP effluent polishing and reliability of operational parameters have already been tested for 2.5 years on a pilot scale. The first full-scale demonstration site will be built in 2017 at the WWTP Rheinbach in North Rhine-Westphalia (Germany).

Keywords: combined sewer overflow, micropollutants, nutrients, retention soil filter, wastewater treatment plant

1 INTRODUCTION

With the adoption by the European Union of the Water Framework Directive in 2000, regulations were established on surface water quality [1]. Water courses that are mostly fed by combined sewer overflows (CSOs) and separate sewer outlets (SSOs) are of special concern. To reduce pollutant loads during event-specific inflows, especially in North Rhine-Westphalia retention soil filters (RSFs) have been installed below stormwater basins for the treatment of rainwater and CSOs. Besides performing mechanical and biological treatment, RSFs are recognized for reducing hydraulic stress on receiving water bodies [2]. The addition of a list of priority substances to the Water Framework Directive has helped to bring chemical water quality into focus. Studies have shown that conventional WWTPs are not able to reduce micropollutants, which has led to an increased demand for additional treatment steps.

RSFs have already proven their effectiveness in treating CSOs and SSOs. Results from studies of filter material and water samples in the Erft catchment area will be presented in this article. According to Scheurer et al. [3], chemical cleaning capacities of RSFs for certain substances are comparable to those of activated sludge treatment. Moreover, nutrient retention rates of more than 75% have been reported [4]. Good elimination rates for heavy metals, numerous micropollutants such as pharmaceuticals, pesticides and industrial wastes as well as faecal bacteria have been shown [2, 3, 5].

DOI: 10.2495/EI-V1-N3-364-374

2 CONVENTIONAL RETENTION SOIL FILTERS

Since 1997, 32 RSFs have been installed by the Erftverband water management association in the Erft catchment area. All of these RSFs are still in use.

The majority of the Erftverband RSFs were built between 2000 and 2005 and are used for the treatment of CSOs. Only 5 of the 32 RSFs are fed by SSOs. A schematic view of the structure of RSFs is shown in Fig. 1.

RSFs consist of sand as filter material to which 20–30 vol.-% of $CaCO_3$ is added. A drainage layer consisting of gravel and drainage pipes to channel the treated water to the receiving river is installed below the filter body. Beneath the drainage layer the RSF is sealed by a sealing membrane to protect the soil layers below the filter body and the groundwater from contamination. RSFs are planted with reeds to avoid clogging. Only five RSFs have either lawn or yarrow vegetation.

RSFs in the Erft catchment area vary in size with filter areas spanning from 80 m^2 to more than 5,000 m^2 depending on the connected urban drainage area they serve. Dimensions are determined by the RSF guideline of North Rhine-Westphalia DWA-M-178 [6]. RSFs differ in their distribution of CSO over the filter surface. This difference has major impact on the loading of the filter material. While some RSFs have only one overspill at the inflow (Fig. 2A), there are others with distribution channels across the filter area to guarantee homogeneous feeding (Fig. 2B).

2.1 Sampling strategy

To comply with relevant permits soil samples have to be taken every 5 years to measure the contamination of the filter material. To measure the purification efficiency, water samples have been analysed in the course of studies funded by the state government for promotion of research and innovation projects [8].

Filter material samples from 24 RSFs constructed between 1999 and 2011 were taken to examine the degree of contamination and to estimate their remaining lifetime. The remaining eight RSFs were too new for filter material analysis. Sampling took place close to and at a

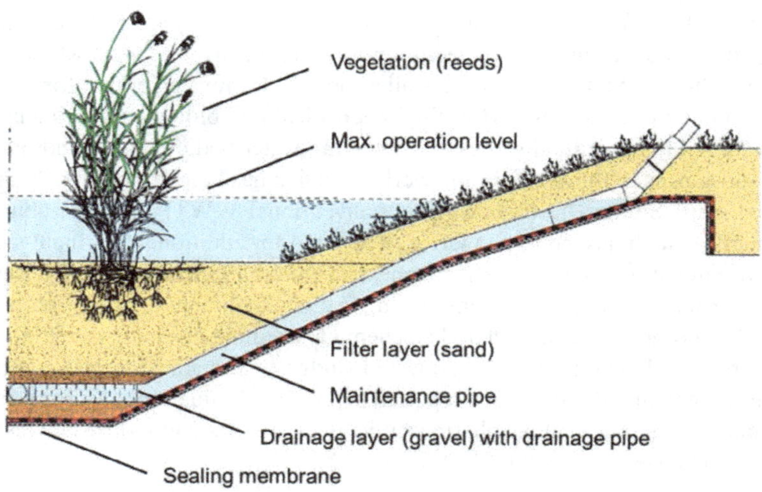

Figure 1: Structure of a conventional RSF [7].

Figure 2: Distribution systems of RSFs.

greater distance from the inflow. The filter material was sampled from different depths (0–5, 5–10, 10–30, 30–60 cm), with six samples taken at each depth. Composite samples were then obtained from each depth. For samples of the upper two layers a spade was used whereas for depths greater than 10 cm an Edelman auger for sandy soil was employed. Besides nutrients and heavy metals, organic matter, carbonate concentrations and micropollutants were analysed.

Water samples were taken at the inflow and outflow of two representative RSFs, in Altendorf and Kaster (North Rhine-Westphalia). Both have been in use since 2005. For this purpose, autosamplers were installed at the inflow and outflow to collect 12-min composite samples of 2.9 L over a maximum of 132 min at RSF Altendorf. At RSF Kaster, the maximum sampling time at the outflow was raised to 150 min. At the inflow, three composite samples were taken in the periods of 0–15 min, 15–60 min and 60–90 min. Physico-chemical as well as microbiological analysis was performed according to Christoffels et al. [2].

2.1.1 Results

Depending on the age of RSFs and the frequency of feeding, organic matter accumulated mostly in the upper layer. Pollutants clearly accumulated within this upper humous layer as shown in Fig. 3 with the example of zinc. The amount of organic matter as well as the amount of pollutant load within the filter medium decreased with depth.

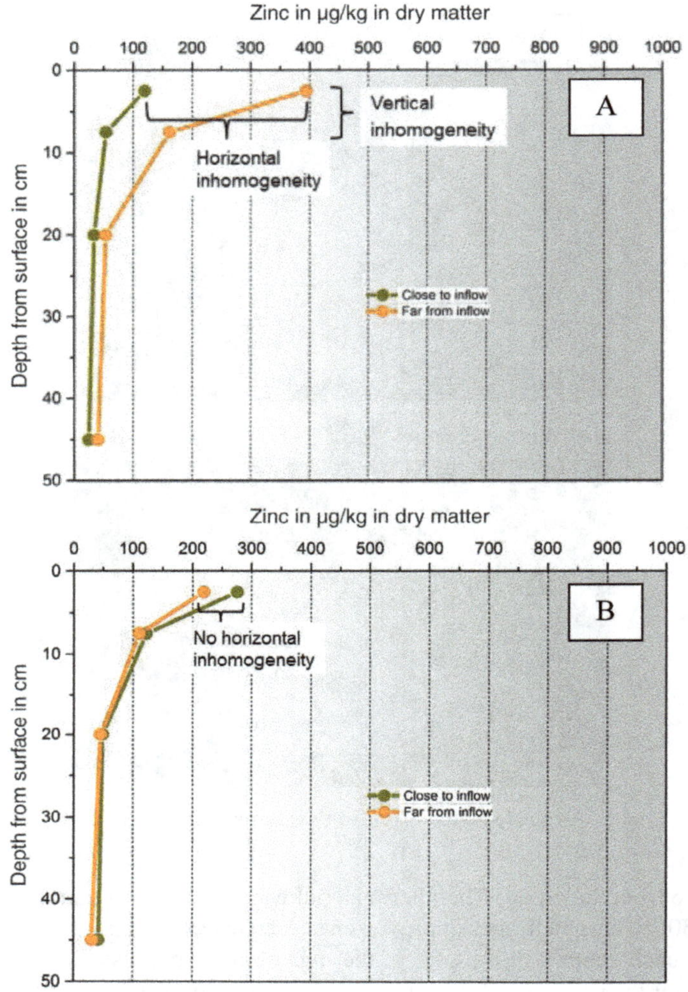

Figure 3: Substrate analysis of zinc picturing vertical and horizontal inhomogeneities of the filter medium.

Besides the vertical inhomogeneity of pollutant concentration, an unequal distribution of the horizontal burden can be seen (Fig. 3A). The filter sand close to the inflow is more contaminated than the material further away from the inflow. This effect is clearly seen for RSFs with only one overspill (compare Fig. 2A). RSFs with distribution channels (compare Fig. 2B) showed less difference in their horizontal pollution profiles (Fig. 3B). In this case, the filter medium is evenly loaded.

$CaCO_3$ concentrations were found to be a suitable factor for evaluating the total burden on the RSF. Carbonate stabilizes the pH during nitrification processes. A depletion in $CaCO_3$ leads to acidification of the soil and hence the remobilization of heavy metals. The pH value was stable in all filter material samples, though a reduction in $CaCO_3$ was noticeable.

Critical depletion of $CaCO_3$ was detected in some RSFs with a reduction of $CaCO_3$ concentrations down to 2.5% in the upper layer. This corresponds with the relatively high concentrations of organic matter and pollutants in this layer. RSFs with higher pollution loads due to more frequent CSO events or larger amounts of water during CSOs are more likely to deplete in $CaCO_3$ than RSFs with low pollution loads. So far, a remobilization of pollutants has not occurred in any of the 24 studied RSFs, which leads to the conclusion that cleaning performance is guaranteed for an operational life of at least 10 years.

Water samples show complementary results. In total, 35 CSO events were analysed with 354 samples from both the inflow and outflow at RSF Altendorf. At RSF Kaster, 23 CSO events with 69 samples from both the inflow and outflow were assessed.

Filtration reduced suspended solids by median concentrations of 98% at RSF Altendorf and 91% at RSF Kaster. Along with the reduction of solid material, pollutants sorbed to the solid material were reduced. Nitrogen compounds such as ammonia were restrained by 98% in median value at RSF Altendorf and by 94% at RSF Kaster. Concentrations of heavy metals were also reduced. For example, median retention rates for zinc were 82% and 71% at RSF Altendorf and RSF Kaster, respectively. Similar results have been reported by Frechen [4]. Retention rates of more than 55% and of greater than 20% as a median are achieved for phosphorus and orthophosphate-phosphorus, respectively (Fig. 4). These divergent results suggest that especially dissolved phosphorus is not retained very well. Total organic carbon (TOC) concentrations are reduced by more than 70% at RBF Altendorf and 50% at RBF Kaster as a median.

High reduction rates for some micropollutants were detected. In contrast to the above-mentioned substances, micropollutants showed a higher variance in retention rates based on their substance-specific characteristics. Diclofenac and Ibuprofen were reduced by more than 65% (Fig. 5). Median reduction rates of more than 90% for caffeine and Metformin were

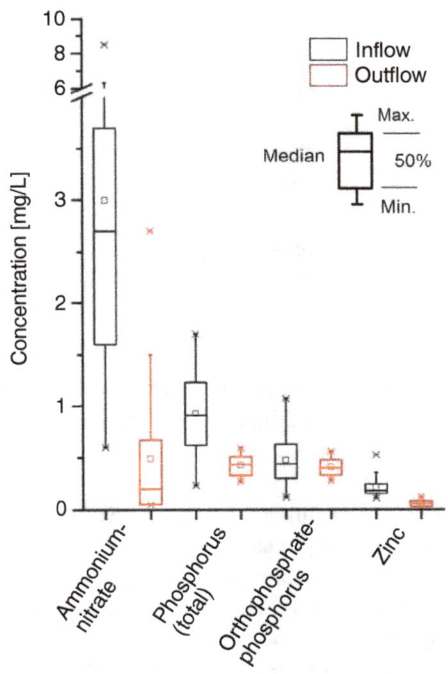

Figure 4: Results of chemical–physical analysis of water samples of RSF Kaster.

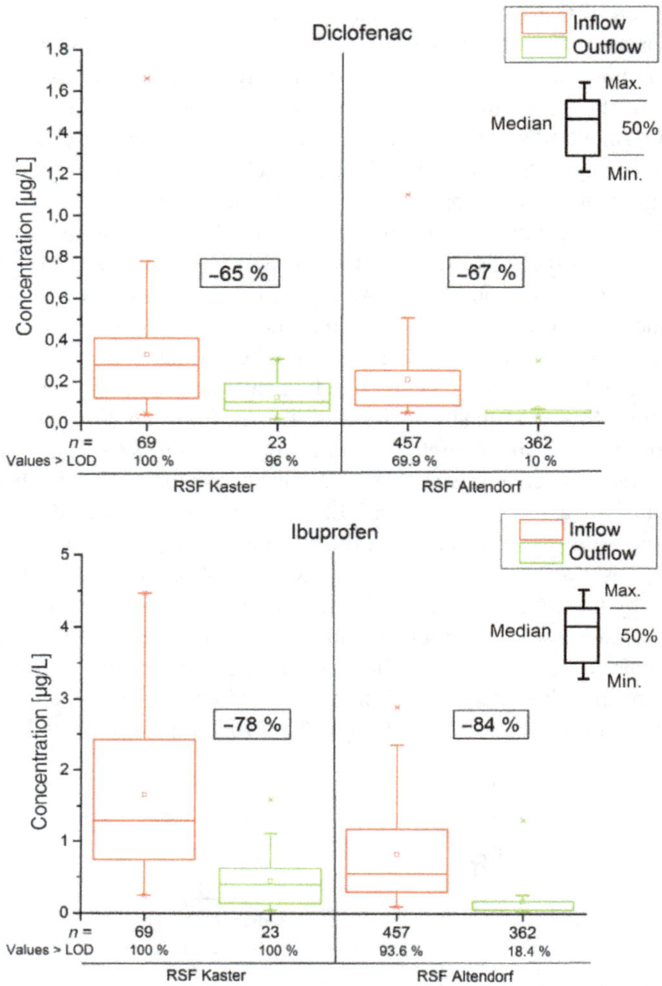

Figure 5: Concentrations of Diclofenac and Ibuprofen at inflow and outflow of RSF Kaster and RSF Altendorf (LOD = limit of detection), respectively.

detected. Metoprolol and Bisphenol-A were reduced by more than 80% as a median. Tondera et al. [5] and Scheurer et al. [3] have obtained similar results.

2.1.2 Assessment of results

The results obtained from the water samples show that RSFs can provide very good chemical cleaning performance with regard to physico-chemical parameters and that good retention rates for micropollutants can also be achieved. In this study, the cleaning capacity of RSFs was only measured for CSOs and SSOs. However, Scheurer et al. [3] report that removal capacities of RSFs are comparable to those achieved by activated sludge treatment.

Results obtained from the filter material sampling indicate long operational lifetimes of RSFs without further investment costs. So far, no breakthrough of pollutants has been detected. However, differences in the degree of contamination of RSFs can be seen, depending on their constructional and operational parameters. RSFs characterized by a point load due to only one overspill show higher contamination close to the inflow than away from the inflow. The study showed that distribution channels have the ability to distribute wastewater evenly, which assures homogeneous loading of the filter medium.

Organic matter accumulated in the upper layer of the filter medium. That way sorption processes of pollutants by the humous layer are improved. A high frequency of overflow events as well as high loads of wastewater during these events led to the largest decrease in $CaCO_3$ concentrations in the upper layer of the filter medium. Nonetheless, both water and filter material samples clearly showed that even after an operational lifetime of 10 years cleaning performance is maintained.

Besides good chemical cleaning capacities and long lifetimes the studied RSFs only needed small amount of maintenance during their entire operational lives. In this respect, they offer an effective tool for low-cost and reliable water purification.

Given the fact that overflow events occur only a few times a year, RSFs could be used for further cleaning purposes during dry weather periods. The idea of advanced treatment of WWTP effluent by RSFs to reduce the micropollutant burden of receiving watercourses during periods with no overflow events is worthy of consideration.

3 PERSPECTIVE: RSF^{PLUS}

Based on the above-mentioned results, the effectiveness of RSFs with substrate additives for WWTP effluent purification will be investigated. A new kind of RSF (RSF^{plus}) for the flexible treatment of CSO and wastewater treatment plant (WWTP) effluent has been designed. The RSF^{plus} is constructed on the site of a WWTP with CSO basin. An additional flocculation filtration stage is mandatory. As shown in Fig. 6, the RSF^{plus} is fed by WWTP effluent during dry weather to serve as a fourth treatment step. During storm events, the CSO is discharged onto the RSF^{plus}. At these times, the WWTP effluent is discharged directly into the receiving river.

3.1 Framework of preliminary studies

The effectiveness of RSFs in performing WWTP effluent polishing was tested at pilot scale. The pilot plant is located at the WWTP in Rheinbach, North Rhine-Westphalia. WWTP

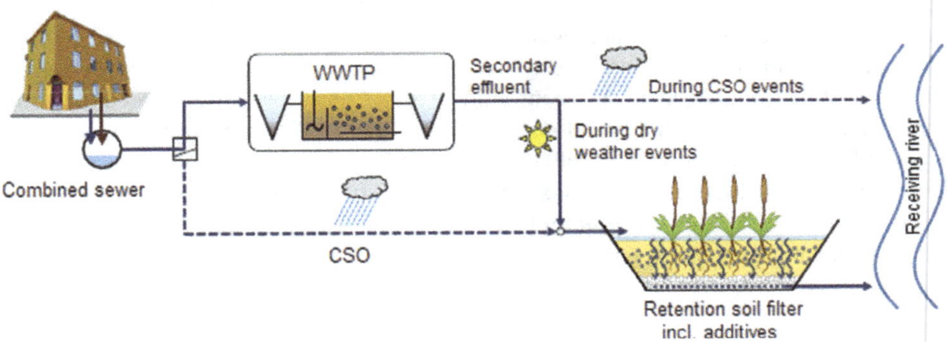

Figure 6: Schematic view of RSF^{plus} system for flexible treatment of CSO and WWTP effluent [9].

Rheinbach is designed for a population equivalent of 27,000 and is equipped with an additional flocculation filtration. Three individual pilot-scale RSFs, each with a volume of 1.39 m^3, were examined over a period of 2.5 years. The pilot system delivered information about the stability of operational parameters, the use of different substrate additives and the lifetime of RSFs utilized for WWTP effluent polishing.

In contrast to conventional RSFs the pilot system is fed periodically with WWTP effluent. The study showed that for a feeding period over 28 h and a dry period of 56 h aerobic conditions are suitable. The corresponding feeding rate was set to 0.03 L/s/m^2 according to the RSF guideline from North Rhine-Westphalia DWA-M-178 [8]. Oxygen conditions remained adequate and the pH remained stable during the study. Hence, periodic feeding of the RSF did not lead to remobilization of heavy metals.

While two of the three pilot-scale RSFs consist of conventional filter material, granular activated carbon (GAC) was added to the filter material of the third pilot-scale RSF to test the influences of GAC on clogging. The studies showed no influence on operational parameters. Clogging did not occur during the investigation.

3.1.1 Demonstration site

A first full-scale demonstration site to deploy RSF^{plus} for flexible treatment of CSO and WWTP effluent will be built in 2017 at the WWTP in Rheinbach. The RSF is designed to treat the CSO during heavy rain events, according to the RSF guideline from North Rhine-Westphalia DWA-M-178 [8]. In this application, the filter medium will cover an area of 5,000 m^2.

During dry weather periods, the average daily inflow amounts to 50 L/s for the demonstration site. At night, it reduces to approximately 17 L/s. CSO events with an inflow of up to 3,000 L/s are possible.

To guarantee aerobic conditions in the filter medium, the RSF^{plus} is separated into three equally sized segments that are fed alternately (Fig. 7). A cycle of 24 h of feeding and 48 h of dry phase appears to be appropriate according to preliminary studies. After feeding of one segment for 24 h, the segment is emptied within a few hours. The remaining time until feeding serves as aeration phase to guarantee aerobic filter conditions.

Distribution channels are necessary for homogeneous loading of RSF^{plus}. Two distribution channels are arranged radially in each segment. At an interval of 4 m, circular flaps are installed on the distribution channels. Depending on the amount of water the distribution channels fill evenly to a certain level. The flaps act as throttle and release nearly the same amount of water close to and far from the inflow onto the filter body. This way the small amount of feeding water during dry periods can be homogeneously distributed over the entire filter area.

As with conventional RSFs, sand is used as filter material. About 20–30 vol.-% of $CaCO_3$ is added to the filter sand. In contrast to conventional RSFs, however, GAC is added to improve micropollutant reduction, and 10 vol.-% is added to the upper 10 cm of the filter. For the purpose of study, different amounts of GAC are added to the lowest 30 cm of the filter. Segments 2 and 3 of the RSF^{plus} contain 30 and 40 vol.-% GAC, respectively. Segment 1 does not contain any GAC in the lowest layer and serves as a reference medium. The structure of the RSF^{plus} is shown in Fig. 8.

3.1.2 Monitoring

A sophisticated monitoring system is to be installed at the RSF^{plus} to observe cleaning performance. Besides autosamplers at the inflow and outflow of the RSF^{plus}, sampling points are

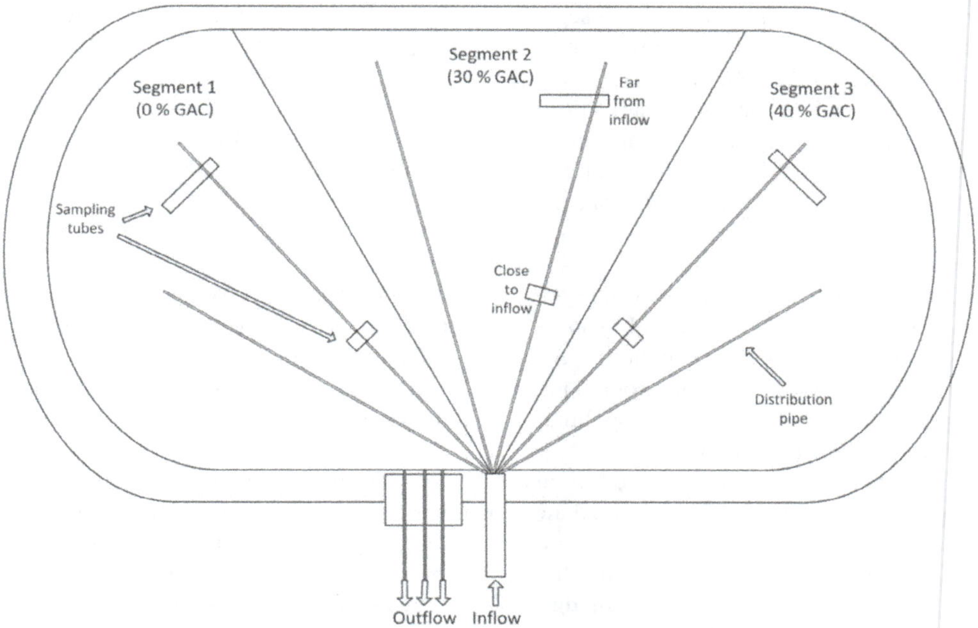

Figure 7: Plan of the RSF^{plus}.

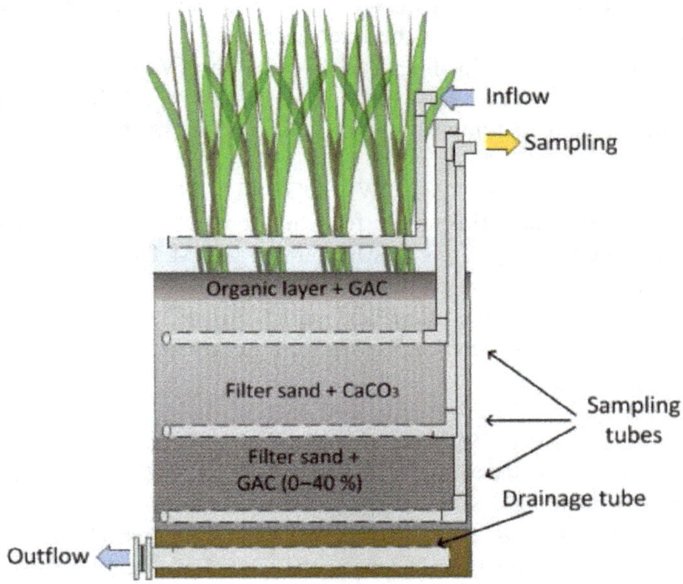

Figure 8: Structure of RSF^{plus} including sampling system.

installed within the filter area. Each segment has a sampling point close to and far from the inflow (Fig. 7). At these sampling points, sampling tubes are connected to horizontal tubes of different depths (Fig. 8). This enables separate sampling of the different layers.

Samples will be taken regularly during dry weather periods and during CSO events. Besides nutrients and heavy metals, great attention is given to the analysis of micropollutants and faecal bacteria. In addition to that monitoring of the receiving river upstream and downstream of the discharge point will take place.

4 SUMMARY

RSFs have been in use in North Rhine-Westphalia for almost 20 years. Water and filter material samples taken from these RSFs show good chemical cleaning capacities, which were also confirmed in further recent studies [2, 3, 5]. Very good retention rates for physico-chemical parameters such as suspended solids, nitrate, phosphorus and heavy metals were detected. Importantly, anthropogenic micropollutants such as pharmaceuticals, pesticides and industrial wastes are also restrained.

Filter material analysis discussed in this article showed that the amount of filter material contamination depends on operational and constructional parameters. Distribution channels clearly prevent inhomogeneous contamination by assuring even loading of the filter material. Organic matter that accumulates during the operational life of the RSF increases sorption processes of pollutants. Even though $CaCO_3$ concentrations decrease depending on the amount and frequency of feeding, remobilization of heavy metals has not been detected so far. Hence, a lifetime of much more than 10 years can be assumed.

Considering the fact that RSFs are only used during overflow events and considering their good cleaning performance, the use of RSFs for purification of WWTP effluent during dry weather periods is worth of consideration. The RSF^{plus} is a new type of RSF with improved structural and operational parameters. Substrate additives such as GAC are expected to improve cleaning capacities. Segmentation of the filter area and periodic feeding guarantee aerobic filter conditions. Homogeneous loading is achieved by improved distribution channels.

Therefore, the RSF^{plus} offers a good tool for cost-effective, reliable and low maintenance WWTP effluent polishing in combination with CSO or SSO treatment.

ACKNOWLEDGEMENTS

The study was funded by the Ministerium für Klimaschutz, Umwelt, Landwirtschaft, Natur- und Verbraucherschutz 'Überprüfung und Bewertung von Maßnahmen zur Reduzierung der chemisch-physikalischen und hygienisch-mokrobiologischen Belastung von Fließgewässern am Beispiel der Swist' and by the European Union Horizon 2020, 'Demonstrating synergies in combined natural and engineered processes for water treatment systems (AquaNES)'.

REFERENCES

[1] European Commission (2000/60/EC), Directive 2000/60/EC of the European Parliament and of the Council establishing a framework for the Community action in the field of water policy, Official Journal L 327, 2000.

[2] Christoffels, E., Mertens, F.M., Kistemann, T. & Schreiber, C., Retention of pharmaceutical residues and microorganisms at the Altendorf retention soil filter, *Water Science & Technology*, **70**(9), pp. 1503–1509, 2014. DOI: 10.2166/wst.2014.404.

[3] Scheurer, M., Heß, S., Lüddeke, F., Sacher, F., Güde, H., Löffler, H. & Gallert, C., Removal of micropollutants, facultative pathogenic and antibiotic resistant bacteria in

a full-scale retention soil filter receiving combined sewer overflow. *Environmental Science, Processes & Impacts*, **17**, p. 186, 2015. DOI: 10.1016/j.watres.2010.08.007.

[4] Frechen, F.B., *Neue Erkenntnisse zur Mischwasserbehandlung in Retentionsbodenfiltern: Leistungsfähigkeit, Nährstoffe, Hygiene*. Planung, Bau und Betrieb: Hennef, Düsseldorf, 2012.

[5] Tondera, K., Koenen, S. & Pinnekamp, J., Survey monitoring results on the reduction of micropollutants, bacteria, bacteriophages and TSS in retention soil filters. *Water Science & Technology*, **68**(5), pp. 1004–1012, 2013. DOI: 10.2166/wst.2013.340.

[6] DWA (Deutsche Vereinigung für Wasserwirtschaft, Abwasser und Abfall e.V.), Merkblatt DWA-M 178, Empfehlungen für Planung, Bau und Betrieb von Retentionsbodenfiltern zur weitergehenden Regenwasserbehandlung im Misch- und Trennsystem. ISBN 3-937758-37-2, Hennef, Deutschland, 2005.

[7] MUNLV (Ministerium für Umwelt und Naturschutz, Landwirtschaft und Verbraucherschutz des Landes Nordrhein-Westphalen), Retentionsbodenfilter. Handbuch für Planung, Bau und Betrieb (Retention Soil Filter. Reference Book for Planning, Building and Operation), Düsseldorf, 2003.

[8] MUNLV (Ministerium für Umwelt und Naturschutz, Landwirtschaft und Verbraucherschutz des Landes Nordrhein-Westphalen), Überprüfung und Bewertung von Maßnahmen zur Reduzierung der chemisch-physikalischen und hygienisch-mikrobiologischen Belastung von Fließgewässern am Beispiel der Swist (Examination and Evaluation of Actions for the Reduction of Chemico-Physical and Hygienic-Microbiological Pollution of Watercourses on the Example of the River Swist), Düsseldorf, 2012.

[9] ERC, European Research Council under the European Union's Horizon 2020 research and innovation programme, Demonstrating synergies in combined natural and engineered processes for water treatment systems (AquaNES), 2015.

RELATIONS BETWEEN THE MICROSCALE RIVERBED MORPHOLOGY AND THE MACROZOOBENTHOS – IMPLICATIONS FOR THE ECOLOGICAL QUALITY ASSESSMENT AND THE DEFINITION OF REFERENCE CONDITIONS

MICHAEL GROLL
Department of Geography, Philipps-Universität Marburg, Germany

ABSTRACT

European rivers have been constantly affected by anthropogenic impairments throughout the course of human history. Since 2000 the European Water Framework Directive (WFD) accelerated the process of recognizing and reclaiming rivers as valuable ecosystems, and the restoration of river stretches has become an essential tool for the improvement of watercourses in Central Europe. These restoration measures need to be complemented by in-depth scientific evaluations to ensure the cost-effectiveness of the extensive rehabilitation programme. In this study, two restored river stretches in the middle reaches of the River Lahn (Germany) were analysed in great detail between 2006 and 2008, with a special focus on the macrozoobenthos (MZB) and the small-scale riverbed morphology. The MZB is one of four quality components used for the evaluation of rivers within the WFD and despite the more than a century-long history of limnological research focused on benthic invertebrates, the central question of which morphological processes and structures have to be improved in order to reach the 'good ecological status' is still largely unanswered as detailed research of the faunistic-morphological connections on the micro scale (choriotopes) is still few and far between. Both areas – the MZB and the hydromorphology – were covered in an extensive field work programme. The riverbed was mapped using the TRiSHa method ('Typology of Riverbed Structures and Habitats') and a detailed survey of the MZB (165 samples) was carried out in accordance with the PERLODES method. Both data sets were analysed for their spatial diversity and for their interrelations. Furthermore the 'ecological status' was assessed in accordance with the WFD. This evaluation revealed gaps in the German river typology and the assessment method that should ignite a broader scientific discussion about the benefits of nationwide unified evaluation methods versus regionalized approaches.

Keywords: ecological quality assessment, European water framework directive, Germany, Lahn, macrozoobenthos, microhabitats, monitoring, river restoration.

1 INTRODUCTION

The intensive use of the European rivers during the last hundreds of years has led to profound changes in the physicochemical properties, river morphology and the aquatic fauna [1, 2]. As a result, the quality of the German freshwater resources deteriorated so much during the 20th century that many rivers could no longer fulfil their multi-purpose functions within the ecological and economic systems. This and several incidents which gained considerable media coverage (like the Sandoz hazard in 1986) led to a programmatic water quality improvement of the German rivers during the last 50 years. Because of enforced water quality regulations and wastewater treatment processes, the biological state of most German rivers increased to a 'good' state [3]. However, the hydromorphology did not improve so that the structural degradation remains and 60% of all German surface water bodies are considered to morphologically heavily modified [3]. Rectifying these deficits is the biggest unsolved challenge for the German rivers and streams to date [4]. The most important step towards this goal was the approval of the European Water Framework Directive (WFD) with its central aim of restoring all European surface water bodies to a good chemical and ecological status until 2027 [5]. One of

DOI: 10.2495/EI-V1-N3-375-389

the four ecological quality parameters used for the assessment of this status is the macrozoobenthos (MZB) – the larger invertebrate animal species living on/in the riverbed substrates. These species have been studied for more than a century and are used for decades as indicators for the pollution and the morphological degradation of freshwater ecosystems [1, 6–8]. The implementation of the WFD requires a detailed knowledge about the morphological and ecological deficits which forms the basis for extensive restoration programmes. But this knowledge of how to select suitable measures in a cost-effective way is mostly theoretical and insufficient [9, 10]. There are hardly any detailed data about the relations between the morphological properties of rivers and the colonization patterns of the aquatic fauna and evaluations of implemented restoration measures are still rather the exception than the norm [2, 11–14]. Considering the large investment needed for reaching the aims of the WFD (between 18 and 50 Bln. € in Germany alone [2]), closing this knowledge gap is paramount for the successful improvement of the freshwater resources. To analyse these important relations between the riverbed morphology and the aquatic fauna in revitalized river stretches is the central premise of the research presented here [15].

2 RESEARCH AREA

The river selected for this research was the Lahn near Marburg, a large river of the lower mountain region (type 9.2 [16]) in Hesse (Germany) with – at that point – a catchment size of 1,666.2 km² and an average discharge of 16.7 m³/s. Two restored river section and one control section of the Lahn were selected as the research areas (Fig. 1). These three areas represent different characteristics of the river Lahn. The first area ('Auf der Weide') had been restored in 2002 and is located in the inner city of Marburg, downstream of a weir. It features a high flow variability and in parts a shallow riverbed. The research area 'Cappeler Fischerwiese' is located 6 km downstream of the first area and 1 km upstream of a weir, resulting in a lower flow velocity and a deeper riverbed. This area had been restored in two steps in 1997 and in 2005. A control area ('Gisselberger Straße') is located between the first two river stretches, downstream of the Marburg gauge. It has not yet been restored and is characterized by a straight river course, a shallow riverbed and a high flow velocity. A total of ten study sites were selected within the three research areas in order to cover a wide range of different characteristics – two side channel and two main channel sites in each restored area and two sites in the control stretch (Fig. 1). The river banks in all three research areas are 1.5 m high and fortified by rubble. The restoration measures followed the concept of a braided river which supposedly was the natural state of the Lahn in this region, though a thorough examination of historical maps (ranging back to 1715) could not find any evidence for natural river branches [15]. The restoration measures were implemented following the passive dynamic river development approach which supports processes caused by the natural discharge dynamic while being very inexpensive. This makes this approach the preferred tool for small-scale river restoration projects in Germany, but without a proper post-restoration evaluation, the effectiveness of this approach remains uncertain.

3 METHODS

The field work for this study consisted of two major steps: the detailed mapping of the riverbed morphology and the sampling of the macrozoobenthos. The riverbed substrates, organic structures (like coarse and fine particulate organic matter or macrophytes), the water depth and the flow velocity were mapped using the TRiSHa method ('Typology of Riverbed Structures and Habitats') [15, 17]. This mapping of 1,069 plots (each measuring 1m²) was repeated

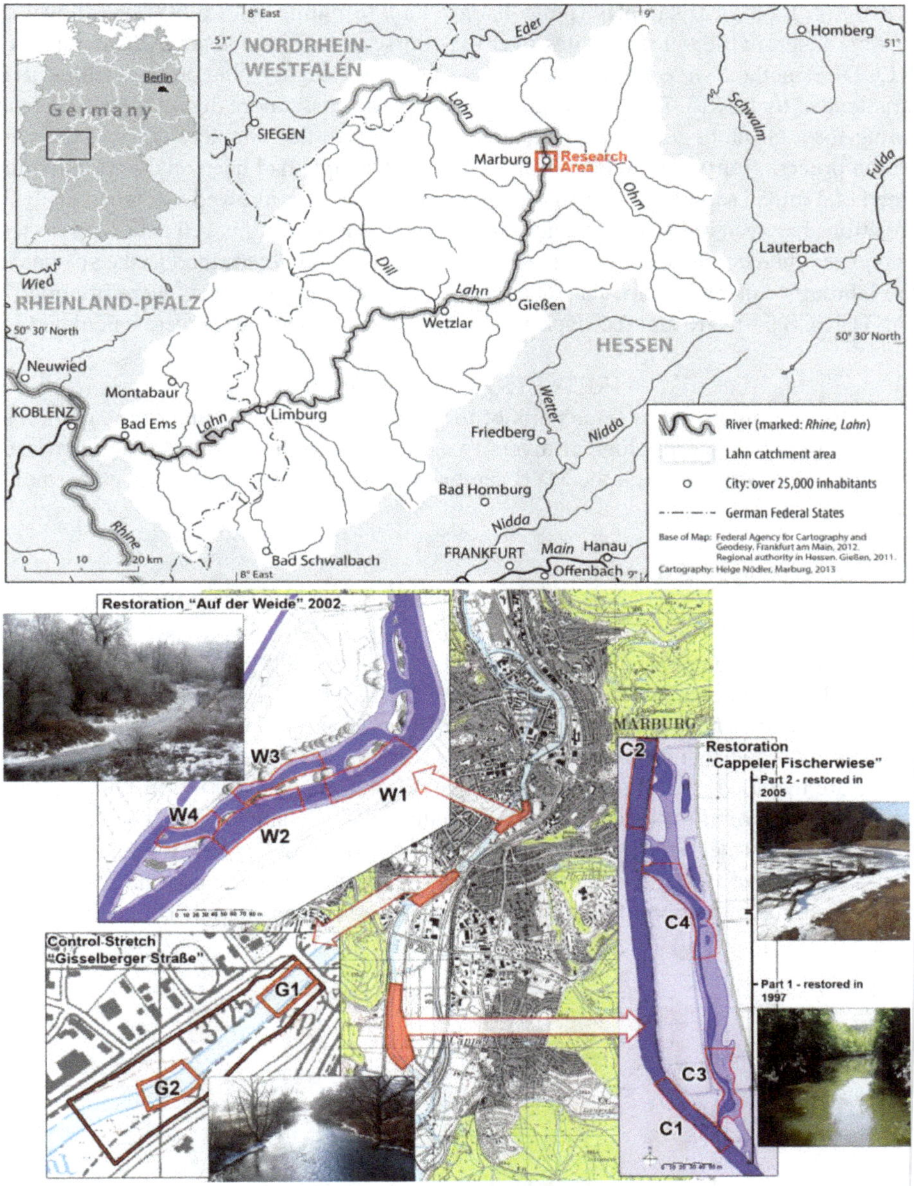

Figure 1: Location of the ten study sites within the three research areas and the Lahn catchment.

annually between 2006 and 2008 for a total of 3,207 data sets with 21 parameters each, resulting in the most detailed hydromorphological database for the analysed river stretches. Based on these morphological results a list of 32 microhabitat types was derived (see [17] for a detailed presentation of the TRiSHa method), which formed the basis for the biological sampling. Following the official WFD method for the assessment of the MZB (PERLODES

[7, 18]), 165 biological samples (multi-habitat kick-sampling with a 500 μm shovel sampler) were taken in the ten study sites over the course of the three years. Adjustments to the PERLODES method were implemented in so far as the highly generalized habitat assessment intended by PERLODES had been replaced by the much more detailed TRiSHa results, allowing for a far more accurate representation of the riverbed characteristics in the sample selection process. Furthermore each PERLODES subsample had been analysed individually (instead of a mixed sample for the whole research area), which allowed a detailed analysis of the relations between individual species and microhabitat conditions. All taxa were sorted on site and specimen copies were preserved in 70% alcohol and determined to the species level in the laboratory using a binocular microscope. The resulting taxa lists were imported into the ASTERICS software and the river type specific evaluation metrics were calculated [18].

4 RESULTS AND DISCUSSION

This results section discusses three parts of the study and presents some of the main results for each – the hydromorphological diversity as a result of the river restoration measures, the faunistic diversity based on the MZB sampling and the ecological quality assessment for the WFD.

4.1 Hydromorphological diversity

The results of the TRiSHa mapping are extensive and have previously been published in [15] and [17] and a detailed third paper has been submitted [19]. That is why only a brief hydromorphological characterization of the restored river stretches is presented here. The restoration measures have led to an increase of the morphological diversity. The total number of habitats recorded increased from 25 detected in the main channel to 32 when the side channels are included. There is also a significant difference ($p < 0.0001$; $Chi^2 = 107.1$) in the dominance structure of these habitat types in main and side channel sites. The main channel of the river is characterized by habitats dominated by rocks and gravel, while sand and clay habitats characterized the side channels (Fig. 2). These differences are even more pronounced when the spatial distribution of the different habitat types is considered. The main channel is

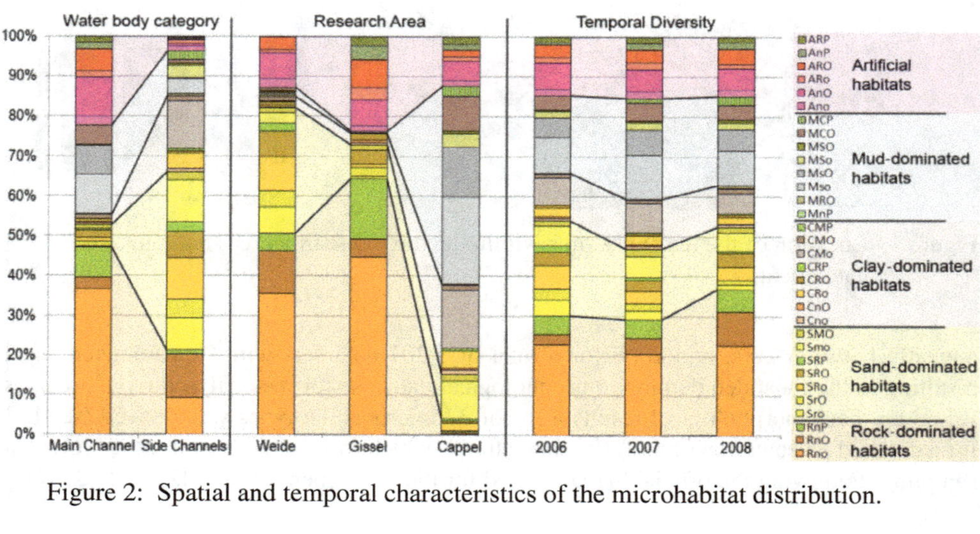

Figure 2: Spatial and temporal characteristics of the microhabitat distribution.

characterized by a mostly homogeneous riverbed dominated (36.8% in all main channel sites and up to 62.2% coverage in individual main channel sites) by just one habitat type (Rno – Rocks and Gravel without organic matter) with most of other habitat types restricted to the riverbanks. The side channels, on the other hand, are characterized by a heterogenic habitat mosaic throughout the riverbed ([15, 17, 19]) and the dominance of the most abundant habitat type is far less pronounced than in the main channel (CMo = 11.9%, Smo = 10.7% and SRo = 10.6% for the three most frequently detected habitat types). There were also significant differences ($p < 0.0001$, $Chi^2 = 142.1$) detected between the free-flowing and the impounded river stretches (Fig. 2, where the research areas 'Weide' and 'Gissel' are free flowing while 'Cappel' is impounded). The reduced flow velocity in the impounded river stretch caused a strong shift towards mud-dominated habitat types. Similar results have been presented by Arscott et al. [20], Jähnig et al. [21] and Dewson et al. [22], although those studies did not focus on the micro scale of the choriotopes.

The temporal dynamic detected in the three research areas was also affected by the impoundment, resulting in a very high similarity of the impounded main channel sites from year to year ($p = 0.871$; $Chi^2 = 8.3$) and significant temporal changes in the free-flowing restored side channels ($p = 0.001$; $Chi^2 = 51.2$). Overall the restored river stretches showed a high flood-driven morphodynamic and especially the side channels were characterized by an ongoing change of their habitat composition so that the restoration measures have been very successful in creating a highly dynamic freshwater ecosystem. The high habitat diversity in river stretches with mixed flow conditions (like the main and side channel study sites analysed here) confirms the findings published by Raven et al. [23] and Lorenz et al. [24], who also could show an increase in the number of substrate and microhabitat types in re-meandering German lowland rivers with variable flow velocities. Rohde et al. [25] detected a heterogeneous mosaic pattern of the microhabitats in restored river stretches comparable to the structures presented here. Interestingly those mosaic structures were even more complex than in near-natural reference sites. Jähnig et al. [26] analysed the riverbed structure of seven German mountain river sites (including the river Lahn) with restored multi-channel beds and could – in unison with the results presented here – show that the restoration led to a decrease of the dominance of the main substrate (from 75% in the single sites to 62% in the multi-channel sites).

4.2 Faunistic diversity

Based on the spatial distribution of the 32 microhabitat types 165 MZB samples were taken in the ten study sites. In order to study the relations between the riverbed morphology and the aquatic fauna and for the assessment of the spatial diversity of the MZB, the 165 taxa lists (one for each sample) were aggregated into 32 taxa lists – one for each habitat type. Two general metrics used of for the comparison of the different habitat types and groups are the animal abundance in individuals per m^2 and the species richness as the number of taxa per habitat type (Fig. 3). The results vary greatly from habitat type to habitat type. The rock- and gravel-dominated habitat types were characterized by the highest species richness (up to 74 taxa in the type Rno and a total 85 taxa for the whole habitat group). This group is followed by the artificial habitats with up to 41 taxa for an individual habitat type (ARP – artificial habitats with rocks and macrophytes) and a total of 81 taxa for the whole habitat group. This high species diversity in the artificial habitats can be explained by the physical structure of the bank fixations found in the Lahn which emulate large rocks and an extensive interstitial, the natural habitat for a mountain river like the Lahn. The clay- and mud-dominated habitat groups, on the other hand, were characterized by a much smaller species richness (41 and 43 taxa) and with just ten taxa three clay and mud habitat types yielded the least species in this study (Fig. 3).

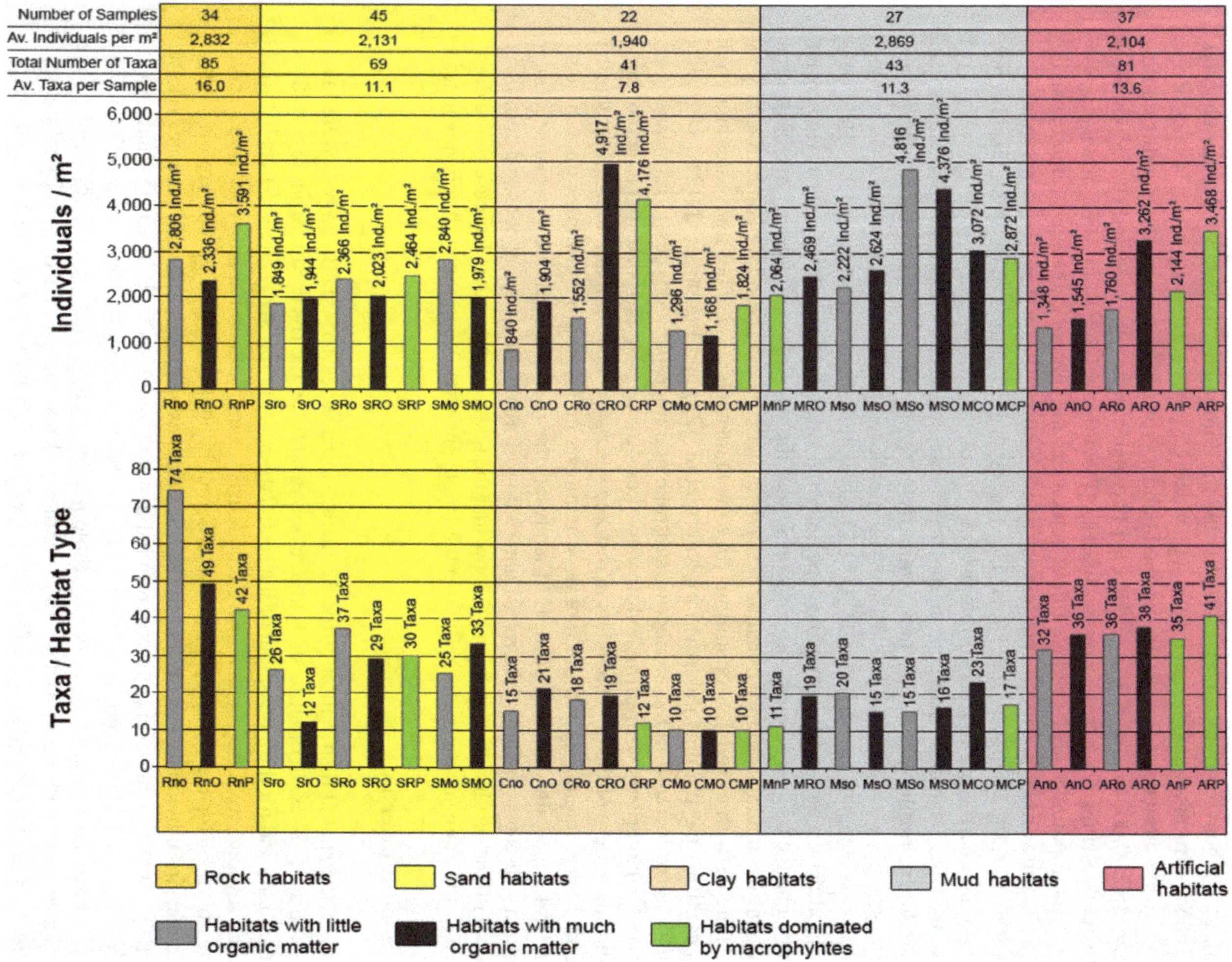

Figure 3: Macrozoobenthos abundance (Ind./m^2) and species richness (taxa/habitat type) for the individual microhabitats and habitat groups.

These results reflect that clay is a substrate that is difficult to colonize by the MZB as it mostly lacks the interstitial and structuring features as rocks or organic material while areas dominated by mud are characterized by extremely low oxygen concentration and even anoxic conditions. The *Trichoptera*, *Diptera* and *Ephemeroptera* provided the majority of the taxa, while the *Diptera* and *Crustacea* showed the highest population densities (with averages up to 1,112 Ind./m² for a single taxon) (Fig. 4). Only three taxa (*Hydracarina sp.*, *Tubificidae Gen. sp.* and *Gammarus roeseli*) had been detected with a high consistency (found in more than 2/3 of all samples) and a high population density (>200 Ind./m²), while 91.6% of all taxa had been found in less than one-third of all samples and with densities of less than 50 Ind./m². The lower half of Fig. 4 exemplifies this specialization for the group of the mayflies (*Ephemeroptera*). *Ephemera danica* showed the highest abundances in sand-dominated habitats, where it could easily burrow itself into the fine sediments, while *Serratella ignita* was closely connected to habitat types characterized by macrophytes and *Cloeon dipterum* dominated the artificial habitats. This high level of specialization reflects the influence of a variety of abiotic parameters – from flow velocity to water depth, microhabitats, channel characteristics and even the research area.

A principal component analysis performed with the TRiSHa and MZB data sets revealed that the flow velocity, the microhabitats and the presence/absence of aquatic macrophytes are the most important environmental parameters responsible for the spatial distribution of the aquatic fauna while the water depth but also the differences between main and side channel were much less important (Fig. 5).

Cross-referencing this result with the autecological profiles of the MZB taxa confirmed the general accuracy of these results, even though the PCA explains only 32.48% of the detected variability. The PCA results were furthermore confirmed by an agglomerated hierarchical clustering (AHC) performed with the faunistic data sets. The nine clusters identified could be differentiated from each other by the flow velocity, the research areas (as a proxy for the effect of the impoundment), the macrophytes and the habitat groups (Fig. 6). These AHC results illustrate very well that each of the research areas has a unique MZB fauna and that the river restoration measures have had a positive effect on the overall species richness and biodiversity of the Lahn river.

4.3 Ecological quality assessment

The previous two subchapters have demonstrated that the river restoration measures have led to an increase of the faunistic diversity, which is a valuable goal in itself. But the question remained if this also led to an improved assessment of the research areas in regard to the WFD. In order to evaluate this, the habitat-specific taxa lists were aggregated into one list for each study area. The ecological quality was then calculated using the ASTERICS software. This evaluation is tailored to the different German river types by comparing the present state to reference conditions [16, 18]. The core metrics for the assessment of the morphological integrity of the Lahn (stream type 9.2) are the German Fauna Index (tolerance metric), the percentage of metarhithral-taxa (functional metric), the percentage of EPT (*Ephemeroptera*, *Plecoptera* and *Trichoptera*) abundance classes (composition metric) and the number of EPTCBO (the EPT plus the *Coleoptera*, *Bivalvia* and *Odonata*) taxa (diversity metric). The results presented in Table 1 show that the overall ecological quality of the three study areas is only 'moderate' to 'unsatisfactory'. While the saprobic quality in all study areas reached a 'good' result (with index values between 1.98 and 2.15), the morphological degradation

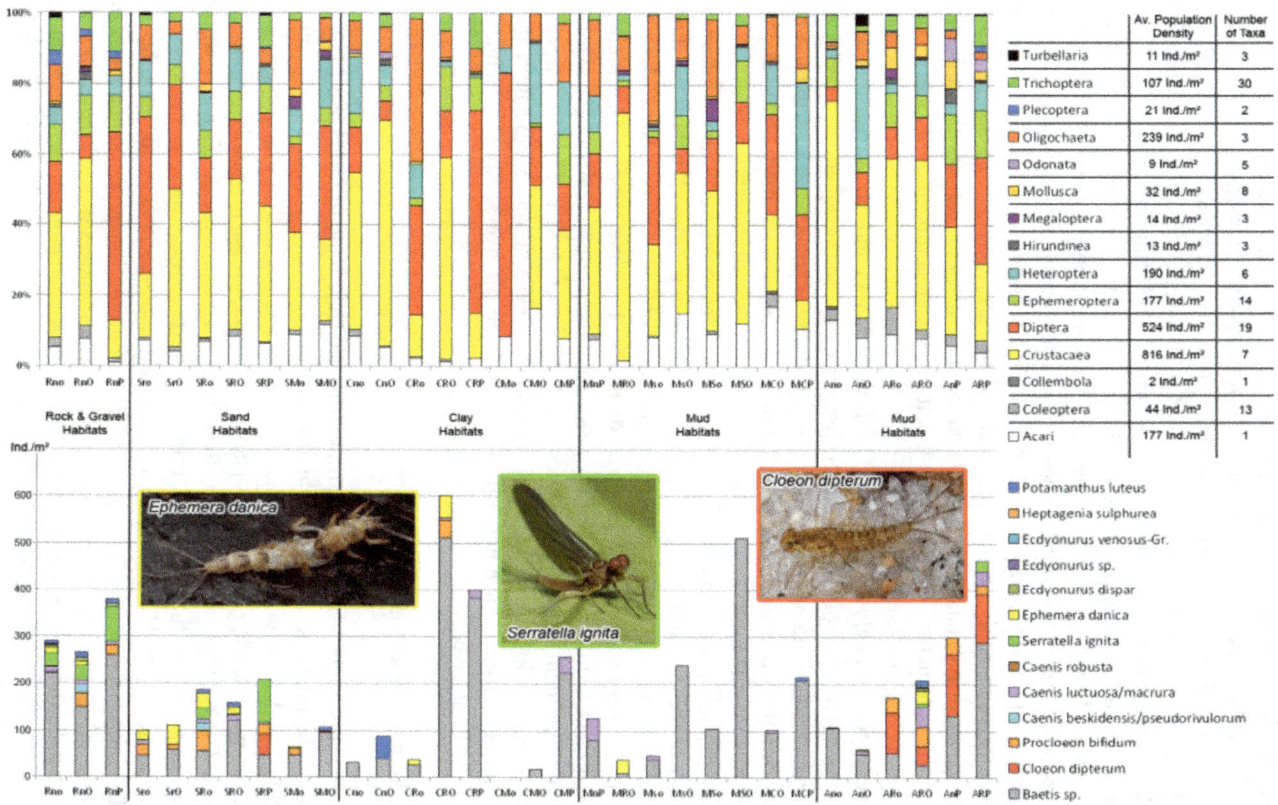

Figure 4: Macrozoobenthos composition per microhabitat type for the taxonomic orders in general and the *Ephemeroptera* in specific.

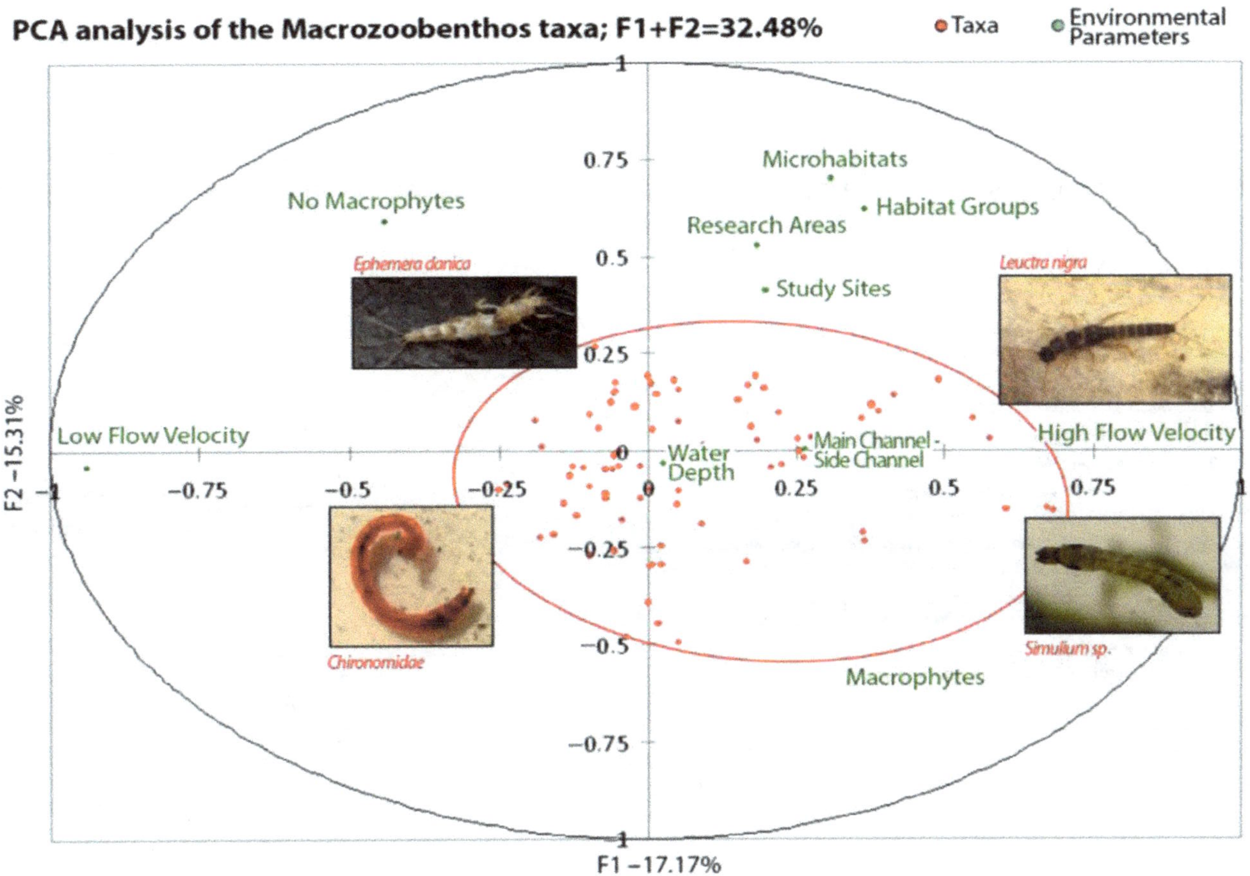

Figure 5: Categorisation of the macrozoobenthos samples using principal component analysis.

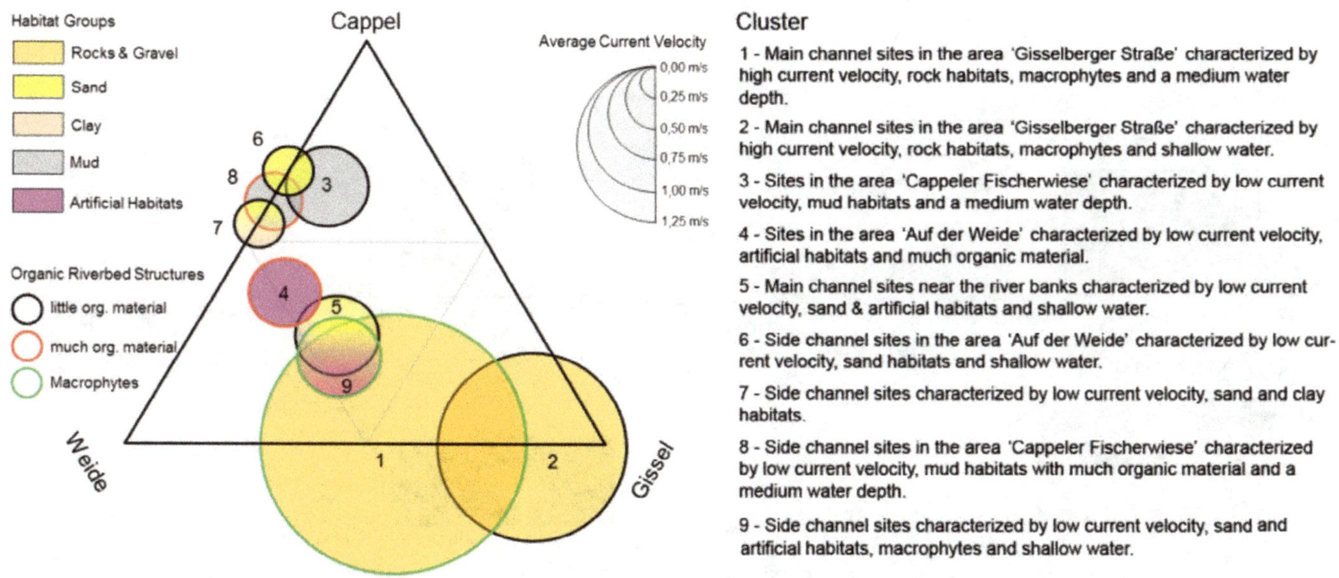

Figure 6: Results of the agglomerative hierarchical clustering of the macrozoobenthos samples with the research area, habitat groups, organic riverbed structures and the current velocity as the main factors.

revealed the main deficits of the three areas. Of the four core metrics, only the diversity metric (the number of EPTCBO taxa) showed no degradation, meaning that the Lahn is characterized by a heterogeneous macroinvertebrate fauna which reflects the structural diversity of the riverbed and confirms the results presented earlier. The functional metric, on the other hand, was consistently rated as 'unsatisfactory'. The detected percentages ranged between 10.7% and 12.9%, while the reference conditions for a type 9.2 river suggest that up to 25% of all taxa should be metarhithral taxa [16]. The low percentages in the study areas cannot be explained as direct consequences of the impoundment as even the free-flowing study areas showed severe deficits. But these results might well be the result of the high degree of impoundment in the Lahn river as a whole, which is a considerable obstacle for the migration of rheophilic taxa. The last two core metrics show a clear distinction between the free-flowing and the impounded study areas and thus are well suited for evaluating the impact of weirs on the ecological quality of the Lahn. The tolerance metric (German Fauna Index type 9.2) scored 0.55 ('moderate' quality) and 0.54 ('unsatisfactory' quality) in the free-flowing study areas, while the score in the impounded study area was only 0.23 ('bad' quality). Even more drastic are the differences between free-flowing and impounded areas for the composition metric (% of EPT abundance classes). This core metric was rated 'good' in the free-flowing study areas ('Auf der Weide': 0.663 (44.9%); 'Gisselberger Straße': 0.689 (45.7%)) which is close to the suggested reference conditions (55% EPT abundance classes), while it was rated 'bad' in the impounded study area (0.123 (28.7%)). In the impounded area, on the other hand, the result is close to the threshold for a complete degradation (25%) [16].

Based on these results the ecological quality of the Lahn still shows deficits. Even worse, the restored areas have not been evaluated with higher scores than the unrestored control stretch. In order to analyse the influence of the restored side channels on the ecological quality in more detail, the taxa lists for the side and main channels were computed separately as well (Table 1). The results showed that there are indeed only very small differences between the main channel evaluation and the scores for the combined system. In the free-flowing study area 'Auf der Weide' the inclusion of the restored side channel study sites leads to a marginally improved saprobic index (2.02 to 1.98) and to a slightly lower score for the morphological degradation (0.55 in comparison to 0.57 for the main channel alone). All the core metrics were also considerably lower in the side channel, but the overall assessment worsened for only the functional and tolerance metrics. In the impounded study area the saprobic index was also improved by the inclusion of the side channel study sites (2.2 to 2.15) while the morphological degradation remained unchanged. Three of the four core metrics showed worse scores in the side channel (with the diversity metric being the exception) which leads to lower scores for the functional and the composition metric. The composition metric for the combined system of main and side channel study sites was even rated 'bad', while the main channel itself had been rated only 'unsatisfactory'. The other two core metrics, however, showed slightly better scores for the combined system so that the overall score for the morphological degradation remained the same. The low scores for the restored side channels are surprising as they were characterized by a very high spatial diversity and temporal dynamic. Similar results, however, were provided by [16, 27–30] who could show that even a high habitat diversity and heterogeneity in restored river stretches did not have a major impact on the development of the benthic communities. Lorenz et al. [16] analysed restoration measures on the upper reaches of the Lahn and concluded that the sites need more time before developing high-quality MZB communities. The restoration measures analysed in the study presented here, however, were in part nine years old before the study began. Palmer et al. [30] conclude that restoration measures which focus solely on increasing the structural

River type 9.2 ("Large river of the lower mountain region")	"Auf der Weide"			"Cappeler Fischerwiese"			"Gisselberger Straße"
Study area characteristic	Free flowing			Impounded			Free flowing
Study site characteristic	Main + Side channel	Main channel	revitalized Side channel	Main + Side channel	Main channel	revitalized Side channel	Main channel control stretch
Modul "Saprobic Quality"	**"good"**	**"good"**	**"good"**	**"good"**	**"good"**	**"good"**	**"good"**
German Saprobix Index	**1.98**	**2.02**	**2.00**	**2.15**	**2.20**	**2.15**	**2.03**
Dispersion	0.059	0.059	0.069	0.054	0.069	0.054	0.053
Sum of abundance classes	92	81	64	64	43	69	85
Modul "Morphological Degradation"	**"moderate"**	**"moderate"**	**"moderate"**	**"unsatisfactory"**	**"unsatisfactory"**	**"unsatisfactory"**	**"moderate"**
Score (0-1)	**0.55**	**0.57**	**0.43**	**0.23**	**0.23**	**0.23**	**0.54**
Tolerance metric (German Fauna Index type 9.2)	"moderate"	"moderate"	"unsatisfactory"	"bad"	"bad"	"bad"	"unsatisfactory"
Score (0-1)	0.449	0.497	0.297	0.008	0.000	0.000	0.390
Functional metric (% metarhithral taxa)	"unsatisfactory"	"unsatisfactory"	"unsatisfactory"	"unsatisfactory"	"unsatisfactory"	"unsatisfactory"	"unsatisfactory"
Score (0-1)	0.317	0.352	0.284	0.287	0.299	0.276	0.396
% of metarhithral taxa	11.3%	12.1%	10.7%	10.7%	11.0%	10.5%	12.9%
Composition metric (% EPT abundance classes)	"good"	"good"	"moderate"	"bad"	"unsatisfactory"	"bad"	"good"
Score (0-1)	0.663	0.603	0.458	0.123	0.237	0.119	0.689
% of EPT abundance classes	44.9%	43.1%	38.7%	28.7%	32.1%	28.6%	45.7%
Diversity metric (EPTCBO taxa)	"very good"	"very good"	"very good"	"very good"	"very good"	"very good"	"very good"
Number of EPTCBO taxa	48	38	33	38	23	32	36
Ecological Quality	**"moderate"**	**"moderate"**	**"moderate"**	**"unsatisfactory"**	**"unsatisfactory"**	**"unsatisfactory"**	**"moderate"**

Table 1: Results of the ecological quality assessment with PERLODES/ASTERICS.

heterogeneity of the riverbed are bound to fail if larger-scale factors like the land use or – in the case of the Lahn – the impoundment – are not addressed as well. Another explanation for the lack of improvement of the ecological quality of the river despite the well-documented morphological diversity and dynamic can also be found in the reference conditions used for the assessment. The stream type 9.2 is characterized by a shallow riverbed dominated by rocks and gravel and a high flow velocity caused by a valley inclination of 3‰. These characteristics are valid for the upper reaches of the Lahn but not for the middle reaches analysed here which was also proven by the analysis of several historical maps of the Lahn catchment [15]. Downstream of the confluence of the rivers Lahn and Ohm, the Lahn valley has an inclination of only 0.76‰ and the riverbed is much deeper. Because of the reduced flow velocity, especially in the restored side channels, the typical coarse sediments are replaced by finer sediments which in turn result in a different MZB fauna [31, 32]. The taxa found in the side channels are indicators for lentic conditions – not because of an anthropogenic degradation but because of the natural landscape features of the Lahn.

5 CONCLUSIONS

The detailed analysis of the riverbed morphology revealed a high diversity and dynamic of the substrates and the aquatic microhabitats in the main and side channel sections of the river Lahn. The diversity was much larger in the restored side channels and the flow velocity turned out to be the most influential parameter for the recorded dynamic. This proves that the approach of the passive dynamic river restoration is well suited to create dynamic and valuable river ecosystems within the limited timeframe of the WFD. At the same time, the results have shown that the impoundment is an impairment that cannot be overcome with a passive dynamic approach. The faunistic assessment, however, does not reflect the detected high morphological diversity and quality. The restored research areas did not show any significant change of their ecological state. Possible reasons for this are catchment-wide effects of the river impoundment as well as the reference conditions used for the ecological assessment which are not adequate for the river Lahn downstream of the confluence with the Ohm. These results either question the value of a nationwide river typology with clearly defined and binding reference conditions or call for a modified or extended assessment tool for a reliable evaluation of rivers like the Lahn, which share characteristics with both mountain rivers and lowland streams and are heavily impaired by anthropogenic flow regulation.

REFERENCES

[1] Barbour, M.T., Gerritsen, J., Snyder, B.D. & Stribling, J.B., *Rapid Bioassessment Protocols for Use in Streams and Wadable Rivers – Periphyton, Benthic Macroinvertebrates, and Fish*, 2nd ed., United States Environmental Protection Agency (US-EPA): Washington, DC, 344pp., 1999.

[2] Dickhaut, W., *Fließgewässerrenaturierung Heute – Forschung zu Effizienz und Umsetzungspraxis – Abschlussbericht*, Hamburg University of Applied Sciences: Hamburg, 120pp., 2005.

[3] BMU (ed.), *Die Wasserrahmenrichtlinie – Ergebnisse der Bestandsaufnahme 2004 in Deutschland*, Bundesministerium für Umwelt, Naturschutz und Reaktorsicherheit (BMU): Berlin, 67pp., 2005.

[4] HMULF (ed.), *Gewässerstrukturgüte in Hessen 1999 – Erläuterungsbericht*, Hessisches Ministerium für Umwelt, Landwirtschaft und Forsten (HMULF): Wiesbaden, 55pp., 2000.

[5] European Commission (ed.), *Directive 2000/60/EC of the European Parliament and of the Council of 23 October 2000 Establishing a Framework for Community Action in the Field of Water Policy (Water Framework Directive).* OJ L 327, 22.12.2000, 93pp., 2000.

[6] Rolauffs, P., Hering, D., Sommerhäuser, M., Rödiger, S. & Jähnig, S., Entwicklung eines Leitbildorientierten Saprobienindexes für die biologische Fließgewässerbewertung. *UBATexte 11/2003*, Umweltbundesamt: Berlin, 137pp., 2003.

[7] Sandin, L., Hering, D., Buffagni, A., Lorenz, A., Moog, O., Rolauffs, P. & Stubauer, I., Experiences with different stream assessment methods and outlines of an integrated method for assessing streams using benthic macroinvertebrates. *AQEM 3rd Deliverable*, 87pp., available at http://www.eugris.info/displayresource.aspx?r=118, 2001.

[8] Schmedtje, U. & Kohmann, F., Bewertung von Fließgewässern- Aussagekraft und Grenzen biologischer und chemischer Indizes. *Wasser und Boden*, 11, Blackwell Verlag: Berlin, pp. 610–612, 1988.

[9] Interwies, E., et al., Grundlagen für die Auswahl der kosteneffizientesten Maßnahmenkombinationen zur Aufnahme in das Maßnahmenprogramm nach Artikel 11 der Wasserrahmenrichtlinie – Handbuch. *UBA-Texte 02/04*, Umweltbundesamt: Berlin, 2004.

[10] Lepori, F., Palm, D., & Malmqvist, B., Effects of stream restoration on ecosystem functioning: Detritus retentiveness and decomposition. *Journal of Applied Ecology*, **42**, pp. 228–238, 2005. DOI: 10.1111/j.1365-2664.2004.00965.x.

[11] Bernhardt, E.S., et al., Synthesizing U.S. river restoration efforts. *Science*, **308**, pp. 636–637, 2005. DOI: 10.1126/science.1109769.

[12] Koenzen, U., Erfolgskontrolle von Maßnahmen zur naturnahen Entwicklung von Fließgewässern – Hinweise für gezielte Maßnahmen zur Kompensation von Strukturdefiziten unter Berücksichtigung der Strahlwirkung. *Schriftenreihe des DRL*, **81**, pp. 35–42, 2008.

[13] Moerke, A.H. & Lamberti, G.A., Restoring stream ecosystems: Lessons from a Midwestern State. *Restoration Ecology*, **12**(**3**), pp. 327–334, 2004. DOI: 10.1577/1548-8675(2002)022<0001:AROSRT>2.0.CO;2.

[14] Purcell, A.H., Friedrich, C. & Resh, V.H., An assessment of a small urban stream restoration project in Northern California. *Restoration Ecology*, **10**(**4**), pp. 685–694, 2002. DOI: 10.1577/1548-8446(1997)022<0012:AEPORA>2.0.CO;2

[15] Groll, M., *Beziehungen zwischen der Gewässermorphologie und dem Makrozoobenthos an renaturierten Abschnitten der Lahn,* Philipps-Universität Marburg: Marburg, 442pp., 2011.

[16] Döbbelt-Grüne, S., Hartmann, C., Zellmer, U., Reuvers, C., Zins, C. & Koenzen, U., Hydromorphologische Steckbriefe der deutschen Fließgewässer-typen – Anhang 1 von Strategien zur Optimierung von Fließgewässer-Renaturierungsmaßnahmen und ihrer Erfolgskontrolle. *UBA-Texte 43/2014*, Umweltbundesamt: Dessau-Roßlau, 288pp., 2013.

[17] Groll, M., Thomas, A., Jungermann, L. & Schäfer. K., Typology of Riverbed Structures and Habitats (TRiSHa) – A new method for a high resolution characterization of the spatial and temporal dynamic of freshwater ecosystems. *Ecological Indicators*, **61**, pp. 219–233, 2016. DOI: 10.1016/j.ecolind.2015.09.019.

[18] Meier, C., et al., *Methodisches Handbuch Fließgewässerbewertung – Handbuch zur Untersuchung und Bewertung von Fließgewässern auf der Basis des Makrozoobenthos vor dem Hintergrund der EG-Wasserrahmenrichtlinie*, Universität Duisburg-Essen: Essen, 79pp., 2006.

[19] Groll, M., The passive river restoration approach as an efficient tool to improve the hydromorphological diversity of rivers – Case study from two river restoration projects in the German lower mountain range. *Geomorphology* (accepted).

[20] Arscott, D.B., Tockner, K., van der Nat, D. & Ward, J.V., Aquatic habitat dynamics along a Braided Alpine River ecosystem (Tagliamento River, Northeast Italy). *Ecosystems*, **5(8)**, pp. 802–814, 2002. DOI: 10.1007/s10021-002-0192-7.

[21] Jähnig, S.C., Lorenz, A.W., Lorenz, R.R.C. & Kail, J., A comparison of habitat diversity and interannual habitat dynamics in actively and passively restored mountain rivers of Germany. *Hydrobiologia*, **712(1)**, pp. 89–104, 2013. DOI: 10.1007/s10750-012-1264-0.

[22] Dewson, Z.S., James, A.B.W. & Death, R.G., A review of the consequences of decreased flow for instream habitat and macroinvertebrates. *Journal of the North American Benthological Society*, **26(3)**, pp. 401–415, 2007. DOI: 10.1899/06-110.1.

[23] Raven, P.J., Holmes, N.T.H., Vaughan, I.P., Dawson, F.H. & Scarlett, P., Benchmarking habitat quality: Observations using River Habitat Survey on near-natural streams and rivers in northern and western Europe. *Aquatic Conservation Marine and Freshwater Ecosystems*, **20**, pp. 13–30, 2010. DOI: 10.1002/aqc.1103.

[24] Lorenz, A.W., Jähnig, S.C. & Hering, D., Re-meandering German lowland streams: Qualitative and quantitative effects of restoration measures on hydromorphology and macroinvertebrates. *Environmental Management*, **44(4)**, pp. 745–754, 2009. DOI: 10.1007/s00267-009-9350-4.

[25] Rohde, S., Kienast, F. & Bürgi, M., Assessing the restoration success of river widenings: A landscape approach. *Environmental Management*, **34(4)**, pp. 574–589, 2004. DOI: 10.1007/s00267-004-0158-y.

[26] Jähnig, S.C., Lorenz, A. & Hering, D., Hydromorphological parameters indicating differences between single- and multiple-channel mountain rivers in Germany, in relation to their modification and recovery. *Aquatic Conservation Marine and Freshwater Ecosystems*, **18**, pp. 1200–1216, 2008. DOI: 10.1002/aqc.875.

[27] Brooks, S.S., Palmer, M.A., Cardinale, C.M. & Ribblett, S., Assessing stream ecosystem rehabilitation – Limitations of community structure data. *Restoration Ecology*, **10(1)**, pp. 156–168, 2002. DOI: 10.1126/science.289.5477.284.

[28] Jähnig, S.C., et al., A comparative analysis of restoration measures and their effects on hydromorphology and benthic invertebrates in 26 central and southern European rivers. *Journal of Applied Ecology*, **47**, pp. 671–680, 2010. DOI: 10.1111/j.1365-2664.2010.01807.x.

[29] Kokes, J., River channel habitat diversity (RCHD) and macroinvertebrate community. *Biologia*, **66(2)**, pp. 328–334, 2011. DOI: 10.2478/s11756-011-0010-9.

[30] Palmer, M.A., Menninger, H.L. & Bernhardt, E., River restoration, habitat heterogeneity and biodiversity – A failure of theory or practice? *Freshwater Biology*, **55**, pp. 205–222, 2010. DOI: 10.1111/j.1365-2427.2009.02372.x.

[31] Brunke, M., Hoffmann, A. & Pusch, M., Use of mesohabitat-specific relationships between flow velocity and river discharge to assess invertebrate minimum flow requirements. *Regulated Rivers – Research & Management*, **17**, pp. 667–676, 2001. DOI: 10.1002/rrr.626.

[32] Downes, B.J., Lake, P.S., Glaister, A. & Bond, N.R., Effects of sand sedimentation on the macroinvertebrate fauna of lowland streams – Are the effects consistent? *Freshwater Biology*, **51**, pp. 144–160, 2006. DOI: 10.1007/s002679900019.

UNCERTAINTY IN MINIMUM INSTREAM FLOW REQUIREMENTS FOR STREAMS IN SEMI-ARID ENVIRONMENTS

MICHAEL BARBER[1] & ROBERT MAHLER[2]
[1]Civil and Environmental Engineering, University of Utah, USA
[2]Soil Science Division, University of Idaho, Moscow, ID, USA

ABSRACT
Increased competition for water resources in arid and semi-arid watersheds is prompting management agencies to adopt instream flow requirements for critical aquatic habitats. For instance, because of increasing concerns for bull trout and salmonid species in the Touchet River system, minimum instream flows are needed to protect several important rearing and spawning reaches. The study included a field reconnaissance of the region with specific emphasis on known migration blockages, spawning and rearing habitat areas, and other areas identified in the project scoping meeting. Eight representative cross sections were selected at each of seven stream segments based on this initial field reconnaissance survey. Care was taken to include pool, spawning, riffle and other unique stream characteristics. Depth and velocity profiles were measured at each of the cross sections at three different water stages: high, medium and low. Substrate grab samples were taken along each of the eight reaches for subsequent analysis. Temperature data along with several other basic water quality parameters (i.e. dissolved oxygen, pH, conductivity and turbidity) and an assessment of vegetative cover were also recorded during sampling. Minimum instream flow modelling and analysis were conducted using the instream flow incremental methodology technique. Habitat suitability indices were assigned to each stream cell. Non-binding recommendations for instream flow values were provided to management agencies. A considerable amount of uncertainty exists in the preference factors assigned to the various life stages of bull trout and salmonid species. Much of this stems from the processes traditionally used to determine factors related to juvenile rearing, spawning and migration. A conclusion of this study was that preference curves developed for wet regions were not applicable to semi-arid river basins where stream flows are often quite small in comparison. This study documents the assumptions, procedures and results of this investigation and demonstrates the potential impacts of uncertainty on the results.
Keywords: aquatic habitat, ecological protection, environmental flows, IFIM

1 INTRODUCTION

Over-appropriation of water in many river basins throughout the world is increasingly causing shortages of domestic, industrial, agricultural and environmental supplies [1]. It has been estimated that 1.4 billion people live in areas where water demand is greater than the supply during at least part of the year [2]. In the semi-arid western United States, where the prior appropriation doctrine governs water allocation through water right systems based on first in time priority, increased competition for water resources for population growth and climate change adaptation is prompting management agencies to adopt instream flow requirements for protection of critical aquatic habitats. The three most commonly used methods for establishing these flow requirements are the (1) Instream Flow Incremental Methodology (IFIM), (2) toe-width and (3) Tennant (aka 'Montana') method [3]. In keeping with the format adopted by the Washington Department of Ecology, the IFIM technique was used in this study. Computational routines for assessing the appropriate level of instream flow requirements have been developed and used based on the IFIM [4], although close analysis of input and interpretation of output results is still more art than science. The simplicity of the weighted usable area (WUA) curves generated by the Physical HABitat Simulation (PHABSIM) model can

DOI: 10.2495/EI-V1-N3-390-400

be deceptive in terms of identifying appropriate flows across the various life stages of each species. A number of studies have critically examined the WUA process with most reaching the conclusion that the nature of the relationship between fish and WUA may be unique for each stream [5]. Furthermore, a considerable amount of uncertainty exists in the preference factors assigned to the various life stages of bull trout and salmonid species. Much of this stems from the processes traditionally used to determine factors related to juvenile rearing, spawning and migration. Preference factors developed for larger rivers are sometimes inappropriately used for smaller stream prevalent in arid and semi-arid regions.

This study used the Touchet River Basin in Southeastern Washington, USA, as the basis for evaluating the hypothesis that understanding uncertainty and variability in the PHABSIM preference curves will result in improved decision-making capabilities. We used an earlier IFIM study by Barber et al. [6] as the starting point for the initial calibrated modelling effort. The original study of instream flow requirements was conducted for Columbia Conservation District in Dayton, Washington, USA, as part of an effort to protect and restore salmonid species in the Touchet River watershed. Modifications to the preference factors were used to demonstrate the impacts of uncertainty on the results.

2 BACKGROUND

The Touchet River watershed (Fig. 1) is a tributary to the Walla Walla River, which contains several cold water stream reaches that are important to anadromous and endangered fish species. The most valuable fish habitat consists of approximately 425 km^2 (163 mi^2) upstream of the Patit Creek confluence [7]. The PHABSIM computer model developed by the United States Geological Survey was used in this investigation [8]. PHABSIM is intended to simulate streamflow and physical habitat relationships for various life stages or recreational activities and is commonly used to calculate the WUA for each cross section over a range of flows. The WUA is computed sequentially using three distinct model subroutines (WSL, velocity and Habtae) and an equation in the form of

$$\mathrm{WUA} = \sum_{i=1}^{N} A_i \times (\mathrm{PF_d PF_v PF_s PF_c})_i \tag{1}$$

where A is the cell area, PF_d is a depth preference or weighting factor, PF_v is a velocity preference factor, PF_s is a substrate preference factor, PF_c is a cover preference factor and N is the number of cells in the cross section.

WSL is the routine used to predict water surface elevations at cross sections based on field data and the total energy as a channel section based on the open channel form of Bernoulli's equation. Velocity is the model that simulates velocity profiles using an empirical approach and an assumption that each cross section can be treated independently. Habtae is the main model that compiles results from WSL, velocity and preference factors into the WUA predictions [8].

Preference factors involve the important aspects of all fish life stages; substrate, cover, water depth and water velocity. Table 1 summarizes the substrate preferences for important fish species in the Touchet River watershed as a function of life stage. Table 2 expresses the values for cover. This information is collected during field investigations at the same time velocity and depth characteristics are being compiled. As indicated in eqn (1), these factors are multiplied together without regard to any special weighting, thus making all factors equally important.

Figure 1: Touchet River Basin in Southeastern Washington State, USA.

Table 1: Substrate preference values [9].

Code	Substrate	Spawning				Rearing		Holding
		Salmon	Steelhead	Trout	Bull trout	Fry	Juvenile	Adult
1	Silt, clay	0.0	0.0	0.0	0.0	0.1	0.1	0.1
2	Sand	0.0	0.0	0.0	0.0	0.1	0.1	0.1
3	Small gravel (0.25–1.25 cm)	0.3	0.5	0.5	1.0	0.1	0.1	0.1
4	Medium gravel (1.25–3.8 cm)	1.0	1.0	1.0	1.0	1.0	0.3	0.3
5	Large gravel (3.8–7.6 cm)	1.0	1.0	1.0	1.0	1.0	0.3	0.3
6	Small cobble (7.6–15.2 cm)	1.0	1.0	1.0	0.7	1.0	0.5	0.3
7	Large cobble (15.2–30 cm)	0.5	0.3	0.3	0.7	1.0	0.7	0.3
8	Boulders (>30 cm)	0.0	0.0	0.0	0.0	1.0	1.0	1.0
9	Bedrock	0.0	0.0	0.0	0.0	0.1	0.3	0.3

Spawning preference curves (depth and velocity curves) have been developed for two types of waterways: (1) large rivers and (2) streams and rivers. The dividing line between types 1 and 2 is a mean annual flow of 28.3 m^3/s (1,000 ft^3/s). However, this divide is somewhat subjective and arguably not appropriate for streams and rivers in arid and semi-arid environments where mean flows are much lower and even peak flows may not reach the threshold. When they exist, preference factors developed for smaller streams are based on a small sample size

Table 2: Cover preference values [9].

Cover type	Rearing		Holding
	Fry	Juvenile	Adult
Undercut bank	1.0	1.0	1.0
Overhanging vegetation	1.0	1.0	1.0
Root wad	1.0	1.0	1.0
Log jam/submerged brush pile	1.0	1.0	1.0
Log(s) parallel to bank	0.3	0.8	0.8
Aquatic vegetation	1.0	0.8	0.8
Short (<30 cm) terrestrial grass	0.4	0.1	0.1
Tall (>90 cm) terrestrial grass	0.7	0.7	0.1
Vegetation beyond the bank-full water edge	0.2	0.2	0.2

or other similar streams which are contrary to studies indicating unique stream characteristics [5]. Rearing preference is considerably more subjective, while spawning preference is relatively straightforward because depths and velocities can be measured at redds (salmon/steelhead spawning beds); rearing surveys are based on visual observations where a fish is spotted, which may be a preference or random.

For example, based on data collected at 25 redds, Figs 2 and 3 illustrate the spawning depth and velocity preference factors for steelhead. In each case, the measured values are shown in dashed lines, whereas the recommended values are shown in blue.

The variability between calculated and observed, which is clearly visible in the adult steelhead curves, is also present in the juvenile depth and velocity preference factors. Similarly, preference curves for salmon species (chinook, coho and sockeye) as well as bull and cutthroat trout demonstrate inherent variability that can have significant impact on interpretation of the results if users are not careful.

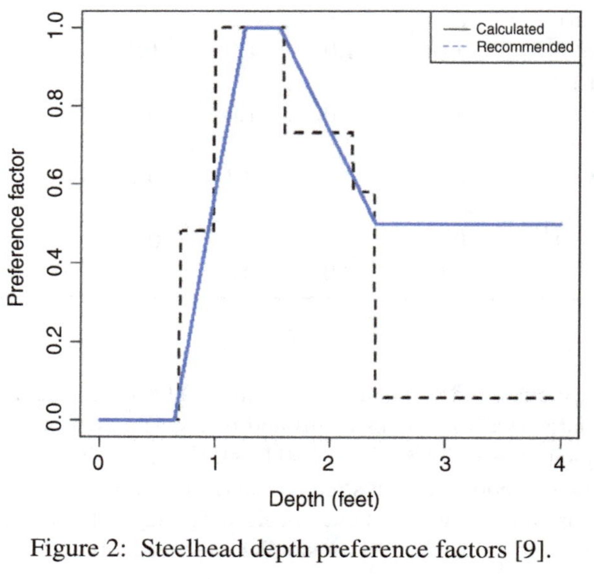

Figure 2: Steelhead depth preference factors [9].

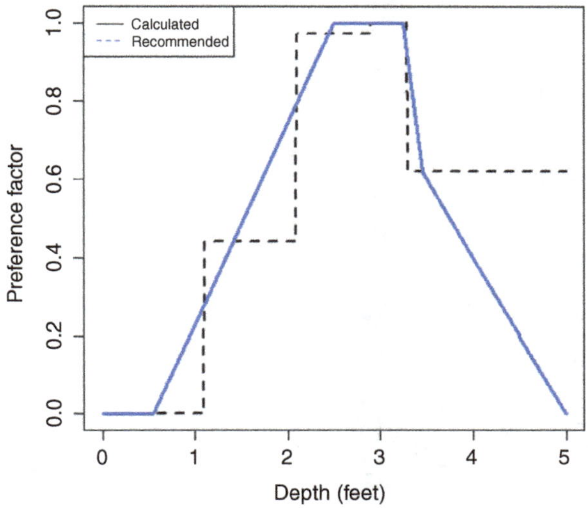

Figure 3: Steelhead velocity preference factor [9].

3 METHODOLOGY

The amount of water needed to protect aquatic fish species is determined by considering existing data, the hydrology of a stream and its natural variations in flow over the year, fish requirements and other related factors. Several procedures are available for determining minimum flow requirements, ranging from standard-setting techniques to incremental techniques. IFIM was originally developed by the U.S. Fish and Wildlife Service [10], primarily as a means of determining flow requirements downstream of hydropower relicensing efforts. The methodology integrated water supply requirements with analytical models from hydraulic and water quality engineering and empirically derived habitat versus flow functions. Over a period of 15 years, IFIM has developed into a river network analysis that incorporates fish habitat, recreational opportunity and woody vegetation response to alternative water management schemes [3].

There are five critical steps to conduct an IFIM analysis:

1) Problem identification
2) Study planning
3) Study implementation
4) Alternative analysis
5) Problem resolution

These steps interact to form the basis for the watershed plan. In this article, we focus on steps 1, 2 and 3 as impetus for the follow-on activities. Using the PHABSIM for Windows Version 1.5.1 model and the procedure and preference curves previously discussed, WUA curves were developed for steelhead and bull trout on three segments: Touchet River, North Fork Touchet River and Wolf Fork Creek. A screenshot of the Habtae user interface showing the range of options associated with WUA calculations is shown in Fig. 4. Clicking the Run button (after calibrating the WSL and velocity routines) generates files and graphs of WUA.

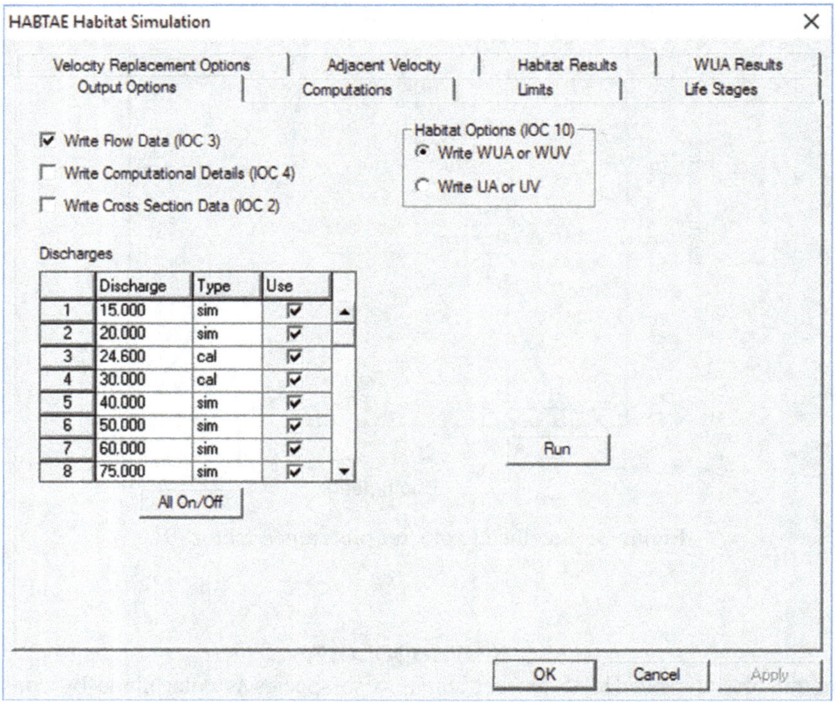

Figure 4: HABTAE user interface for calculating weighted usable area (WUA).

Because the variability in results is linear to the preference factors (eqn (1)), there was no need to run additional simulations to evaluate uncertainty. This was done using Excel and by performing sensitivity analysis of WUA values.

4 RESULTS AND DISCUSSION

WUA values were generated for various life stages of anadromous and resident fish populations threatened or endangered in the Pacific Northwestern USA. Tables 3 and 4 present the WUA values for steelhead and bull trout on the Wolf Fork Creek tributary. By themselves, these values are difficult to interpret. Information regarding existing stream flows, fish use throughout the year and options for mitigating shortages must be considered. Furthermore, determining which is the critical life stage is difficult to do with variable hydrologic conditions and the impacts of harvest, hatcheries, hydropower and downstream habitat factors. The three submodels (WSL, velocity and Habtae) generate copious amounts of data that is important to carefully examine in order to understand and interpret results. Space limitations that dictate only brief summaries can be presented here. More information can be found in the original technical report prepared for the Conservation District [6].

It is considerably easier to understand the meaning of these values by using the graphing function of PHABSIM. Figures 5 and 6 illustrate WUA as functions of flow for steelhead and bull trout on the Wolf Fork Creek tributary. Similar graphs were generated for the other streams in the study area.

Table 3: WUA values for steelhead on Wolf Fork Creek.

Discharge (m^3/s)	Spawning	Fry	Juvenile
0.43	635.5	5,210.0	787.8
0.57	1,279.4	4,598.9	962.2
0.70	2,027.1	4,416.4	1,103.0
0.85	2,950.2	4,377.7	1,241.9
1.13	4,731.0	3,703.6	1,454.1
1.42	6,330.8	2,728.9	1,622.8
1.70	7,530.0	2,302.3	1,683.8
2.13	9,240.9	1,837.2	1,674.8
2.55	10,463.2	1,504.2	1,669.3
2.89	10,571.6	1,100.9	1,669.2
3.12	10,693.7	933.4	1,673.7
3.54	10,611.4	986.9	1,652.9
3.97	10,455.5	1,195.6	1,614.3
4.25	10,037.5	1,331.4	1,593.7

Table 4: WUA values for steelhead on Wolf Fork Creek.

Discharge (m^3/s)	Spawning	Adult
0.43	8,667.1	2,082.1
0.57	9,649.3	2,431.2
0.70	10,162.7	2,810.9
0.85	10,594.9	3,300.4
1.13	11,122.7	4,306.8
1.42	11,395.3	4,911.3
1.70	11,524.0	5,435.2
2.13	11,290.7	5,687.8
2.55	10,930.7	6,025.9
2.89	10,678.7	6,023.0
3.12	10,539.6	6,023.9
3.54	10,260.6	5,707.2
3.97	9,804.1	5,277.6
4.25	9,412.3	4,955.2

Although easier to understand, interpretation for the purposes of establishing instream flow requirements is challenging. Steelhead spawning peaks are above 2.8 m^3/s (100 ft^3/s) while bull trout spawning peaks are close to 1.7 m^3/s (60 ft^3/s). Determining which species should be preferred even if they use the stream at the same time of the year is difficult. The low adult curve for bull trout indicates the adults may actually prefer to migrate elsewhere in the watershed so perhaps their needs could be considered not as important. Furthermore, other life stages peak at considerably different values. However, it is also important to realize the impact of even minor changes in the preference curves on interpretation of results. For example, instead of multiplying depth and velocity preferences of $0.7 \times 0.7 = 0.49$, using values

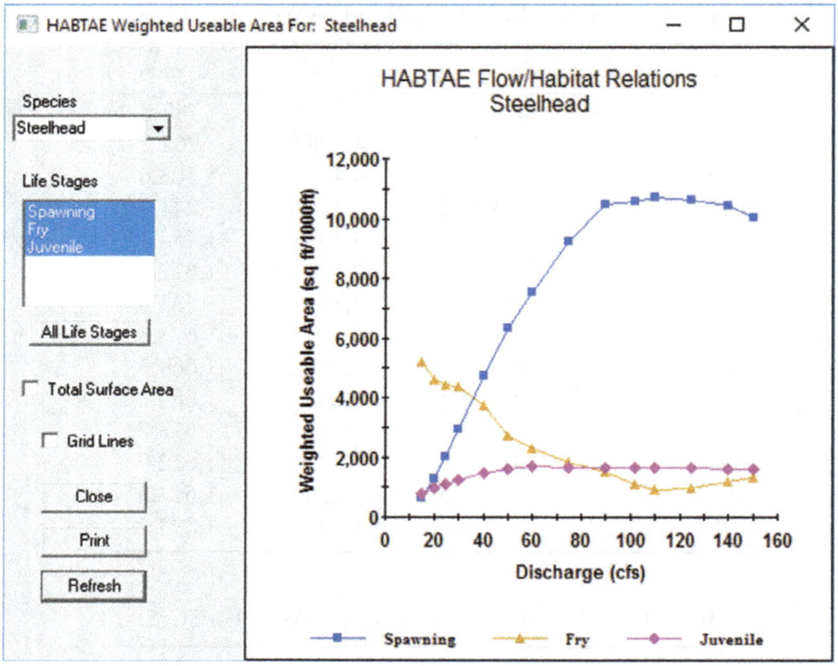

Figure 5: WUA for all life stages of steelhead on Wolf Fork Creek.

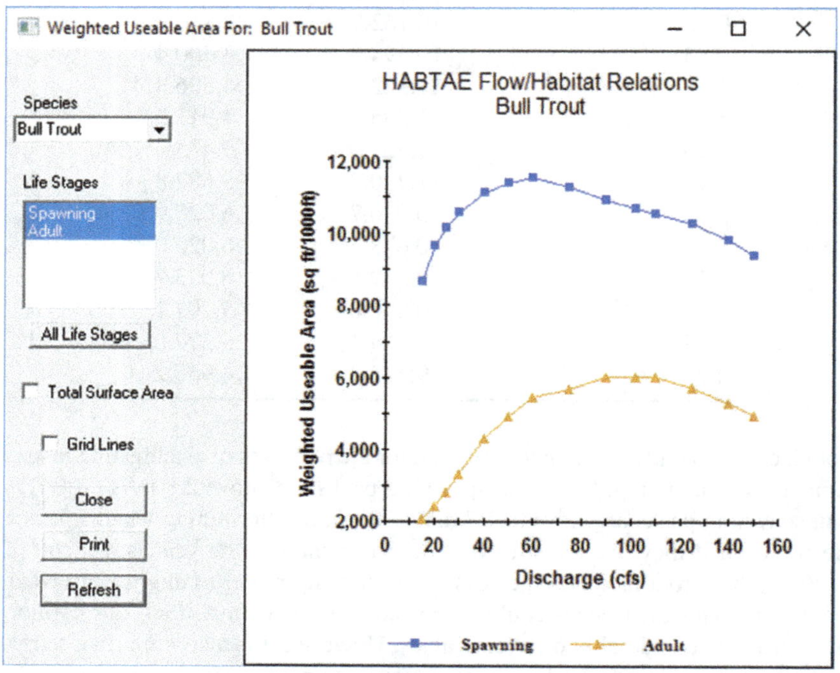

Figure 6: WUA for all life stages of bull trout on Wolf Fork Creek.

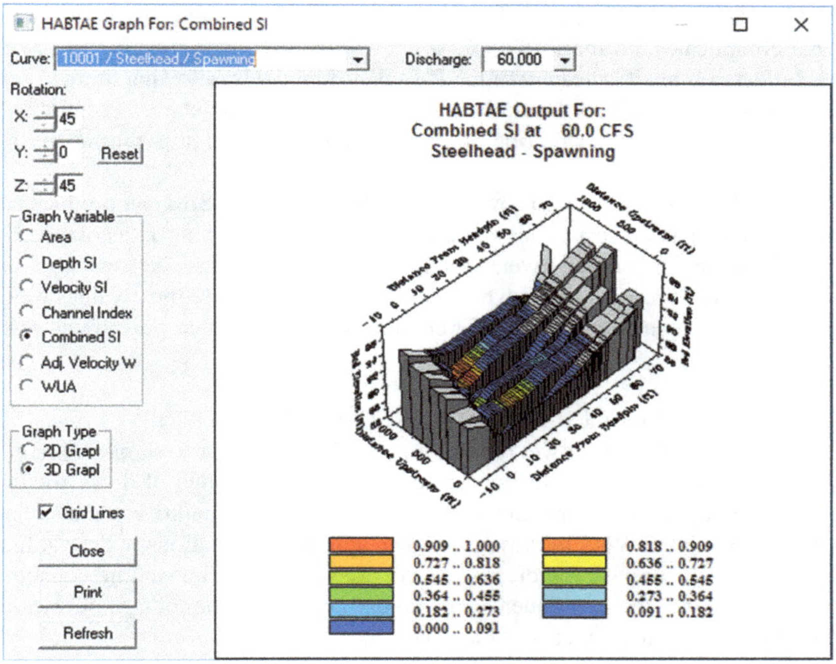

Figure 7: Location of steelhead spawning at 1.7 m³/s (60 ft³/s) on the Touchet River.

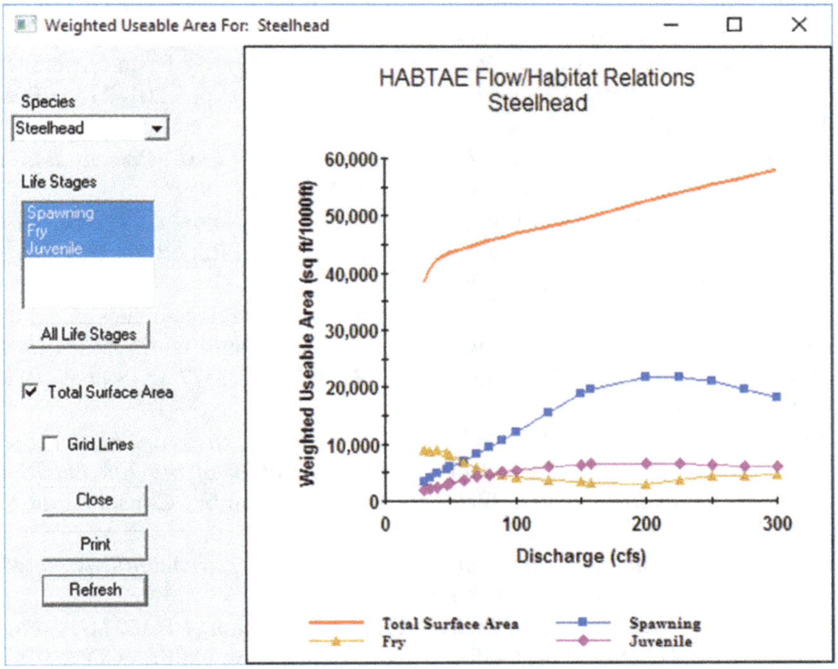

Figure 8: WUA with respect to total stream area for Touchet River Steelhead.

that vary by 10% (e.g., 0.63) reduces WUA by 20%. Compounding cover and substrate factors further complicates the analysis.

Figure 7 illustrates another feature of the PHABSIM model by allowing users to see where the high value habitat is within the study area. Connectivity of habitat is essential to healthy fish populations so habitat restoration activities should consider this during the planning phase.

To understand the options for improving stream habitat, PHABSIM can produce results in relationship to total stream surface area. Figure 8 illustrates the usable area compared to total area for steelhead in the Touchet River. As shown, fish are able to use between 10% and 50% of the total area. While an analysis of what could be done to improve these values was outside the scope of this work, it is very useful for comparison purposes and prioritizing restoration efforts.

5 CONCLUSIONS AND RECOMMENDATIONS

Preference curves developed for wet regions may not be applicable to semi-arid river basins, where stream flows are often quite small in comparison. The range of flows shown in our analysis are typically far below the threshold dividing large and small rivers and streams. The sensitivity of moving the WUA curves up or down with small variations in flow indicates the importance of establishing local preference curves. While this will be a time-consuming and expensive undertaking, the consequences of failure could be extinction of species or overprotection at high costs to residents of the watershed.

REFERENCES

[1] Richter, B., Mathews, R., Harrison, D. & Wigington, R., Ecologically sustainable water management: Managing river flows for ecological integrity. *Ecological Applications*, **13**(1), pp. 206–224, 2003. DOI: 10.2307/1938620.

[2] Falkenmark, M. & Molden, D., Wake up to realities of river basin closure. *International Journal of Water Resources Development*, **24**(2), pp. 201–215, 2008. DOI: 10.1080/07900620701723570.

[3] Rushton, C.D., *Instream Flows in Washington State: Past, Present and Future*, Washington Department of Ecology: Olympia, WA, 2000.

[4] Stalnaker, C., Lamb, B., Henriksen, J., Bovee, K. & Bartholow, J., *The Instream Flow Incremental Methodology: A Primer for IFIM*, Biological Report 29, U.S. Department of the Interior: Washington, DC, 1995.

[5] Conder, A.L. & Annear, T.C., Test of weighted usable area estimates derived from a PHABSIM model for instream flow studies on trout streams. *North American Journal of Fisheries Management*, **7**, pp. 339–350, 1987. DOI: 10.1577/1548-8659(1987)7<339:TOWUAE>2.0.CO;2.

[6] Barber, M., Juul, S., Saul, D., Cichosz, T. & Rabe, C., *Instream Flow Incremental Methodology Analysis and Streamflow Data Collection of the Touchet River System Within Columbia County*. Technical Report, Columbia Conservation District, Dayton, WA, 2001.

[7] GeoEngineers, *Geomorphic Assessment, Touchet River Upstream of Dayton, Washington*, GeoEngineers: Spokane, WA. 2011.

[8] Milhous, R.T. & Waddle, T.J., *Physical Habitat Simulation (PHABSIM) Software for Windows (v.1.5.1)*, USGS Fort Collins Science Center: Fort Collins, CO, 2012.

[9] Beecher, H., Vada, B., Caldwell, B., Pacheco, J. & Marti, J., *Instream Flow Study Guidelines: Technical and Habitat Suitability Issues*, Washington Department of Fish and Wildlife: Olympia, WA, 2008.

[10] Trihey, E.W. & Stalnaker, C.B., Evolution and application of instream flow methodologies to small hydropower development: An overview of the issues. *Proceedings of the Symposium on Small Hydropower and Fisheries*, eds. F.W. Olson, R.G. White & H. Hamre, The American Fisheries Society: Denver, CO, 1985.

Author index

www.ingramcontent.com/pod-product-compliance
Lightning Source LLC
LaVergne TN
LVHW062222200926
848484LV00006B/1453